ENVIRONMENTAL COSTS AND BENEFITS OF TRANSGENIC CROPS

Wageningen UR Frontis Series

VOLUME 7

Series editor:

R.J. Bogers
Frontis – Wageningen International Nucleus for Strategic Expertise, Wageningen University and Research Centre, The Netherlands

The titles published in this series are listed at the end of this volume.

Environmental Costs and Benefits of Transgenic Crops

Edited by

J.H.H. WESSELER

University of Wageningen,
The Netherlands

A C.I.P. Catalogue record for this book is available from the Library of Congress.

ISBN 1-4020-3248-X (PB)
ISBN 1-4020-3247-1 (HB)
ISBN 1-4020-3249-8 (e-book)

Published by Springer,
P.O. Box 17, 3300 AA Dordrecht, The Netherlands.

Sold and distributed in North, Central and South America
by Springer,
101 Philip Drive, Norwell, MA 02061, U.S.A.

In all other countries, sold and distributed
by Springer,
P.O. Box 322, 3300 AH Dordrecht, The Netherlands.

Printed on acid-free paper

Printed in the Netherlands.

Contents

Preface

The impacts of releasing transgenic crops for the environment and the economic consequences are not yet well understood. The level of concern about the environmental impacts differs between Europe and North America and influences not only government response but also international trade and public as well as private incentives for R&D.

To address the complex issues around the environmental impacts of transgenic crops a workshop on Environmental Costs and Benefits of Transgenic Crops was organized at Wageningen in June 2003. The participants were a heterogeneous group of natural and social scientists from North America and Europe. This book contains revised and reviewed papers presented at the workshop.

This book is not intended to cover all the aspects related to environmental costs and benefits of transgenic crops. But every paper presents specific aspects of the problem and contributes to a better understanding of the whole.

The organizers

The workshop was organized by Frontis – Wageningen International Nucleus for Strategic Expertise, together with the Environmental Economics and Natural Resources Group of the Department of Social Sciences, both at Wageningen University and Research Centre, Wageningen, The Netherlands.

Acknowledgments

I would like to thank the workshop participants and the authors for their contributions, the reviewer for his comments, advice and openness while discussing the different issues, Wendelien Ordelman (Frontis), Erik Ansink, Anna Maciejewska and Sara Scatasta for making the organizational arrangements for the workshop, Debbie Kleinbussink for proof-reading and Paulien van Vredendaal for technically editing the manuscripts, Hugo Besemer for facilitating the lay-out process and the Netherlands Ministry of Agriculture, Nature and Food Safety, the Netherlands Ministry for Spatial Planning, Housing and the Environment, the Netherlands Organization for Scientific Research – Social Science Research Council (NWO-MAGW), Wageningen University and Research Centre and the Wageningen Institute for Environment and Climate Research (WIMEK) for financial support, and Rob Bogers of Frontis for stimulating the workshop and the publication of this book.

The editor,

Justus Wesseler

Wageningen, May 2004

1

Environmental benefits and costs of transgenic crops: introduction

Justus Wesseler[#]

Introduction

Concern about the environmental impacts of transgenic crops is one of the major reasons for the EU's quasi moratorium on GMOs (European Environment Council 1999). The contributions in this book show that the economic implications of these concerns are far-reaching and complex. At the centre of the theoretical framework for analysis stands the linear chain of agricultural biotechnology development as depicted in Figure 1. The public and private sector invest resources in the development and use of knowledge to produce agricultural crops with new traits. Those new crops are sold to farmers who plant them and sell the harvest to the downstream sector, which further processes the products until they finally reach the end consumer via the retailers. The major concern about environmental impact at the farm level, where the deliberate release into the environment takes place, is indicated by the circle around the box in the centre of Figure 1.

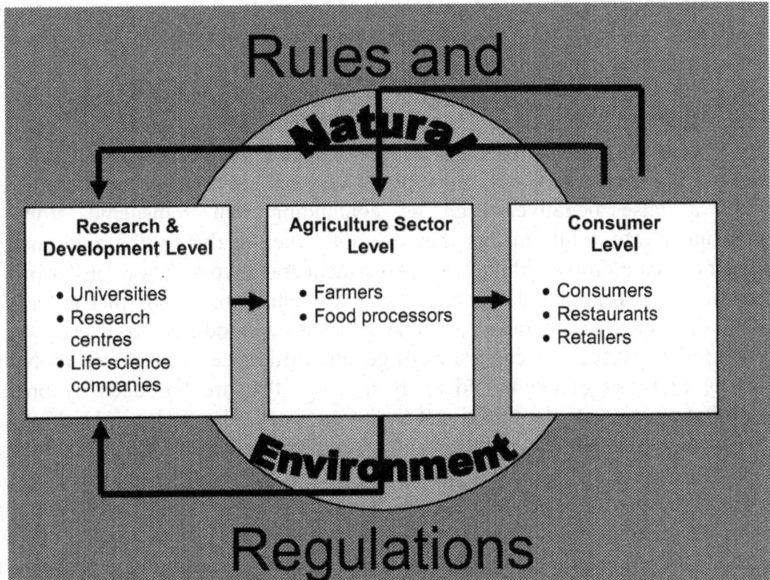

Figure 1. The transgenic-crop development chain

[#] Environmental Economics and Natural Resources Group, Wageningen University, The Netherlands. E-mail: justus.wesseler@wur.nl

J. H. H. Wesseler (ed.), Environmental Costs and Benefits of Transgenic Crops, 1–6.

This linear model of technology development ignores important feedback mechanisms. In reality, consumers send signals about their food preferences back to the farm sector and the farm sector sends them back to the technology provider. Additionally, within each box exchange of information between agents in the chain influences whether and how a new transgenic crop and derived food products will be successfully introduced and, hence, influences their environmental impact.

The rules and regulations that national governments and international organizations use to govern the release of transgenic crops also influence the behaviour of agents within the chain and, consequently, the environmental impact. Those rules and regulations do not appear out of the blue; in fact they are made by humans who act in their own interest. The political economy of deciding about rules and regulations adds another dimension of complexity.

Considering the interactions between the agents, the problem of identifying the economic costs and benefits of environmental impact from transgenic crops appears like the Gordian knot, a problem almost impossible to solve. This book takes a step towards trying to disentangle parts of the knot by examining specific aspects of environmental costs and benefits of transgenic crops from the points of view of the natural and social sciences.

From the various contributions in this book, a structure emerges that, while not yet complete, helps to understand the economic benefits and costs of environmental impact of transgenic crops.

The papers, written by natural and social scientists, cover both environmental and economic perspectives. They were presented at a workshop held in Wageningen in June 2003. The first nine chapters focus on the interaction between the natural environment and the economy at farm level. The last four chapters include the research and development perspective as well as a view from the consumer side. Each chapter in this book includes a comment by a workshop participant that raises issues discussed after the presentation.[1]

Ervin and Welsh start with an overview of the environmental costs and benefits of transgenic crops and then look at the current environmental regulatory process in the US. They argue in line with the National Research Council (NRC) and other researchers that more emphasis must be placed on controlling the type-II error (accepting a false negative, such as concluding that transgenic crops are environmentally safe, while in fact they are not) when analysing the environmental effects of transgenic crops within a risk-assessment framework. They further propose a differentiated environmental-risk assessment for transgenic crops that is based on the genetic difference of the transgene to the genes of the modified crop. They suggest three risk models for the five categories of genetic difference as discussed by Nielsen (2003). The first risk model would apply to crops that are produced by breeding processes that are close to the traditional (non-GM) ones. The ecological risk of those crops is assumed to be similar to the risk of crops produced by traditional methods. In this case the control of the type-I error (accepting the hypothesis of no negative impact on the environment when in fact it should be rejected) is sufficient. Examples are herbicide-tolerant crops. New crops that are produced with methods that go beyond what is possible with traditional breeding methods, such as insect-resistant crops, require a higher standard of ecological-risk assessment. The third category includes synthetic genes and novel proteins. As the genetic distance for this type of crops from those that are bred by traditional methods is greatest, the authors propose even stricter standards of tests. They suggest that tests should be carried out by experts who, in the case of wilful fraudulence will be held liable to avoid biased assessments. The risk assessment is then combined with a benefit-costs analysis that differs according to the category of crop under consideration. With the differentiated risk assessment the authors propose a case-by-case approach that goes beyond the current regulatory approach in the US and supports the view of the EU that the precautionary principle should be applied. The authors also suggest that a proper regulatory system can provide incentives for the private sector to develop transgenic crops with traits that provide ecological benefits of a public good nature. One eminent example would be a crop that is resistant rather than toxic to pests. Developments in this direction will provide a number of environmental benefits that may result in a reduced release of toxic substances.

Kleter and Kuiper address this particular topic in Chapter 3. Changes in pesticide use emerge as one of the most important impacts of transgenic crops on the environment. The measurement of the environmental impact of pesticide use is still not solved. One indicator is the Environmental Impact Quotient (EIQ) that includes impact of pesticides on the environment, on farm workers and on consumers. An application of the EIQ to herbicide-tolerant (HT) soybeans indicates an overall positive environmental impact of HT soybeans over non-HT soybeans. On the other hand, available indicators have several shortcomings addressed by the authors. The EIQ does not consider temporal aspects. Those can be important, e.g. for measuring the effect on water reservoirs of a continuous use of glyphosate on herbicide-tolerant crops. Long-term effects also pose problems for environmental-risk assessment, which should include an economic assessment as indicated in the previous chapter.

The paper by Laxminarayan and Simpson addresses the inter-temporal aspects of toxin-producing transgenic crops for pest management. They use a bio-economic model that combines a pest-population model with a social-welfare model. The resistance of pests to the toxin is modelled as a renewable resource that can be controlled by refuge areas. The social welfare depends on the net yield from agriculture and the amount of land used as a refuge area. The model is fairly simple and is introduced to illustrate possibilities for modelling transgenic crops. The contribution by Laxminarayan and Simpson indicates that in general a monopolistic technology provider may have incentives providing stacked varieties, but this will depend on the protection of IPRs, the market structure and the life-span of the single trait and stacked varieties. As Soregaroli points out in his comments on the paper, the model will become more complex if instead of insect pests weeds are considered.

A direct link from the previous chapter can be made to the contribution of Schubert, Matoušek and Supp in Chapter 5. The authors present results from research on virus-resistant potatoes and discuss the potential of private-sector investment for virus-resistant potato varieties. The private sector has low incentives to develop virus-resistant potato varieties as other traits like taste or colour are far more important. As virus-resistant potatoes can lead to reduced pesticide use, which provides a public benefit, public-sector research in that area can be justified. The net benefits of this research depend on the research costs, which can increase significantly with an increase in bio-safety regulation, and hence reduce potential public benefits. In addition, virus-resistant potatoes may require managed planting of potatoes to sustain the benefits from-virus resistant potatoes over several years. The paper illustrates two important problems for pest management as mentioned in the comments by Laxminarayan. First, potato viruses replicate fast and with a high mutation rate, overcoming the defence mechanism of the resistance gene, and second, wide use of the resistance gene can result in a more virulent strain of the virus. The comment again highlights the need for managed planting of the virus-resistant potato and the consideration of economic and biological factors for a successful development and application of the technology.

The importance of combined biological and economic assessment for transgenic crops directly leads to the contribution by Hurley in Chapter 6. The paper illustrates that the economic efficiency of the regulations for planting *Bt* corn in the US can be improved by considering the economic environment for planting *Bt* corn. If economists had been involved from the beginning in preparing the regulations, the economic efficiency of the adopted regulations could have been improved. The problem of expert panels appears again in the presentation by Hurley. It is also a good example of the need for interdisciplinarity in designing cost-effective regulatory policies.

Assessing the environmental costs and benefits of transgenic crops and designing appropriate regulatory systems is difficult, even if the potential implications are known. In Chapter 7 van de Wiel, Groot and den Nijs present the high variations of results in studies on gene flows, which make it difficult to get a consistent view about the implications for the environment. Regional aspects seem to be very important in quantifying the magnitude of gene flow. Scatasta stresses this point in her comment and asks whether the observed low risk of gene flow in the US would also be observable in Europe with its mosaic structure of fields.

Demont, Wesseler and Tollens in Chapter 8 demonstrate that those implications can be important. They discuss the difference between irreversibilities from a biological and economic point of view. While resistances of pests to the *Bt* toxin can be reversible from a biological perspective, occurrence of resistance will nevertheless result in irreversible costs. The economic implications of irreversible costs are explained using the case of transgenic sugar beets. The expected benefits for farmers are compared with environmental concerns of consumers. The results show that, if a household in Europe were willing to pay about 1,00 Euro per year for not having transgenic sugar beets introduced, based e.g. on concerns about the impact on biodiversity, this amount would from a social point of view justify not releasing transgenic sugar beets. As one Euro is a very small amount, this can justify the decision of the European Union not to release herbicide-tolerant sugar beets. Moreover, the results differ considerably by member state. In combination with regional differences in gene flow, there will be regions where the planting of transgenic crops will be more important from an environmental point of view. The regional differences are of relevance for Europe, as European agriculture is more heterogeneous than agriculture in the US and Canada. The importance of heterogeneity for the spatial adoption of transgenic crops is illustrated in the case study on ht-soybeans by Weaver in Chapter 9.

In Chapter 10 Gilligan, Claessen and van den Bosch provide further evidence about the spatial importance of planting transgenic crops. The theoretical framework they present allows consideration of the spatial and temporal dynamics of gene movements. By using the case of oilseed rape they show that stochastic models are far more important than deterministic models of gene movement. The resulting probability distributions about local persistence of novel genes provide important information for an environmental-risk assessment of transgenic crops. The model indicates the scope for reducing the environmental risk by introducing novel genes that are spatially explicit. This provides further support for regional management of transgenic crops.

Soregaroli and Wesseler further substantiate the importance of regional aspects in Chapter 11, where spatial implications of coexistence are analysed. The authors show

how *ex-ante* regulations and *ex-post* liability rules can effect the decision of a farmer to adopt transgenic crops and, importantly, that the choice of the regulatory system has an impact on adoption that will not be independent of the farm size. Decision-making bodies can influence the adoption rate by the governance rules for transgenic crops. While Soregaroli and Wesseler only analyse the adaptation to the regulatory framework within the farm, Beckmann in his comments adds another view that looks into the possibilities for co-operation among farmers. The possibility for co-operation results in a different cost structure that suggests the existence of a threshold between choosing a solution based on co-operation with neighbours and choosing a solution within the farm. These results again stress the importance of the regional impact for adoption of transgenic crops and consequently for environmental impacts as well.

The planting of transgenic crops and the environmental implications do not only depend on the decisions made at the farm level. There is a feedback to the upstream supply side, the R&D sector and a feedback from the downstream demand side. The private-sector incentives for developing technologies depend to a large extent on the regulatory framework. In Europe, the precautionary principle guides the decisions on the release of transgenic crops. Van den Belt provides an overview of the implications of a strong and weak interpretation of the principle and its implications for international trade, specifically the dispute between the US and the EU.

The discussion about different regulatory systems directly leads to the question of intellectual property rights (IPRs) and transgenic crops. Goeschl in Chapter 13 looks into the implications of different IPR systems for incentives to invest in environmentally friendly transgenic crops such as virus-resistant potatoes. The incentives will basically depend on the lifetime of the IPR-protected technology. The lifetime depends on the IPR system itself but also on the impact of environmental factors such as the build-up of pest resistance. In general, it cannot be concluded that the current IPR system unilaterally supports crop developments that harm the environment. However, the appearance of transgenic crops offers the opportunity to re-assess the existing IPR system. Further research is warranted before conclusions about superior systems can be drawn. Hogeveen and Michalopoulos in their comment provide a very critical view about IPRs in the context of transgenic crops and raise the ethical issues that are associated with patenting living organisms.

Graff, Roland-Holst and Zilberman discuss the implications of IPRs on the efficiency of research. IPRs limit the access of others to knowledge that is important for R&D. The authors present implications of different IPR systems for public and private-sector research priorities and suggest a clearinghouse for IPRs, where information on IPRs is traded. The authors show that a clearinghouse can increase the social benefits from modern biotechnology. De Wit provides a cautious view about the potential benefits from biotechnology. He argues that, in the end, competition will eat-up the windfall profits of early technology adopters and farmers will not be better off.

But, who in the end will benefit from the introduction of transgenic crops? Hobbs and Kerr in Chapter 15 build the link between consumers and technology adopters. They highlight the problem that without labelling – either voluntary or mandatory – consumers cannot identify GM food. In the case without labelling, consumers who prefer GM-free food will be worse off. In the case of labelling, markets can be segregated in GM and non-GM food and consumers can express their preferences by selective buying behaviour. Whether the welfare of consumers will be enhanced in the case of labelling depends on the additional segregation and labelling costs and consumer preferences for GM and non-GM food. Generalizations about gains are difficult to make and the impacts need to be analysed on a case-by-case basis as stressed in the comments by Scatasta.

The book ends with the major conclusions drawn from the contributions and research priorities for further investigation of the environmental costs and benefits from transgenic crops.

I hope that readers will enjoy the chapters in this volume with the same enthusiasm as they were discussed during the workshop. The interested audience can find further updates on the topic at: http://www.sls.wau.nl/enr/frontisworkshop/.

References

European Environment Council, 1999. *2194th Council meeting: environment, Luxembourg, 24/25 June 1999.* [http://www.eel.nl/council/2194.pdf]

Nielsen, K.M., 2003. Transgenic organisms - time for conceptual diversification? *Nature Biotechnology,* 21 (3), 227-228.

Zaid, A., Hughes, H.G., Porceddu, E., et al., 2001. *Glossary of biotechnology for food and agriculture : a revised and augmented edition of the Glossary of biotechnology and genetic engineering.* FAO, Rome. FAO Research and Technology Paper no. 9. [http://www.fao.org/DOCREP/004/Y2775E/Y2775E00.HTM]

---

[1] The terminology within this book is in line with the glossary published by the Zaid et al. (2001).

2a

Environmental effects of genetically modified crops: differentiated risk assessment and management

David E. Ervin[#] *and Rick Welsh*[##]

Abstract

A review of the literature shows that the environmental risks and benefits of genetically modified crops have varying degrees of certainty. For example, field studies have documented growing resistance to highly used pesticides. However, the risks of gene flow and deleterious effects on non-target organisms have not been evaluated at large field scales. Similarly, reduced pesticide use and toxicity have been estimated for some transgenic crops in some regions. Yet, the effects of herbicide-resistant crops on erosion, carbon loss and supplemental water use generally have not been evaluated. Recent assessments have concluded that inadequate monitoring and evaluation of the ecological risks are being conducted. Among other limitations, the US regulatory system must rely on the small science base to assess the biophysical risks of transgenic crops. The system evaluates the occurrence of a suite of hazards for all such crops and applies the standard science protocol of minimizing type-I error (i.e., rejecting the null hypothesis of no environmental risk, when in fact the null is true). However, genetically modified crops vary widely in their potential for environmental risks, some with minor and others with major possible ecological disruptions. We illustrate a differentiated risk-assessment process based on the 'novelty' of the genetically modified organism, as measured by the genetic distance from its source of variation. As 'novelty' increases, information about hazards and their probabilities generally diminishes and more precautionary risk-assessment standards would be invoked. Three different models are illustrated: (1) the current US approach that controls type-I error for crops that are close to conventionally bred crops; (2) a model for transgenic and similar crops that minimizes type-II error (i.e., accepting the null hypothesis when the alternative of significant ecological effect is true) at a moderate power of test standard, and (3) a model for the most novel and complex genetically modified crops that imposes a very high power of test standard. The discussion then develops parallel risk-management approaches that include economic costs for the first and second models. The paper concludes with a discussion of how a biosafety regulatory system that effectively distinguishes the relative risks of genetically modified organisms can stimulate public and private research into a new generation of biotechnology crops that reduce unwanted environmental risks and perhaps provide ecological benefits.

Keywords: environmental; risk; hazard; transgenic; genetically modified; precautionary; type-I error; type-II error

---

[#] Portland State University, P.O. Box 751, Portland, OR 97207, USA. E-mail: dervin@pdx.edu

[##] Clarkson University, P.O. Box 5750, Potsdam, NY 13699-5750, USA. E-mail: welshjr@clarkson.edu

J. H. H. Wesseler (ed.), Environmental Costs and Benefits of Transgenic Crops, 7–29.

Introduction

The growth in transgenic-crop plantings is the most rapid technology revolution in the recent history of United States (US) agriculture. Beginning from zero in 1996, data from the US Department of Agriculture (USDA) show farmers intend to plant approximately 80 percent of soybean acreage, 70 percent of cotton, and 38 percent of corn to transgenic varieties (NASS 2003). Barring a serious environmental or human-health problem linked to the crops, these plantings likely will grow and spread across ecosystems throughout the US over the next decade.

Given the rapid pace of adoption, perhaps it should not be surprising that knowledge of the environmental risks and benefits of the crops is immature. Independent appraisals have concluded that the science on the environmental risks of transgenic crops is small and incomplete (Ervin et al. 2001; Wolfenbarger and Phifer 2000). Estimates of the benefits also are crude, mostly aggregate changes in pesticide use. A root cause of the science deficiencies is inadequate monitoring of environmental effects at field or ecosystem scales (Ervin et al. 2001; National Research Council NRC 2002).

The central question addressed in this paper is how to make sound regulatory decisions about releasing transgenic crops under such information deficiencies. We suggest the development of risk assessment and management approaches that are tailored to the nature of the ecological risks posed by the genetically modified (GM) plant. Two reasons underpin the need for such a differentiated approach. First, the organisms inserted into transgenic crops vary and expose the environment to quite different hazards. The distance between the engineered organism and the source of the genetic variation may be a useful measure for assessing the novelty of the introduced genetic changes and risks (Nielsen 2003). A second related reason is the varying amount of information about the environmental risks and benefits of transgenic crops. For example, field studies have documented growing resistance to highly used pesticides. In contrast, the risks of gene flow and deleterious effects on non-target organisms mostly have not been evaluated at large field scales. Reduced pesticide use and toxicity have been estimated for some transgenic crops in some areas, albeit not in relation to ecological conditions. But, the effects of herbicide-resistant crops on yields, soil erosion, carbon loss and supplemental water use have not been measured or estimated.

We begin with an interpretative summary of the latest evidence on the environmental risks and benefits of transgenic crops in the US. After a brief review of the current US regulatory process and its limitations, we develop the framework for a differentiated risk-assessment approach. We close with a discussion of the implications of more effective regulation on private and public R&D for GM crops.

Environmental risks[1]

Transgenic crops do not present new categories of environmental risk compared to conventional methods of crop improvement. "However, with the long-term trend toward increased capacity to introduce complex novel traits into the plants, the associated potential hazards, and risks, while not different in kind, may nonetheless be novel" (National Research Council NRC 2002, p. 63). The nature of the risks vary depending on the characteristics of the crop, the ecological system in which it is grown, the way it is managed, and the private and public rules governing its use. Three categories of hazard emerge from the interaction of these factors[2]. Table 1 shows often-mentioned environmental concerns for herbicide-tolerant, virus-resistant and insect-resistant crops.

Table 1. Selected transgenic traits and environmental concerns

Genotype	Environmental concerns
Herbicide tolerance (HT)	• Increased weediness of wild relatives of crops through gene flow • Development of HT weed populations through avoidance and selection • Development of HT 'volunteer' crop populations • Negative impact on animal populations through reduction of food supplies
Insect resistance (IR)	• Increased weediness of wild relatives of crops through gene flow • Development of IR populations • Toxicity to non-target and beneficial insect and soil micro-organism populations
Virus resistance (VR)	• Increased weediness of wild relatives of crops through gene flow • Disease promotion among plant neighbours of VR crops through plant alteration • Development of more virulent and difficult to control viruses through virus alteration

Resistance evolution

Current commercial transgenic crops emphasize effective pest control via the increased use of certain pesticides, such as *Bt*. Crops bred to resist herbicides, viruses and insects have the potential to change agricultural practices dramatically. The lack of long-term studies poses a serious obstacle to performing an adequate assessment of the potential environmental effects. Nevertheless, some studies have assessed the impacts.

Herbicide-tolerant crops

The primary environmental concern from herbicide-tolerant (HT) crops is the development of weed populations that are resistant to particular herbicides. This resistance can occur from the flow of herbicide-resistant transgenes to wild relatives or to other crops or from the development of feral populations of herbicide-resistant crops (National Research Council NRC 2002). Also, if farmers rely on only one or a few herbicides, weed populations can develop that can tolerate or 'avoid' certain herbicides, which enables them to out-compete weeds that do not manifest such tolerance. Weed scientists find the latter development likely. In fact, Owen (1997) reports that in Iowa "[c]ommon waterhemp (*Amaranthus rudis*) populations demonstrated delayed germination and have 'avoided' planned glyphosate applications. Velvetleaf (*Abutilon theophrasti*) demonstrates greater tolerance to glyphosate and farmers are reporting problems controlling this weed". And VanGessel (2001) reports that horseweed (*Conyza canadensis*) has been found to be resistant to glyphosate through experiments conducted in a farmer's field in Kent County, Delaware.

Virus-resistant crops

There is a relative dearth of research on the ecological risks associated with these crops. However, scientists have voiced several environmental concerns related to virus-resistant (VR) transgenic crops. First, these bio-engineered varieties may promote disease in neighbouring plants by altering such plants so they become hosts for particular viruses, when such plants were not previously susceptible to infection by the viruses of concern. Second, VR transgenic crops may alter the methods through which viruses are transmitted (Rissler and Mellon 1996; Royal Society 1998). These changes could result in the development of stronger viruses (Hails 2000; Rissler and Mellon 1996; Royal Society 1998). Scientists are also concerned that the genome in VR crops may recombine with the plant-virus genome (which is comprised of RNA in most/all plant viruses) during viral replication (Rissler and Mellon 1996; Royal Society 1998). Researchers believe that such recombination could lead to genetically unique viruses that may be difficult to control (Greene and Allison 1994). Third, the flow of VR transgenes may enhance the weediness of wild relatives of VR crops (see section Transfer of genes – gene flow on the next page). A National Research Council (NRC) assessment (2000) found that the USDA's assumption that transgenic resistance to viruses engineered in cultivated squash will not result in enhanced weediness of wild squash through gene flow, needs verification through longer-term studies. The NRC study also concluded that the USDA's assessment of the potential for virus-protective transgenes in cultivated squash to affect wild populations of squash "is not well supported by scientific studies", especially for transgenic squash engineered to be resistant to several viruses instead of three or fewer (2000, p. 124). In a new report, the NRC (2002, p. 134) argued that the evidence collected to date is "scientifically inadequate" to support the conclusion of USDA's Animal and Plant Health Inspection Service (APHIS) that gene flow from VR squash would not result in increased weediness of free-living *Cucurbita pepo*.

Insect-resistant crops

The innate ability of insect populations to adapt rapidly to pest-protection mechanisms poses a serious threat to the long-term efficacy of insect-resistant (IR) biotechnologies. Such adaptations can have environmental impacts. For example, adaptation by insect populations to a more environmentally benign pest-control technique, such as *Bt*, could result in the use of higher toxicity pesticides (National Research Council NRC 2000). The Canadian Expert Panel on the Future of Food Biotechnology (Royal Society of Canada 2001) finds that it is important to account for insect movement when devising resistance-management plans. Regional or interregional-scale plans, rather than local, are needed if the insect of concern is highly mobile (Gould et al. 2002; Hails 2000). Field outbreaks of resistance to *Bt* crops have not yet been documented (Morin et al. 2003). The body of science to inform resistance management is limited to laboratory studies of specific insect pests (Ervin et al. 2001; Morin et al. 2003). Such studies show the potential for resistance to develop. Indeed, the NRC (2002, p. 76) finds that the evolution of "insect resistance to *Bt* crops is considered inevitable". Similarly, Tabashnik and colleagues write that eventually insects will develop resistance to IR crops, and therefore, "...any particular transgenic crop is not a permanent solution to pest problems" (Tabashnik et al. 2001, p. 1). A recent laboratory study found that *resistant* populations of diamondback-moth larvae may be able to use a toxin derived from *Bt* "...as a supplementary food protein, and that this may account for the observed faster development rate of *Bt* resistant insects in the presence of the *Bt* toxin" (Sayyed, Cerda and Wright 2003).

One of the few field studies by Tabashnik and colleagues (2000; 2001) found that in 1997 approximately 3.2 percent of pink-bollworm larvae collected from Arizona *Bt* cotton fields exhibited resistance. This level was far above what was expected, raising fears that rapid resistance development would occur. However, data collected in 1998 and 1999 showed no increase in resistant populations of pink bollworm. Tabashnik and colleagues (2001) conclude that there might be high fitness costs for insects to develop resistance to *Bt*. In addition, Carrière and colleagues (2003, p. 1523) found that widespread and sustained use of *Bt* cotton can suppress regional pink-bollworm populations and thus, "...*Bt* cotton could reduce the need for pink bollworm control, thereby facilitating deployment of larger refuges and reducing the risk of resistance". In addition, a recent greenhouse study found that stacking or pyramiding two unrelated *Bt* toxins in a plant can slow the rate of resistance development (Gould 2003; Zhaio et al. 2003).

The potential for insect resistance implies that integrating transgenic crops into a multiple tactic pest-management regime may prove to be a more effective long-term strategy. The exact path of the emergence of insect resistance is yet to be characterized. However, progress has been made on identifying resistant alleles in pests of certain crops, such as the pink bollworm in cotton (Morin et al. 2003). Therefore research is needed to better define the parameters of resistance development, as well as to design crops that minimize the opportunities for resistance to develop in the first place. This latter point on fostering precautionary technology development is discussed in the concluding section.

Transfer of genes – gene flow

There is little doubt in the scientific community that genes will move from crops into the wild (Hails 2000; National Research Council NRC 2000; Snow and Palma 1997). The relevant research questions are whether transgenes will thrive in the wild, and how they might convey a fitness advantage to wild plants that makes them more difficult to control in areas (Hails 2000; Keeler, Turner and Bolick 1996; National Research Council NRC 2000; Royal Society of Canada 2001; Snow and Palma 1997).

Generally, crops with wild relatives in close proximity to the areas where the crops are grown, pose higher risk for gene flow to wild relatives. US examples include sunflower (*Helianthus annuus L*) and oilseed rape (*Brassica napus*) (Hails 2000; Keeler, Turner and Bolick 1996; Snow and Palma 1997). Gene transfer could become a problem if the transferred genes do not have deleterious effects on the crop–wild hybrids, but instead confer an ecological advantage (Hails 2000; Royal Society of Canada 2001; Snow and Palma 1997). Gene flow from classically bred crops to wild plants has been documented. Ellstrand (2001) finds that classically bred crop-to-wild gene flow has enhanced the 'weediness' of weeds for seven of the world's thirteen most important crops (e.g. Johnson grass (*Sorghum halepense*) from cultivated sorghum (*Sorghum bicolor*)).

Snow and Palma (1997) argue that widespread cultivation of transgenic crops could exacerbate the problem of gene flow from cultivated to wild crops, enhancing the fitness of sexually compatible wild relatives. Traditional breeding typically results in the inclusion of deleterious alleles (i.e., alternative forms of a gene at a given site on the chromosome) linked to the desired beneficial genes. The inclusion of the deleterious genes decreases the likelihood that crop-to-wild outcrossing will result in enhanced weediness of the wild plants. In contrast, biotechnological methods enable solitary genes to be selected without including neutral or deleterious genes (Snow and Palma 1997; National Research Council NRC 2000, p. 85).

Scientists generally expect that herbicide-resistant transgenes will not result in increased weediness of wild relatives, as such genes tend to impose a cost, or are neutral, to wild relatives. Nonetheless, in situations where herbicides are typically used to control weedy plants, herbicide resistance could confer a competitive advantage to unwanted volunteer crops (Keeler, Turner and Bolick 1996). Indeed, the flow of herbicide-resistant transgenes has already become a problem regarding within-crop gene flow. Hall, Huffman and Topinka (2000) reported the presence in a Canadian farmer's field of volunteer oilseed rape resistant to three herbicides: glyphosate, imidazolinone, and glufosinate. The 'triple-resistant' oilseed rape developed from gene flow among three oilseed rape varieties designed to resist each of the herbicides, which were planted in close proximity to each other. The Canadian expert panel finds that "...herbicide-resistant volunteer canola plants are beginning to develop into a major weed problem in some parts of the Prairie Provinces of Canada" (Royal Society of Canada 2001, p. 122). They expressed special concern about the potential for 'stacked' resistance to multiple herbicides, which could force farmers to employ older herbicides that are often more environmentally harmful than newer classes (Royal Society of Canada 2001).

In general though, ecologists tend to be more concerned about potential fitness advantages of insect- and virus-resistant transgenes (Hails 2000). For example recent research has shown that the *Bt* gene for lepidopteran resistance can increase seed production in wild sunflowers (Snow 2002; Snow et al. 2003). Another study showed that crossing *Bt* oilseed rape with a wild relative (*Brassica rapa*) did not enhance the weediness of the resulting plant relative to unmodified *Brassica rapa* (Adam 2003). To understand the contrasting findings, more studies along with monitoring and testing are needed to detect potential ecological problems. However, actions by Pioneer Hi-Bred International and Dow AgroSciences to the Snow and colleagues (2003) study may impede such research. The firms blocked a follow-up study by denying access to the materials they controlled that were needed to conduct further investigation (Dalton 2002). The firms denied three requests to continue studying *Bt* sunflower using the scientists' research funding. Therefore, Snow and her colleagues are legally barred from continuing their investigations (personal e-mail communication, April 25, 2003).

Impacts on non-target animals and plants

While crops bred to resist pests may suffer less damage and lead farmers to use less insecticide, there is concern that the toxins these plants produce may harm non-target organisms, including animals and plants that are not pests (Royal Society 1998;

National Research Council NRC 2002). Laboratory research confirms that transgenic insecticidal crops can have negative impacts on potentially beneficial non-target organisms, including lacewings (Hilbeck et al. 1998a; Hilbeck et al. 1998b), ladybird beetles (Birch et al. 1997), monarch-butterfly larvae (Losey, Raynor and Carter 1999) and soil biota (Watrud and Seidler 1998).

Tabashnik (1994) asserts that reductions in pest populations due to transgenic crops may negatively affect available numbers of desirable natural predators. Similarly, the NRC finds that "Herbicide tolerant crops might cause indirect reductions on beneficial species that rely on food resources associated with the weeds killed by the herbicides" (National Research Council NRC 2002, p. 70). In this vein, Watkinson and colleagues (2000) modelled the potential impacts on skylarks (*Alauda arvensis*) from a reduction in seeds of a weed of sugar beet (primary food supply of skylarks) from the introduction of HR sugar beet. They found that the weed populations could be almost completely eradicated depending on the conditions surrounding adoption of such transgenic sugar beets, such as the management practices. Severe reductions in weed populations could significantly affect the skylark's use of fields as a food source. Conversely, some bird populations may increase if farmers replace broad-spectrum synthetic herbicides, which have cut into the birds' food supply, with transgenic crops (National Research Council NRC 2000, p. 80).

Recent farm trials in the UK confirm that enhanced weed-control efficacy from using herbicide-tolerant crops can reduce food supplies and lower the populations of non-target species such as bees, butterflies and seed-eating beetles. However the results varied considerably by the type of herbicide employed in the system. Additional findings from the data collected from these trials should be forthcoming and help to shed further light on these complex interactions (Andow 2003).

Other potential neutral and positive impacts on non-target species of transgenic crops have also been found. For example, research sponsored by the European Commission on the safety of genetically modified organisms (GMOs) found no negative impacts on honeybees from transformed oilseed-rape plants. Also, *Galanthus nivalis* agglutinin (GNA) lectin accumulation in aphids did not result in acute toxicity to ladybird beetles or prevent *Eulophus pennicornis* from successfully parasitizing tomato-moth larvae (Kessler and Economidis 2001; Pham-Delègue et al. 2000). A field study in Wisconsin found that populations of predators and parasites were higher in *Bt* potato fields than in conventional potato fields where conventional insecticides were used. Non-chemical or less-intensive chemical treatments were not evaluated (Hoy et al. 1998). This finding points to the need to evaluate the impacts of transgenic crops relative to conventional chemically intensive practices and alternative systems (Dale, Clarke and Fontes 2002; National Research Council NRC 2002).

Given that research results on potential impacts on non-target organisms point to negative, neutral and positive effects, generalizations may well be inappropriate as to the impact on non-target organisms, with each crop and region requiring specific research. For example, in a widely publicized laboratory study, Losey, Raynor and Carter (1999) found a 44% mortality rate in monarch-butterfly larvae fed on milkweed leaves dusted with *Bt* corn pollen. No mortality occurred in monarchs fed on leaves with non-*Bt* corn pollen. These laboratory finding on the toxicity of *Bt* corn to monarch butterflies generated significant controversy and prompted responses as to the applicability of the finding to field settings (Beringer 1999), follow-up research supporting the original findings (Hansen-Jesse and Obrycki 2000), and a risk-assessment study finding that monarchs are not at risk from *Bt* corn since "overall exposure of monarch larvae to *Bt* pollen is low" (Sears et al. 2001). In its turn, the NRC panel asserted "In the upper Midwest, herbicide-tolerant soybeans might cause indirect reductions of monarch populations because their milkweed host plants are killed by the herbicides" (National Research Council NRC 2002, p. 71). Though many consider the debate over monarchs and *Bt* corn closed, there are still questions being raised about the effects of long-term and low-level exposure to *Bt* in corn pollen on monarch-larvae survival and fitness (Stanley-Horn et al. 2001).

Insects and other animals are not the only organisms potentially affected by transgenic crops. The Canadian expert panel found the cultivation of transgenic crops could impact the diversity and abundance of soil microflora, however the impacts are "...minor relative to the natural variability..." (Royal Society of Canada 2001). They observed that transgenic manipulation aimed at modifying biogeochemical cycles should receive more scrutiny. The NRC (2002) largely concurs by arguing that no effects on soil organisms have been found to date, though it has been discovered that *Bt* toxin 'leaks out' of corn roots and can persist in the soil for months (Saxena and Stotzky 2000; Dale, Clarke and Fontes 2002).

Risk summary

In general, the environmental hazards associated with transgenic crops are potential risks. However, research results provide emerging parameters to evaluate the relative magnitude of the potential risk. For example, good evidence is emerging that the combination of natural promiscuity regarding gene flow among crop varieties and engineered herbicide resistance is a serious concern. Likewise, it is becoming clearer that HR crops will probably not create 'superweeds' through crop–wild flow of genes that enable plants to tolerate particular herbicides. Rather weed problems will be enhanced by the selection of resistant weed populations through increased use of herbicides tied to particular transgenic crops, such as glyphosate-resistant soybeans. Research efforts should concentrate on the latter potential risks.

Also of concern is the enhanced weediness of wild relatives of crops from the flow of genes enabling plants to resist insects and viruses. However, the research to evaluate the extent of these risks is incomplete. More study is needed to assess the potential for the widespread adoption by farmers of IR sunflower and VR squash to promote the development of wild plants with improved fitness relative to other wild plants. The improved fitness of particular plants in wild populations could alter plant and animal ecosystems. The controversy over the potential for *Bt* corn to harm monarch butterfly populations also illustrates the need to move beyond laboratory studies to comprehensive field scale when assessing the potential negative impact on susceptible but beneficial populations. That is, studies that account for the temporal and spatial interaction between the introduced technology and the organism of interest.

That field outbreaks of resistance to *Bt* crops have not yet been documented despite widespread adoption of such crops deserves more investigation. Potential questions include:

- Have the refugia plans prevented the development of such resistance?
- If so, can such plans be developed for herbicide-resistant technologies to delay the development of weed populations resistant to herbicides linked with HR crop varieties?

Environmental benefits

Reduced pesticide use and toxicity

Data to assess the effects of transgenic crops on pesticide use should capture the full range of climate, pest and economic conditions. The data should also be linked to environmental conditions to estimate changes in acute and chronic toxicity on ecological systems (Antle and Capalbo 1998). The impacts of changes in pesticide use for transgenic crops on the environment can be determined only by comparing the fate, transport and toxicity of the full array of compounds available to farmers, and how they are applied. The following estimates for three major US transgenic crops do not yet measure up to these standards.

Bt cotton

Results from farm surveys generally indicate that farmers who plant *Bt* cotton apply fewer insecticides than on conventional cotton (Carlson, Marra and Hubbell 1998; Hubbell, Marra and Carlson 2000; Economic Research Service USDA 1999a; 1999b; Fernandez-Cornejo and McBride 2002). USDA analysts recently estimated that *Bt* cotton plantings in 1997-98 reduced insecticide use by approximately 250,000 pounds of active ingredients (a.i.) (Fernandez-Cornejo and McBride 2002). Other studies have found that the reductions vary by area and year depending on pest pressures and other factors. For example, Carlson, Marra and Hubbell (1998) report that the average number of insecticide applications by farmers who adopted *Bt* cotton in 1996 was 3.29 on their traditional acres in the upper South, but only 2.58 in the lower South, a difference that likely reflects different insect conditions in the two regions. Farmers who plant *Bt* cotton likely use more conventional insecticides in the first place and can save more money than farmers who applied lower levels. The long-term effects of *Bt* cotton on insecticide use may require analyses of 10 years or more to cover the cycles of pest, climate and economic variations.

Herbicide-tolerant soybeans

The latest national analysis estimated that planting of HT soybeans in 1997-98 increased overall pesticide use by approximately 2.5 million pounds a.i. (Fernandez-Cornejo and McBride 2002). The increase was largely in the form of glyphosate, which is anywhere from 3 to 16 times less toxic than the herbicides it has replaced and 1.6 to 1.9 times less likely to persist in the environment (Heimlich et al. 2000). Whether and how long these shifts in herbicide composition on soybeans will persist depend on how fast the resistance problems discussed above unfold. If glyphosate becomes ineffective, farmers will use other herbicides to control weeds that develop resistance to it, and the environmental implications depend on the substitute compound, location and other conditioning factors.

Herbicide-tolerant corn

Over 1997-98, US farmers who planted HT corn were estimated to decrease their herbicide use by approximately 5 million pounds a.i. (Fernandez-Cornejo and McBride 2002).

Net effect

The overall effect for the three crops was a net reduction of approximately 3 million pounds of pesticide a.i. in 1997-98. This is characterized as "…a small but statistically significant effect…" (Fernandez-Cornejo and McBride 2002, p. 36). Since nearly 80% of the reduction in pesticide treatments is attributed to switching to glyphosate, the overall toxicity quotient decreased by more than the simple volume reduction. Since 1997-98, the adoption of transgenic crops has spread and the decrease in pesticide volume and toxicity likely has increased as well, but likely still remains a small percentage of overall pesticide use.

Reduced tillage, erosion, carbon loss and water savings

Several other potential environmental benefits of transgenic crops have been hypothesized. For example, manufacturers and advocates of transgenic crops have asserted that HT varieties will increase use of conservation tillage. However, the USDA analysis could not support this hypothesis, concluding that farmers already using conservation tillage were more likely to adopt HT crops (Fernandez-Cornejo and McBride 2002). Evidence on the other potential environmental benefits has been nil.

The current environmental regulatory process

The US Environmental Protection Agency (EPA) and USDA's Animal, Plant, and Health Inspection Service (APHIS) share responsibility for assessing the environmental risks of transgenic crops before their release for testing and commercialization. EPA evaluates plant-incorporated protectants, such as *Bt* in cotton and corn, and regulates them in the same way it regulates conventional chemical pesticides. The APHIS evaluation, depending upon the particular plant line, covers a broad range of potential environmental effects: (1) the potential for creating plant-pest risk; (2) disease and pest susceptibilities; (3) the expression of gene products, new enzymes or changes to plant metabolism; (4) weediness, and impact on sexually compatible plants; (5) agricultural or cultivation practices; (6) effects on non-target organisms, including humans; (7) effects on other agricultural products, and (8) the potential for gene transfer to other types of organisms (McCammon 2001).

Each agency conducts a risk analysis of the biophysical effects of the crop. We discuss the more comprehensive APHIS process here. The risk analysis includes three stages – hazard identification (as covered in the previous section Environmental risks), risk assessment and risk management. The risk assessment stage is the focus of this section. One or more of several techniques may be used to perform the assessment: (a) epidemiological analysis; (b) theoretical models; (c) experimental studies; (d) expert judgments, and (e) expert regulatory judgments. APHIS generally uses expert regulatory judgment, a less rigorous technique than (a), (b) or (c) according to the NRC (2002, p. 60).

APHIS uses a two-part model in which a transgenic plant is divided into (1) the unmodified crop and (2) the transgene and its product (National Research Council NRC 2002, p. 90). The theoretical reasoning behind this choice is that the transgene is a small genetic change that is likely to have only a small phenotypic effect. Invoking that principle accepts the simple linear model of 'precise' single gene modifications that do not significantly alter other plant processes. Experts have reservations about this rationale: "...unanticipated changes can be induced by expression of a novel gene, and their phenotypic consequences need to be assessed empirically across time and environments" (Royal Society of Canada 2001, p. 185). The NRC panel also noted the assumption that single gene changes have small ecological effects is not always true (National Research Council NRC 2002, p. 91).

The APHIS risk-assessment process evaluates the risk of the unmodified crop separate from the risks of the transgene and its products. The test for biophysical risk basically attempts to control the type-I error, α, of rejecting the null hypothesis that the transgene has *no* effect on the environment when in fact the null is true[3]. Minimizing the frequency of type I-error (a 'false positive') by requiring a high level of certainty, most often 95%, is the dominant approach used for hypothesis testing in science. It is also considered to be a conservative approach in detecting an ecological effect, because it is difficult to construct studies with sufficient power to reject the null hypothesis under $\alpha = 0.05$. This is especially true for studies dealing with environmental issues where natural variation, both in space and time, is typically great (Buhl-Mortensen 1996; National Research Council NRC 2000). The effect of this high variation is to spread out the distribution of the random variable for testing the null hypothesis and increase the range of values for which the sample test statistic is interpreted as not rejecting the null hypothesis.

The NRC panel (2002), noting several limitations of the two-part model, recommended more evaluation of 'fault-tree' and 'event-tree' risk analyses that systematically search for potential ecological risks. These and other techniques place more emphasis on understanding and controlling type-II error, β, the error of failing to reject the null hypothesis when it is in fact false, i.e., a 'false negative' when the transgene actually causes an ecological hazard. Lemons, Shrader-Frechette and Cranor (1997) argue that scientists and decision-makers should be more willing to minimize type-II errors (and accept higher risk of committing type-I errors) because of the pervasiveness of scientific uncertainty in complex ecological processes. They explain that the normal argument to minimize type-I error when adding new scientific knowledge does not apply equally to environmental regulatory decisions. Type-I error with a 95% confidence level is appropriate for evaluating new scientific results in order to prevent or minimize the inclusion of "speculative knowledge to our body of [scientific] knowledge" (Lemons, Shrader-Frechette and Cranor 1997, p. 228). However, knowledge generation to support environmental regulatory decisions is different in kind as it is based on finding whether negative environmental or health outcomes are likely, or unlikely, to occur, and not generating new scientific results per se (Lemons, Shrader-Frechette and Cranor 1997, p. 224-230). Buhl-Mortensen (1996) also notes that since ecosystem models, with a few exceptions, have low predictive power, it is prudent to control for type-II error when evaluating the potential ecological impacts of industrial processes or technologies. In a similar vein, Jasanoff concludes that the current US system has "...biased the assessment exercise away from large, holistic questions..." (Jasanoff 2000, p. 279), instead focusing on relatively precise genetic manipulations. Part of the reason for the bias may be the lack of post-market monitoring and testing of biotechnology crops to provide adequate data for examining the larger ecological questions (National Research Council NRC 2002; Taylor and Tick 2003).

Figure 1 illustrates the differences in risk-assessment tests under control of type-I versus type-II errors. Assume for purposes of illustration that the horizontal axis measures the difference in seed production by wild relatives containing a transgene compared to seed production in wild plants without the transgene. The hypothetical distributions measure the probability density of various values. The distributions centred over $\mu = 0$ and $\mu = SD_a$ are the null and alternative population distributions. If a value of SD* is observed from the sample test, the APHIS risk assessment would fail to reject the null hypothesis of no significant gene flow, because SD* lies to the left of the critical test statistic for a one tailed test, assuming $\alpha = 0.05$ (and $\beta = 0.20$).

However, under a criterion to minimize type-II error by setting β at 0.10 (i.e., the power of the test = 0.90), the alternative hypothesis of a positive effect on seed production in wild relatives would fail to be rejected. Note also that if the natural processes and the assumed distribution become more variable or flatter, the range under which the alternative hypothesis is not rejected would expand. That is, the probability of a 'false negative' increases, *ceteris paribus*.

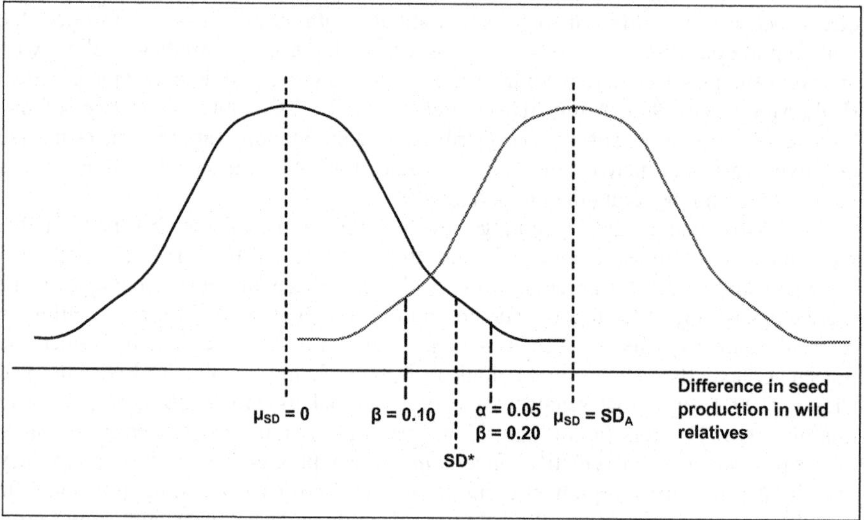

Figure 1. Differences in statistical tests to control type-I and type-II errors

The NRC analysis repeatedly emphasized that "For purposes of decision support, risks must be assessed according to the organism, trait and environment" (National Research Council NRC 2002, p. 63). The review of potential environmental risks also suggests information deficiencies could be added to this list. The recognition that GM crops vary in their potential for environmental risk invites consideration of a risk-assessment process that uses different methods and different standards of proof for different types of genetic modification.

A differentiated risk-assessment process

A differentiated risk-assessment process captures the novelty of the ecological hazard(s) from the GM crop and the information quality about the potential hazards and their occurrences. Different models might cover a range, from controlling type-I error for GM crops using high-quality information that shows little potential ecological risk to a very high standard of avoiding type-II errors for crops judged to pose serious potential ecological disruptions with little scientific evidence to assess the nature of the risk(s).

To differentiate the risk assessment, a robust method for characterizing the nature of the ecological risks of GM crops is needed. However, scientific or regulatory consensus on such a robust method does not exist. To facilitate our analysis, we adopt an approach suggested by Nielsen (2003), who argues for conceptual diversification in discussing and regulating genetically engineered organisms. She claims the current process-based categorization is imprecise and does not adequately convey the sources, extent and novelty of the genetically modified organism. For example, GM crops with simple nucleotide changes are unlikely to generate serious ecological concerns beyond those of their traditional counterparts. In contrast, species-foreign genes, synthetic genes and some other changes in GM crops deviate substantially from what classical selective-based breeding has achieved. The latter organisms have genetic compositions that do not reflect evolutionary processes occurring under natural conditions. She cites *Bt* corn, derived by genetic engineering of several unrelated DNA segments, as an example of an organism that cannot be replicated by natural processes within the same time scale. Thus, the 'genetic distance' between the engineered organism and the source of the new genetic variation could serve as a functional criterion for determining the type of risk assessment conducted.

This approach dovetails with earlier arguments by Snow and Palma (1997) and the NRC (2000, p. 85), which imply that the differences between classical breeding approaches and transgenic methods can justify differences in risk assessment. However, historically the NRC (2002) has emphasized regulating the products of genetic manipulation over the processes (Snow 2003). Snow (2003) argues that focusing on the phenotype over the process used to engineer it into a plant is appropriate "...given that many GEOs [genetically engineered organisms] have truly new characteristics relative to what can be created by conventional breeding". Therefore, as very novel traits are engineered into crops, regulating such traits in essence means regulating the process by which the traits were engineered. Classical approaches to breeding could never result in the production of crops with such truly novel traits.

We argue that as novel processes are developed, it could be in the public interest to consider the process in regulating the organism, in addition to the trait or phenotype expressed in the plant. For example, it may be appropriate to treat differently, two plants with the same engineered phenotype that was produced through different engineering processes. And that, as the 'genetic distance' (Nielsen 2003) from conventional techniques increases, the regulatory regime becomes increasingly precautionary in its approach and conceptualization.

It is important to note that we are not asserting that classical breeding inherently produces safer products and that deviations from this approach produce more dangerous products. Rather our argument is subtler. We believe that our long experience and familiarity with classical breeding techniques makes it reasonable to assume that as ecological or other problems potentially develop, we are more likely to recognize such problems and take corrective measures. It follows that our relatively brief experience with transgenic and other recently developed techniques and the scant science base makes it more likely that if a biosafety problem develops with these new crops, we may not recognize the problem as quickly because our ability to discern potential problems is primarily based on our experience with traditional breeding techniques. We believe our rationale and approach are consistent with the finding of the NRC stated at the beginning of section Environmental risks that "...the associated potential hazards, and risks [of transgenic crops], while not different in kind, may nonetheless be novel" (National Research Council NRC 2002, p. 63).

Nielsen (2003) proposes five genetic-distance categories that vary from low to high: (1) intragenic (within genome); (2) famigenic (species in the same family); (3) linegenic (species in the same lineage); (4) transgenic (unrelated species), and (5) xenogenic (laboratory-designed genes). We combine categories 1 and 2 and categories 3 and 4 along with category 5 to develop three risk models based on increasing genetic distance[4].

1. Intragenic and famigenic – These two categories of genetic modification respectively include those from directed mutations or recombinations including those arising in classical, selection-based breeding, and from the taxonomic family, including those arising from applying cellular techniques in classical breeding. The risk-assessment process for these cases could reasonably presume no substantial ecological risks from releasing the crops beyond those from conventional breeding. Thus, it would include a straightforward review of evidence submitted by the applying entity and application of the 'probability rule' criterion for all relevant effects reviewed by APHIS (Mooney and Klein 1999). A test of the null hypothesis of no significant effect would be conducted to control type-I error, i.e., failing to accept the null when it is in fact true, at the standard 0.05 level of significance.
2. Linegenic and transgenic – These two categories include organisms that contain genetic variability *beyond that* possible with conventional breeding. Linegenic includes species in the same lineage and the recombination of genetic material beyond what can be achieved by classical breeding methods. Transgenic covers those plants that contain DNA from unrelated organisms, and include most of the GM plants commercialized today. For these plants, the test shifts the framing hypothesis to one that assumes a significant environmental effect because of the increased novelty of the crop and less information. The standard of proof would be set at a specified power of test, for example 0.90 or $\beta = 0.10$. The decision of setting the standard of proof moves beyond science into the realm of public input and political decisions because the standard reflects society's general preference for avoiding such risks, i.e., the degree of precaution (Van den Belt 2003).
3. Xenogenic – This category includes laboratory-designed genes for which no naturally evolved genetic counterpart can be found or expected, e.g., synthetic genes and novel combinations of protein domains. This class is the furthest of the three from natural genetic variability, and therefore poses the greatest potential for ecological hazard and risk. For the hazards that can be characterized with objective or subjective probability distributions, the bar for approval to release would be highest for such plant organisms. Since the genetic distance from classically bred crops is greatest for this category, a higher standard of proof would be applied to control type-II error than for category 2, for example $\beta = 0.05$.

To implement this risk-assessment framework for GM crops, a group of experts with sufficient breadth across ecological and other relevant sciences would be assembled. The composition and independence of the groups is critical if reliable risk assessments are to be completed. To counter criticisms that the expert panels used by USDA inappropriately favour releases, both government and university scientists would be involved and each would face sanctions if their contributions were subsequently determined to be biased, e.g., a bonding mechanism for liability. Due to the scant knowledge that exists for many GM crops, especially new transgenic varieties, the groups would at first conduct case-by-case assessments. However, over time with the accumulation of more systematic knowledge on the potential environmental risks due to the search for type-II errors, the assessments likely would shift to broader categories and become more routine and cost-efficient over time.

Differentiated risk management

The final stage of risk analysis is the management decision taken, including commercial release and regulatory measures that may accompany the releases, such as refugia requirements and post-commercialization monitoring and testing. The differentiation of the risk-management process parallels that taken for risk assessment in that increasing degrees of precaution are imposed on more novel organisms. However, risk-management decisions would involve weighing environmental and economic considerations for the organisms that do not pose potential catastrophic and irreversible hazards. The answers to six questions in Table 2 summarize the differences for the APHIS-like process, a risk–benefit evaluation, and the precautionary approach for the three risk-management categories.

Table 2. Comparison of APHIS, risk–benefit and precautionary risk management

Questions	**APHIS-like approach**	**Risk–benefit approach**	**Precautionary approach**
1. What is the framing hypothesis?	No significant environmental hazard (the null hypothesis).	Significant environmental effect (alternative hypothesis)	Significant environmental effect (alternative hypothesis)
2. What rule is used to test the hypothesis?	Probability of rejecting null hypothesis when it is true (type-I error) is less than critical value, e.g., $\alpha = 0.05$	Power of test to correctly reject null hypothesis, (1 - type-II error) is high, e.g., 0.90.	Power of test to correctly reject null hypothesis, (1 - type-II error) is very high, e.g., 0.95.
3. What party is responsible for the burden of proof?	US Government	Shared between the US Government and entity introducing crop	Entity introducing the transgenic crop (as certified by independent party)
4. What costs are considered?	Lost production, environmental and health benefits from not releasing the crop	Lost production, environmental and health benefits from not releasing the crop	Potential ecological risks from releasing the organism outweigh economic considerations
5. What is the general rule for making release decisions?	Permit release if test to minimize type-I error at standard $\alpha =$ 0.05 level indicates no significant ecological risk, or if net benefits exceed the ecological risks/costs.	Permit release if test to minimize type-II error at high power level does not indicate significant ecological risk, or if net benefits exceed the ecological risks/costs.	Permit release if test to minimize type-II error at very high power level is passed, but avoid irreversible risks until information is available to assure adequate ecological safety.
6. Will compensation be provided to negatively affected parties?	Collect some of the net benefits to compensate 'losers' or remediate damages	Collect some of the net benefits to compensate 'losers' or remediate damages	Not applicable

There is a substantial body of science on the potential environmental effects of intragenic and famigenic GMOs or products from similar techniques. Therefore, the evaluation would be the least precautionary by controlling type-I error using the standard 95% confidence level. If no significant effects were detected, the organism would be approved for release. However, if evidence is found to support the hypothesis of a significant ecological risk, the crop would not be rejected automatically for release, but passed through a risk–benefit test. The estimated value of the ecological damages would be compared to the potential net benefits of releasing the crop, including production, human health and any positive environmental effects, such as pesticide toxicity reductions. Note the production benefits must incorporate the relevant social value of added production due to lowering the supply curve, which for the US and EU countries may be negative if excess supplies are creating deadweight losses. The decision to release is made by a comparison of the estimated ecological risks/costs against the potential social benefits. Non-market valuation methods would be applied to those ecological effects for which reliable monetary values could be estimated. It is doubtful that all effects could be reliably monetized. Thus, expert scientific and policy judgments would be necessary to compare order-of-magnitude effects and implement the decision rule.

If the estimated benefits outweigh the potential costs, then release would be permitted. To turn the cost–benefit decision rule into a real rather than potential Pareto improvement, a portion of the net benefits would be used to compensate for associated losses, such as contamination from genetic drift. The burden of proof lies with the government using information provided by the applying entity for this least precautionary category. Because novel risks are unlikely and good quality information about the ecological risks is likely available for these familiar crops, this risk assessment based on minimizing type-I error will result in more commercialization decisions than the following models. However, APHIS currently uses this probability rule to approve the field testing of approximately 99% of most transgenic crops, which would fall into the next model in our differentiated approach.

For linegenic and transgenic organisms, more stringent tests for ecological risk would be applied. Because of our relative lack of experience with these crops, they conceivably could introduce serious ecological risks. An independent scientific panel would first screen the crops for potentially serious irreversible impacts, and any such organisms would move to the precautionary risk-assessment process (model 3) with higher standards for release. For the remaining crops in this category, the framing hypothesis that the crop causes significant ecological risks would be tested to control type-II error by specifying a minimum power of the test, e.g., 90 percent ($\beta = 0.10$)[5]. Adequate ecological risk information is a prime requirement to frame and test the alternative hypotheses. However, this task presents a conundrum. Small-scale field trials before commercialization can detect order-of-magnitude differences in ecological effects, but low-probability and low-magnitude effects likely will escape detection (National Research Council NRC 2002). Evidence collected from large-scale field trials would be required. Thus, the test may have to be conducted in progressive stages of field experiments, followed by limited releases to gather sufficient data to assess all potential ecological impacts. This process would address a weakness in current ecological monitoring of GM crops (National Research Council NRC 2000, p. 19). The USDA and the entity proposing release would share responsibility in gathering the ecological risk data under scientifically certified protocols. The entity requesting permission to release could conduct the tests if the experimental design and measurement were independently certified. Alternatively, the tests may be conducted by an independent certifying body.

If the test indicates the crop does not cause significant ecological risk, its release would be permitted. Further monitoring of crops that pose unknown long-term effects would be conducted, such as cases of uncertain resistance development. Just as for intragenic and famigenic crops, if evidence is found to support the hypothesis of a significant ecological hazard, the estimated production, environmental and health benefits of the linegenic or transgenic crop would be compared to its estimated ecological damages to decide upon release. Expert and diverse scientific panels would be used to evaluate the ecological effects and their impacts because of less familiarity with these organisms than with the first category. In cases where the science and evidence are not robust, the regulating authority may choose to permit release, but require periodic review with new monitoring data to improve the analysis of risks and renew or revoke commercialization. As for the first model, some of the net benefits would be used to compensate for associated losses, such as transgenic contamination of organic fields from genetic drift.

The final model applies to transgenic crops judged to hold the potential to cause serious irreversible ecological effects and to xenogenic crops. As for linegenic and transgenic crops, the framing hypothesis is for significant environmental effects and type-II error is controlled. However, the standard for approving release of these crops is extremely high. For example, the required power level could be increased over model 2 to 95% ($\beta = 0.05$). The entity applying for release must prove beyond scientific doubt that the organism is safe. Expert scientific panels with representation from all relevant ecological sciences would be used to implement the model and make decisions concerning release. The potential social benefits would not be considered until minimum levels of safety are assured for all ecological hazards.

It is important to note that a well-designed environmental regulatory process does more than minimize the potential for unwanted environmental hazards from new technologies. If implemented properly, environmental regulations can provide incentives and disincentives to influence the research and technology development process beneficially. Under the differentiated risk-assessment framework, imposing higher regulatory costs on organisms that potentially pose higher ecological risks stimulates research and development of GM varieties with traits that provide production benefits with acceptable environmental risk and perhaps ecological benefit. To realize this outcome, a new set of bio-engineered traits would be developed and inserted into important agronomic crops: traits that are less likely, for example, to result in resistant insect populations or harm non-target organisms. Increased and targeted involvement of the public-sector agricultural research and regulatory branches also is necessary to achieve these types of outcomes (Ervin et al. 2003).

In addition, the proposed framework may shift resources toward engineering processes that have less genetic distance from conventional techniques, at least in the short to medium term. However, it would not forestall innovation and even commercialization of crops developed through novel techniques. Rather, consistent with a precautionary approach, our proposed framework brings with it additional safeguards for engineering techniques and their products about which we are less familiar and have had less experience as consumers, regulators and scientists.

Conclusions

There is a substantial need for increased public research funding on the environmental effects of transgenic crops and for research of a different character. It is natural to ask why more public research is needed when private research on transgenic crops has increased so dramatically. Under current US biosafety regulatory policies, private industries have scant incentive to invest in the research to understand the environmental impacts of transgenic crops, especially the ecosystem effects beyond the farm boundary. Most environmental risks stem from missing markets; there are few or no market incentives for reducing the environmental risks of transgenic crops. Thus, private research to control the full range of negatively affected environmental services will not be triggered by current market and regulatory signals (Batie and Ervin 2001).

Evidence in support of this argument is provided by the recent decision by Pioneer Hi-Bred and Dow AgroEvo to deny access to the proprietary materials required by independent scientists to conduct biosafety analysis of *Bt* sunflower (Dalton 2002). A decision made even more problematic by the fact that it was made after the firms initially co-operated with Snow and colleagues. Permission to access the material was withdrawn only after the scientists' preliminary findings indicated potential biosafety risks from *Bt* sunflower (Snow 2002; Snow et al. 2003).

Likewise, it is unrealistic to assume that most private firms will develop transgenic crops that provide ecological benefits and minimize potential risks in line with social preferences. The development of such crops would suffer from the same missing-markets dilemma since the environmental benefits would not merely accrue to the farmer that purchased the transgenic seed. Rather, other farmers, the general public, and even future generations would enjoy the benefits from such crops. For example, a vehicle for addressing many of the identified potential risks from insect-pest-resistant crops is to develop crops that are pest-damage-tolerant rather than toxic to the pest, as are *Bt* crops (Hubbell and Welsh 1998; Pedigo 2002). The difference between tolerance of damage and resistance to pests is fundamental. Tolerance does not rely on toxicity to kill pests and therefore does not negatively impact non-target organisms or promote resistance development (Welsh et al. 2002). Pedigo (2002) finds that certain crops display tolerance to pest damage. This characteristic has been used commercially with great success for decades with no public controversy. For example, cucumbers with stable tolerance to *Cucumber mosaic virus* have dominated the industry since the 1960s. Genetic modification could be used to amplify these types of properties in several other important crops (Pedigo 2002). The publicness of the environmental benefits potentially derived from such crops dampens private-sector enthusiasm to develop and commercialize them[6].

However, if regulatory polices effectively control type-II error for transgenic crops, the private sector would receive signals and incentives to assess environmental risks more fully and to develop crops that cause less risk while providing production, health and other potential market benefits. If, for example, governments assigned liability for the deleterious environmental effects to the biotechnology company, perhaps through the posting of a significant bond upon commercialization, more private R&D resources would likely be devoted to controlling adverse effects either through risk-assessment research or developing technologies such as damage-tolerant crops. In essence, this approach forces firms to take into account the shadow price of environmental risks when making decisions about attempting to commercialize a transgenic technology or investing in the development of crops with particular sets of characteristics or traits.

Acknowledgments

Senior authorship is shared equally. The authors are grateful for the helpful comments and suggestions by Willem Stiekema and other Frontis workshop participants, and an anonymous reviewer. We also express our sincere appreciation to Elizabeth Minor for her excellent editorial assistance in preparing the paper.

References

Adam, D., 2003. Transgenic crop trial's gene flow turns weeds into wimps. *Nature,* 421 (6922), 462.

Andow, D.A., 2003. UK farm-scale evaluations of transgenic herbicide tolerant crops. *Nature Biotechnology,* 21 (12), 1453-1454.

Antle, J.M. and Capalbo, S.M., 1998. Quantifying agriculture environment tradeoffs to assess environmental impacts of domestic and trade policies. *In:* Antle, J.M., Lekakis, J.N. and Zanias, G.P. eds. *Agriculture, trade and the environment: the impact of liberalization on sustainable development.* Edward Elgar, Cheltenham, 25-51.

Batie, S.S. and Ervin, D.E., 2001. Transgenic crops and the environment: missing markets and public roles. *Environment and Development Economics,* 6 (4), 435-457.

Beringer, J.E., 1999. Cautionary tale on safety of GM crops. *Nature,* 399 (6735), 405.

Birch, A.N.E., Geoghegan, I.E., Majerus, M.E.N., et al., 1997. Interactions between plant resistance genes, pest aphid populations and beneficial aphid predators. *In:* Scottish Crop Research Institute ed. *Annual Report 1996/97.* Scottish Crop Research Institute, Dundee, 68-72.

Bishop, R.C. and Scott, A., 1999. The safe minimum standard of conservation and environmental economics. *Aestimum,* 37, 11-40.

Buhl-Mortensen, L., 1996. Type-II statistical errors in environmental science and the Precautionary Principle. *Marine Pollution Bulletin,* 32 (7), 528-531.

Carlson, G.A., Marra, M.C. and Hubbell, B.J., 1998. Yield, insecticide use, and profit changes from adoption of *Bt* cotton in the Southeast. *In: Proceedings Beltwide Cotton Conferences 2, San Diego, California, USA, 5-9 January 1998.* 973-974.

Carrière, Y., Ellers-Kirk, C., Sisterson, M., et al., 2003. Long-term regional suppression of pink bollworm by *Bacillus thuringiensis* cotton. *Proceedings of the National Academy of Sciences of the United States of America,* 100 (4), 1519-1523.

Dale, P.J., Clarke, B. and Fontes, E.M.G., 2002. Potential for the environmental impact of transgenic crops. *Nature Biotechnology,* 20 (6), 567-574.

Dalton, R., 2002. Superweed study falters as seed firms deny access to transgene. *Nature,* 419 (6908), 655.

Economic Research Service USDA, 1999a. *Genetically engineered crops for pest management.* Economic Research Service, US Department of Agriculture, Washington. [http://www.biotechknowledge.com/biotech/knowcenter.nsf/ID/C217513C7EDCB48886256AFB0067D273]

Economic Research Service USDA, 1999b. *Impacts of adopting genetically engineered crops in the United States.* Economic Research Service, US Department of Agriculture, Washington. [http://www.ers.usda.gov/emphases/harmony/issues/genengcrops/genengcrops.htm]

EDEN Bioscience Corporation, 2002a. *What is Harpin?* Available: [http://www.edenbio.com/newdocuments/wp_1.pdf] (26 June 2003).

EDEN Bioscience Corporation, 2002b. *What is Messenger?* Available: [http://www.edenbio.com/documents/TechBul.pdf] (26 June 2003).

Ellstrand, N.C., 2001. When transgenes wander, should we worry? *Plant Physiology,* 125 (4), 1543-1545.

Ervin, D.E., Welsh, R., Batie, S.S., et al., 2001. *Public research for environmental regulation of transgenic crops*. Unpublished paper. Environmental Sciences and Resources Program, Portland State University, Portland. [http://www.esr.pdx.edu/ESR/docs/papers/PRofRTC.pdf]

Ervin, D.E., Welsh, R., Batie, S.S., et al., 2003. Towards an ecological systems approach in public research for environmental regulation of transgenic crops. *Agriculture, Ecosystems and Environment,* 99 (1/3), 1-14.

Fernandez-Cornejo, J. and McBride, W.D., 2002. *Adoption of bioengineered crops*. Economic Research Service, US Department of Agriculture, Washington. Agricultural Economic Report no. 810. [http://www.ers.usda.gov/publications/aer810/aer810.pdf]

Gould, F., 2003. Bt-resistance management-theory meets data. *Nature Biotechnology,* 21 (12), 1450 - 1451.

Gould, F., Blair, N., Reid, M., et al., 2002. *Bacillus thuringiensis*-toxin resistance management: stable isotope assessment of alternate host use by Helicoverpa zea. *Proceedings of the National Academy of Sciences of the United States of America,* 99 (26), 16581-16586.

Greene, A.E. and Allison, R.F., 1994. Recombination between viral RNA and transgenic plant transcripts. *Science,* 263 (5152), 1423-1425.

Hails, R.S., 2000. Genetically modified plants: the debate continues. *Trends in Ecology and Evolution,* 15 (1), 14-18.

Hall, L.M., Huffman, J. and Topinka, K., 2000. *Pollen flow between herbicide tolerant canola (Brassica napus) is the cause of multiple resistant canola volunteers*. Weed Science Society of America Abstracts no. 40.

Hansen-Jesse, L.C. and Obrycki, J.J., 2000. Field deposition of *Bt* transgenic corn pollen: lethal effects on the monarch butterfly. *Oecologia,* 125 (2), 241-248.

Heimlich, R.E., Fernandez-Cornejo, J., McBride, W., et al., 2000. Adoption of genetically engineered seed in U.S. agriculture: implications for pesticide use. *In:* Fairbairn, C., Scoles, G. and McHughen, A. eds. *Proceedings of the 6th international symposium on the biosafety of genetically modified organisms*. University of Saskatchewan, University Extension Press, Saskatoon. [http://www.ag.usask.ca/isbr/Symposium/Proceedings/Section0.htm]

Hilbeck, A., Baumgartner, M., Fried, P.M., et al., 1998a. Effects of transgenic *Bacillus thuringiensis* corn-fed prey on mortality and development time of immature *Chrysoperla carnea* (Neuroptera: Chrysopidae). *Environmental Entomology,* 27 (2), 480-487.

Hilbeck, A., Moar, W.J., Pusztai-Carey, M., et al., 1998b. Toxicity of *Bacillus thuringiensis* Cry1AB toxin to the predator *Chrysoperla Carnea* (Neuroptera: Chrysopidae). *Environmental Entomology,* 27 (5), 1255-1263.

Hoy, C.W., Feldman, J., Gould, F., et al., 1998. Naturally occurring biological controls in genetically engineered crops. *In:* Barbosa, P. ed. *Conservation biological control*. Academic Press, New York, 185-205.

Hubbell, B.J., Marra, M.C. and Carlson, G.A., 2000. Estimating the demand for a new technology: *Bt* cotton and insecticide policies. *American Journal of Agricultural Economics,* 82 (1), 118-132.

Hubbell, B.J. and Welsh, R., 1998. Transgenic crops: engineering a more sustainable agriculture? *Agriculture and Human Values,* 15 (1), 43-56.

Jasanoff, S., 2000. Between risk and precaution: reassessing the future of GM crops. *Journal of Risk Research,* 3 (3), 227-282.

Keeler, K.H., Turner, C.E. and Bolick, M.R., 1996. Movement of crop transgenes into wild plants. *In:* Duke, S.O. ed. *Herbicide resistant crops: agricultural, environmental, economic, regulatory, and technical aspects.* CRC, Boca Raton, 303-330.

Kessler, C. and Economidis, I., 2001. *EC-sponsored research on safety of genetically modified organisms: a review of results.* European Commission, Brussels. [http://europa.eu.int/comm/research/quality-of-life/gmo/index.html]

Knudsen, O.K. and Scandizzo, P.L., 2001. Evaluating risks of biotechnology: the Precautionary Principle and the social standard. *In: 5th international conference on biotechnology, science and modern agriculture: a new industry at the dawn of the century, Ravello, Italy, June 15-18, 2001.* International Consortium on Agricultural Biotechnology Research ICABR, Ravello. [http://www.economia.uniroma2.it/conferenze/icabr01/nontechabsrtact2001/Scandizzo.htm]

Lemons, J., Shrader-Frechette, K. and Cranor, C., 1997. The Precautionary Principle: scientific uncertainty and type I and type II errors. *Foundations of Science,* 2 (2), 207-236.

Losey, J.E., Raynor, L.S. and Carter, M.E., 1999. Transgenic pollen harms monarch larvae. *Nature,* 399 (6733), 214.

McCammon, S.L., 2001. *APHIS' review of biotechnology products.* Animal and Plant Health Inspection Service, US Department of Agriculture, Washington. [http://www.usda.gov/gipsa/psp/issues/millennium/mccammon.htm]

Mooney, S. and Klein, K., 1999. Environmental concerns and risks of genetically modified crops: economic contributions to the debate. *Canadian Journal of Agricultural Economics,* 47 (4), 437-444.

Morin, S., Biggs, R.W., Sisterson, M.S., et al., 2003. Three cadherin alleles associated with resistance to *Bacillus thuringiensis* in pink bollworm. *Proceedings of the National Academy of Sciences of the United States of America,* 100 (9), 5004-5009.

NASS, USDA, 2003. *Prospective plantings, 03.31.03.* National Agricultural Statistics Service, Agricultural Statistics Board, US Department of Agriculture, Washington. [http://usda.mannlib.cornell.edu/reports/nassr/field/pcp-bbp/pspl0303.pdf]

National Research Council NRC, 2000. *Ecological monitoring of genetically modified crops: a workshop.* National Academy Press, Washington.

National Research Council NRC, 2002. *Environmental effects of transgenic plants: the scope and adequacy of regulation.* National Academy Press, Washington.

Nielsen, K.M., 2003. Transgenic organisms: time for conceptual diversification? *Nature Biotechnology,* 21 (3), 227-228.

Owen, M., 1997. *North American developments in herbicide tolerant crops.* Paper presented at the British crop protection conference, Brighton, England. [http://www.weeds.iastate.edu/weednews/Brighton.htm]

Pedigo, L.P., 2002. *Entomology and pest management.* 4th edn. Prentice Hall, Upper Saddle River.

Pham-Delègue, M.H., Wadhams, L.J., Gatehouse, A.M.R., et al., 2000. Environmental impact of transgenic plants on beneficial insects. *In:* Kessler, C. and Economidis, I. eds. *EC-sponsored research on safety of genetically modified organisms: a review of results.* European Commission, Brussels. [http://europa.eu.int/comm/research/quality-of-life/gmo/01-plants/01-07-project.html]

Rissler, J. and Mellon, M., 1996. *The ecological risks of engineered crops.* MIT Press, Cambridge.

Royal Society, 1998. *Genetically modified plants for food use.* The Royal Society, London. [http://www.royalsoc.ac.uk/files/statfiles/document-56.pdf]

Royal Society of Canada, 2001. *Elements of precaution: recommendations for the regulation of food biotechnology in Canada.* The Royal Society of Canada, Ottawa. [http://www.rsc.ca/foodbiotechnology/GMreportEN.pdf]

Saxena, D. and Stotzky, G., 2000. Insecticidal toxin is released from roots of transgenic *Bt* corn in vitro and in situ. *FEMS Microbiology Ecology,* 33 (1), 35-39.

Sayyed, A.H., Cerda, H. and Wright, D.J., 2003. Could *Bt* transgenic crops have nutritionally favourable effects on resistant insects? *Ecology Letters,* 6 (3), 167-169.

Sears, M.K., Hellmich, R.L., Stanley-Horn, D.E., et al., 2001. Impact of *Bt* corn pollen on monarch butterfly populations: a risk assessment. *Proceedings of the National Academy of Sciences of the United States of America,* 98 (21), 11937-11942.

Snow, A., 2002. Transgenic crops: why gene flow matters. *Nature Biotechnology,* 20 (6), 542.

Snow, A., 2003. Genetic engineering: unnatural selection. *Nature,* 424 (6949), 619.

Snow, A., Pilson, D., Rieseberg, L.H., et al., 2003. A *Bt* transgene reduces herbivory and enhances fecundity in wild sunflowers. *Ecological Applications,* 13 (2), 279-286.

Snow, A.A. and Palma, P.M., 1997. Commercialization of transgenic plants: potential ecological risks. *BioScience,* 47 (2), 86-96.

Stanley-Horn, D.E., Dively, G.P., Hellmich, R.L., et al., 2001. Assessing the impact of Cry1Ab-expressing corn pollen on monarch butterfly larvae in field studies. *Proceedings of the National Academy of Sciences of the United States of America,* 98 (21), 11931-11936.

Tabashnik, B.E., 1994. Evolution of resistance to *Bacillus thuringiensis. Annual Review of Entomology,* 39, 47-79.

Tabashnik, B.E., Dennehy, T.J., Carrière, Y., et al., 2001. Resistance management: slowing pest adaptation to transgenic crops. *Acta Agriculturae Scandinavica B,* 53 (Suppl. 1), 57-59.

Tabashnik, B.E., Patin, A.L., Dennehy, T.J., et al., 2000. Frequency of resistance to *Bacillus thuringiensis* in field populations of pink bollworm. *Proceedings of the National Academy of Sciences of the United States of America,* 97 (24), 12980–12984.

Taylor, M.R. and Tick, J.S., 2003. *Postmarket oversight of biotech foods: is the system prepared?* Pew Initiative on Food and Biotechnology, Washington.

Van den Belt, H., 2003. Debating the Precautionary Principle: "guilty until proven innocent" or "innocent until proven guilty"? *Plant Physiology,* 132 (3), 1122-1126.

VanGessel, M.J., 2001. Glyphosphate-resistant horseweed from Delaware. *Weed Science,* 49 (6), 703-705.

Watkinson, A.R., Freckleton, R.P., Robinson, R.A., et al., 2000. Predictions of biodiversity response to genetically modified herbicide-tolerant crops. *Science,* 289 (5484), 1554-1557.

Watrud, L.S. and Seidler, R.J., 1998. Nontarget ecological effects of plant, microbial, and chemical introductions to terrestrial systems. *In:* Huang, P.M. ed. *Soil chemistry and ecosystem health.* Soil Science Society of America, Madison, 313-340.

Wei, Z.M., Laby, R.J., Zumoff, C.H., et al., 1992. Harpin elicitor of the hypersensitive response produced by the plant pathogen *Erwinia amylovora. Science,* 257 (5066), 85-88.

Welsh, R., Hubbell, B., Ervin, D., et al., 2002. GM crops and the pesticide paradigm. *Nature Biotechnology,* 20 (6), 548-549.

Wolfenbarger, L.L. and Phifer, P.R., 2000. Biotechnology and ecology: the ecological risks and benefits of genetically engineered plants. *Science,* 290 (5499), 2088-2093.

Zhaio, J.Z., Cao, J., Li, Y., et al., 2003. Transgenic plants expressing two *Bacillus thuringiensis* toxins delay insect resistance evolution. *Nature Biotechnology,* 21 (12), 1493-1497.

---

[1] Risk is used here to convey the combination of the probability of occurrence of some environmental hazard and the harm associated with that hazard (National Research Council NRC 2002, p. 54). This section updates a similar review in a forthcoming article in *Agriculture, Ecosystems and Environment* (Ervin et al. 2003). For more complete discussions of the risks, readers may consult Ervin et al. 2001, Wolfenbarger and Phifer 2000, National Research Council NRC 2002, and Royal Society of Canada 2001.

[2] The National Research Council (2002) defines four categories of hazards: (1) resistance evolution; (2) movement of genes; (3) whole plants; and (4) non-target effects. Categories (2) and (3) are combined here.

[3] We assume for ease of exposition that the environmental hazards can be described by reliable probability distributions to illustrate the differences in the tests. For some hazards, probability distributions cannot be estimated, either objectively or subjectively, i.e., true uncertainty exists. For those cases, nonparametric analytical techniques must be used. For still other hazards, the outcome space is undefined, i.e., surprises may occur, and game-theory techniques can be used (Bishop and Scott 1999).

[4] Because transgenic crops vary considerably in traits and biological makeup, a disaggregated taxonomy of transgenic crops would improve the analytical power of the differentiated assessment.

[5] Knudsen and Scandizzo (2001) offer a similar approach by reversing the null hypothesis from one of no significant effect to one of presuming a significant effect.

[6] Some new products follow this approach, but they are clearly the exception rather than the rule. The biopesticide *Messenger* is an interesting example of this approach to technology development. *Messenger* is a biopesticide that acts as a non-hormone growth regulator for a wide variety of plants. The active ingredient is the naturally occurring Harpin bacterial protein. Applying *Messenger* topically, as a spray, essentially signals plants to activate their natural plant defenses against a variety of diseases and boosts plant development and growth. The primary mode of action is not toxicity to the invading pathogen, but rather entrapment or the creation of a physical barrier to the movement of the pathogen through localized cell death (see EDEN Bioscience Corporation 2002a, 2002b; Wei et al. 1992).

2b

Comment on Ervin and Welsh: Environmental effects of genetically modified crops: differentiated risk assessment and management

Willem J. Stiekema[#]

The main statement of this paper is the plea to use genetic distance between donor and receiver as criterion for assessing the novelty of the genetic changes and thus the potential risk of such a modification. This so-called differentiated approach leads the authors to the conclusion that species-foreign and synthetic genes potentially pose more risk to the environment compared to genes coming from the same species. This point of view is debatable because of a number of observations:

1. Small changes in genes can lead to major changes in phenotype and possible effects on the environment. An example is the presence of an active or inactive pathogen-resistance gene.
2. Introduction of modified transcription factors of the same species may result in changing whole sets of genes, which may result in the production of a plethora of unexpected metabolites that may harm the environment.
3. After 6 years of growing crops in the US that contain the bacterial *Bt*-toxin gene, *Bt*-resistant insects have not been detected in the fields in which those crops grow.
4. Plant genes conferring resistance to pathogens are often present in multiple copies in selected regions of the genome. In certain cases it has been shown that those multiple copies have risen by recombination between two of these genes, resulting in novel additional genes sometimes having an effect on a different plant pathogen.

Thus, a decrease in genetic distance does not necessarily decrease the potential risk. A case-by-case approach is preferable, bearing in mind that by doing so a knowledge database can be built containing data of crop–transgene combinations that can be used to formulate efficiently the right questions to determine the potential risk of each new introduction of a genetically modified crop into the environment. The experience gained may be used to set up transparent assessment procedures.

The authors suggest that information on the ecological risks of the introduction of a certain transgenic crop can be obtained by a step-by-step approach. This is a good approach except that it will be very difficult to determine how many steps have to be taken for how many years and in how many fields with how many different types of soils and under how many different conditions. Current practices used for registration of new plant varieties may be of help here. Also the input of knowledge of breeders and farmers in addition to that of ecologists is essential.

[#] Wageningen UR, Plant Research International, Wageningen, The Netherlands. E-mail: willem.stiekema@wur.nl

J. H. H. Wesseler (ed.), Environmental Costs and Benefits of Transgenic Crops, 31.

3a

Assessing the environmental impact of changes in pesticide use on transgenic crops

Gijs A. Kleter[#] *and Harry A. Kuiper*[#]

Abstract

Two main traits that have been introduced into genetically modified crops that are currently on the market, viz., herbicide and insect resistance, likely affect pesticide use on these crops. Various surveys have been carried out, such as those of the USDA-ERS and NCFAP, comparing the pesticide use on genetically modified versus conventional crops. Environmental indicators for pesticides may aid in comparing the outcomes of such assays in terms of environmental impact. Previously we applied one indicator, the Environmental Impact Quotient, to pesticide-use data for commercial biotech crops from a recent survey by NCFAP and found that, by this method, the impact paralleled the decreased use of pesticides. The output of many environmental indicators, while lending themselves to comparison of pesticides, is abstract and there may be a need for specific indicators that lend themselves for comparison with other agricultural factors or that are expressed in more tangible terms, e.g., monetary indicators. IUPAC recently initiated a project on the assessment of the environmental impact of altered pesticide use on transgenic crops, with the aim of providing input for risk–benefit analysis of the adoption of genetically modified crops. In conclusion, the use of appropriate environmental indicators enables the assessment of the economic and environmental effects of agricultural biotechnology, including that of altered pesticide use.

Keywords: plant biotechnology; genetically modified crops; pesticides; environmental impact; pesticide-use surveys; environmental indicators; risk-benefit analysis

Introduction

Modern biotechnology has enabled the transfer of genes from biologically unrelated species, opening up avenues for genetic modifications that were hitherto inaccessible by conventional methods of genetic amelioration. The large-scale commercial cultivation of genetically modified (GM) crops started in 1996 and since then has increased at a rapid pace in terms of cultivated acreage, amounting to a global 58.7 million hectares in the year 2002, an area somewhat larger than the total area of France (James 2002). Main countries for GM crop cultivation are the United States of America (US), Argentina, Canada and China.

The most important GM crops are soybean, maize, cotton and canola, while the most important traits that have been conferred by genetic modification are herbicide

[#] RIKILT Institute for Food Safety, PO Box 230, 6700 AE Wageningen, The Netherlands. E-mail: gijs.kleter@wur.nl

J. H. H. Wesseler (ed.), Environmental Costs and Benefits of Transgenic Crops, 33–43.

tolerance and insect resistance. Herbicide tolerance allows for over-the-top application of weed-control agents (herbicides), which would otherwise harm the crop, hence facilitating weed management. Many insect-resistant crops have been modified to express small amounts of insecticidal proteins (*Bt*) from the soil bacterium *Bacillus thuringiensis*, which itself has been used over decades as a biological insect-control agent. The presence of the *Bt* proteins thus substitutes for the application of insecticides aimed at the insect pests against which these proteins afford protection.

Given the fact that both herbicide tolerance and insect resistance deal with pest management, and, in particular, with the way that pesticides are used in crops, changes in the types and amounts of pesticides that are used on GM crops can be anticipated. For herbicide-tolerant crops, for example, a shift towards the herbicides that can be used on the herbicide-tolerant GM crop is likely to occur. An example is provided by the pesticide use on soybeans in the US from 1995 till 2002. As shown in Figure 1, the fraction of GM soybeans (predominantly herbicide-tolerant) among the American soybean acreage has steadily increased from 1996 onwards, accounting for 75% of soybean acreage in 2002. Figure 2 shows that the percentage of soybean acreage to which glyphosate is applied has also increased, whereas that of other herbicides has decreased. In terms of percentage of the total amounts of pesticides used, glyphosate has also expanded in these years (Figure 3). It is likely that the adoption of glyphosate-resistant GM soybeans has contributed to this enlarged market share of the glyphosate herbicide. In this paper, we wish to elaborate on the issue of the changed pesticide use on GM crops and the potential environmental implications of this change.

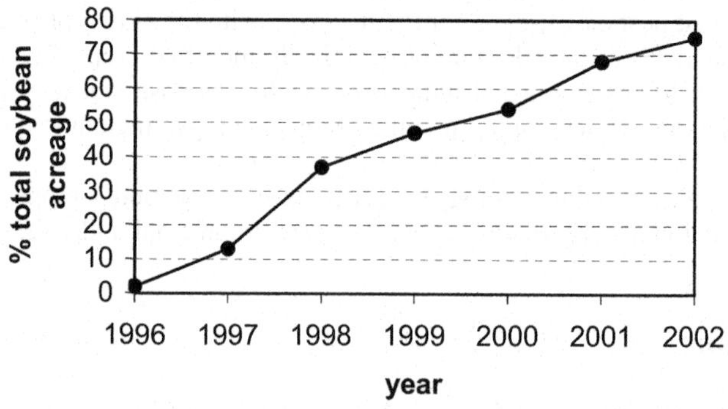

Figure 1. Adoption of GM soybeans in the US, percent of total acreage (data from James (2001) and NASS (2003b))

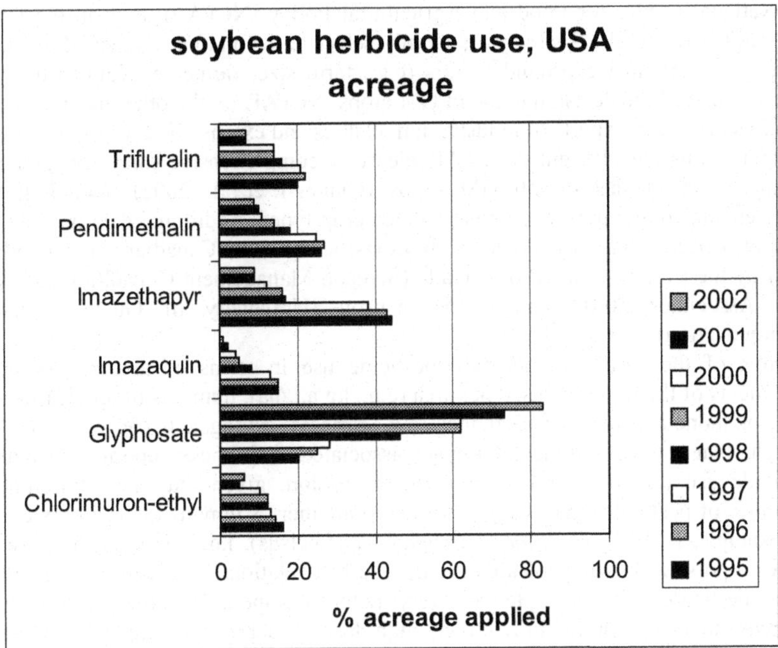

Figure 2. Use of selected herbicides on soybeans in the USA, percent of total acreage (data from NASS (2003a), herbicides selected with minimally 10% acreage in 1995)

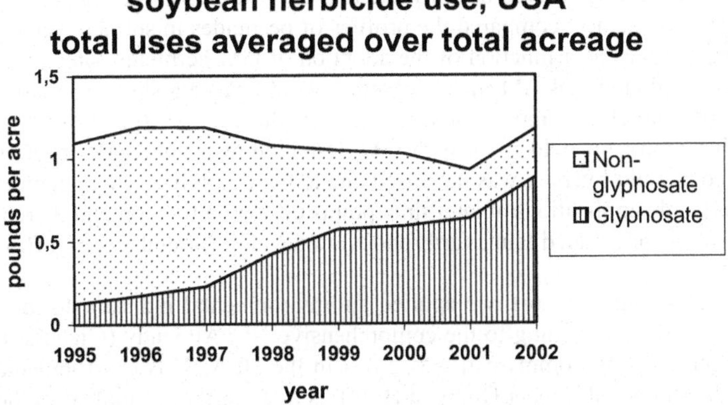

Figure 3. Herbicide use on soybeans in the USA, average active ingredient per area, 1995-2002 (data from NASS (2003a))

Reports on pesticide use on GM crops

The topic of pesticide-use surveys in GM-crop cultivation has been more extensively reviewed in our previous report (Kleter and Kuiper 2003). Organizations that periodically survey the use of pesticides in GM crops include the Economic Research Service of the United States Department of Agriculture (USDA-ERS) and the National Center for Food and Agricultural Policy (NCFAP), both in the US. In general, the USDA-ERS surveys randomly select farms and use statistical models to correct for certain background factors (e.g., farm size, farmer education) that may have influenced the decision to adopt GM crops. NCFAP, on the other hand, generally employs data from experts of industry, universities and extension services on pesticide use and alternatives. Recently, NCFAP released a comprehensive survey of (potential) savings of pesticides on 40 GM crops (Gianessi et al. 2002), including both commercialized and not yet commercialized crop types. Additional data on this issue have also been released by other institutions, such as on Canadian canola (oilseed rape) by the Canola Council of Canada (Serecon Management Consulting and Koch Paul Associates 2001) and G. Stephenson (University of Guelph; personal communication).

Most of these surveys compare pesticide use in terms of amounts of active ingredients of the pesticide used per area (e.g., kg a.i./ha), numbers of applications per crop, or monetary costs of pesticides (e.g., farmers' savings). It can be envisioned, though, that the environmental impact associated with these reported changes in pesticide use do not correlate in a linear fashion, given that the environmental attributes of pesticides may vary from one to another. A famous quote in toxicology, for example, is "*sola dosis facit venenum*" (Paracelsus), i.e. it is only the dose that makes a poison, pointing towards the dose–effect relationships of toxic compounds. Some pesticides, for example, are very potent compounds acting at low doses, whereas others are applied in relatively high doses to achieve a comparable degree of pest control. Also, in their toxicological action on non-target organisms and their environmental behaviour, pesticides may differ widely from each other. The survey outcomes in terms of amounts of active ingredients per area may therefore not reflect the true environmental effect of the changed pesticide use on GM crops.

A limited number of reports have appeared dealing with the environmental impacts associated with the altered use of pesticides on GM crops. Wauchope and co-workers (2002), for instance, compared the profiles of pesticides in surface water in American watershed areas as a function of the adoption of GM herbicide-tolerant maize. These authors predicted that GM-maize adoption would cause a shift to low or no residues of comparatively benign herbicides used on these crops (glyphosate, glufosinate). Another report on field trials with GM herbicide-resistant fodder beets in Denmark showed that biodiversity was increased due to the flexibility in weed control, allowing for herbicide applications at later time points during cultivation and thereby for weeds to grow and associated fauna (arthropods) to develop (Strandberg and Bruus Pedersen 2002).

In our previous report, we applied an environmental indicator to the amounts of pesticides that, according to the comprehensive NCFAP study (Gianessi et al. 2002), had been used on commercial GM crops in the USA. This environmental indicator, the Environmental Impact Quotient (EIQ), is a factor that is applied to the amount of active ingredient used per acre, yielding outcomes that are predictive of the impacts of different pesticides. These outcomes allow for comparison of these pesticides with each other on their environmental effects. Below we will discuss indicators in more detail. The data calculated thus show that the predicted reduction in environmental impact of the pesticides used on GM crops more or less paralleled the reduction in the active amounts of ingredients per area (Table 1, Kleter and Kuiper 2003).

Table 1. Changes in amounts versus environmental impact of pesticides on GM crops [1]

Crop	Difference transgenic versus non-transgenic, %		Method of comparison
	Pesticide use, lbs a.i./A	Environmental impact, EI/A	
Canola, HR	-55	-48	transgenic versus alternative spray programmes for non-transgenic canola
Cotton, HR	-18	-11	Year-2000 figures (transgenic) versus those of base year (non transgenic)
Cotton, IR	-78	-78	statewise determined number of sprays to control pink bollworm / bollworm (non transgenic) times the average amount of pesticides per spray
Maize, HR	-30	-32	substitution scenarios for spray programmes
Maize, IR	-100	-100	no pesticides (transgenic) versus average pesticide use to control European corn borer (non transgenic)
Soybean, HR	-38	-25	glyphosate (transgenic) versus statewise differentiated, equally effective programmes

[1] adapted from Kleter and Kuiper (2003), pesticide use and method of comparison from Gianessi et al. (2002)

Indicators for the environmental impact of pesticides

As mentioned above, an environmental indicator, the EIQ, has been used in our previous work to predict the impact of alterations in pesticide use on GM crops. Whereas some indicators may serve very specific purposes, e.g. avoidance of toxicity to non-target organisms (honeybees), drinking-water contamination, or local environmental impact, the EIQ is a comparatively general indicator. The EIQ, for instance, incorporates the impacts on farm worker (applicator and harvest worker), consumer and ecology (non-target organisms: fish, birds, honeybees and other beneficial insects) (Kovach et al. 1992). Calculation of these separate impacts, which are united in the final EIQ, is based on inherent properties of a given pesticide, e.g. toxicity towards certain organisms and exposure of these organisms due to environmental behaviour (among others: residues on plants, binding to soil, leaching). These inherent properties are assigned ratings, i.e. from 1 to 3 or 1 to 5, with increasing degree of toxicity or harmfulness, based on whether they are within or beyond predefined boundary values (Table 2). Given that zero is not included in the rating, environmentally neutral or benign substances would still attain a non-zero minimum score, which limits the distinction between these and more hazardous substances (Levitan, Merwin and Kovach 1995). The EIQ (total, farm worker, consumer or ecology) is multiplied with the amount of active ingredient used per area (e.g. pounds per acre) to obtain the field-use rating (EI/A). The field ratings for different pesticides lend themselves for comparison of the environmental impact. An example of this is shown in Table 3, where the EI/As, both composed and for subcomponents (e.g. farm worker), for glyphosate use and use of an alternative herbicide mix on GM soybeans in South Carolina are compared. Input data on the amounts of active ingredients have been obtained from Gianessi et al. (2002), who compares the application rate of glyphosate with that of alternative mixes selected for each US state that are equally effective for weed control. The values of the alternative mix in South Carolina are the lowest obtained, which indicates that glyphosate treatment may not always be optimal in terms of calculated environmental impact. However, overall comparison of the total EI/A and its individual components for glyphosate versus alternative mixes in all states of the USA yields favourable figures for the use of glyphosate. Most of the improvement by glyphosate use is to the benefit of the consumer and farm-worker components of the EI/A, which are almost reduced by half (Table 4).

In addition, as noted above, the outcomes of these calculations show that the decrease in environmental impact of pesticides used on GM crops parallels the decrease in amounts of pesticides per standard area used on these crops. In other words, the replacement pesticides do not show particularly more benign environmental properties, which in theory would have decreased the impact of these pesticides beyond the observed reduction in applied amounts. This is also evident from the similarity of the apparent EIQs, i.e. the weighted, aggregate EIQs for the pesticides in both groups that underlie the comparison (Table 5).

In addition to the EIQ, a number of other indicators have also been developed for predicting pesticide impact on the environment. These indicators have been reviewed by several authors (Levitan, Merwin and Kovach 1995; Reus and Middleton 1999) and may serve various purposes, such as a decision-making tool for farmers and extension workers, or a guide for policy decisions by governments. In addition, the input required for performing the calculation of the environmental impact may differ, such as the environmental diffusion of pesticides, toxicity of pesticides to humans or wildlife, and the actual environmental exposure to these pesticides. The EU-sponsored project CAPER, for example, reviewed and compared eight European environmental indicators for pesticide use, i.e. the Environmental Yardstick, the Hasse Diagram, SYNOPS, Ipest, p-EMA, EPRIP, SyPep and PERI. While all these indicators include water contamination by pesticides, they incorporate other environmental compartments and organisms in varying ways. In addition, some of these indicators take actual exposure into account (Environmental Yardstick, SYNOPS, SyPep, EPRIP), while others focus only on inherent pesticide properties (p-EMA, PERI) and the remaining indicators rank intermediate between these two groups (Hasse Diagram, Ipest) (Reus and Middleton 1999).

It should be noted that the outcomes of environmental-impact calculations, such as the one in the example described above, yield abstract numbers, which allow for comparison of pesticides with each other. However, this may not allow for comparison with other non-pesticidal impacts within the same agricultural systems, or may not be tangible for other stakeholders' understanding. One alternative that both provides an avenue for comparison with other impacts and yields outcomes that are tangible for other stakeholders, is to calculate the environmental impact in monetary terms. In their review of pesticide environmental indicators, Levitan, Merwin and Kovach (1995) discerned three types of monetary indicators for environmental impacts of pesticides:

- Single-index systems, where monetary costs, environmental impacts and other scores are added into a composite score.
- Imputing monetary values to environmental impacts, for example as costs for remediation or the costs that farmers would be willing to pay for avoiding these risks.
- Separate indices for environmental and economic impacts, by using an xy-graph with each impact index on one axis.

In addition, it may also be interesting to discern reversible and irreversible effects, both in economic and (eco)toxicological terms, especially with a view on long-term effects of GM-crop adoption. To our knowledge, such a distinction in economic terms would still need to be made, while in some environmental indicators, the toxicological parameter of chronic toxicity to humans and animals has been incorporated into the risk calculation, although not outstanding as a separate feature of the outcome.

Table 2. The EIQ equation

Component	Equation	Input variables (ratings)
Farm worker	C(DT*5)+(DT*P)	C = chronic toxicity (1-3-5) DT = dermal toxicity (1-3-5) P = plant surface residue half-life (1-3-5)
Consumer	(C*(S+P)/2*SY)+(L)	C = chronic toxicity (1-3-5) S = soil half-life (1-3-5) P = plant surface residue half-life (1-3-5) SY = systemicity (1-2-3) L = leaching potential (1-2-3)
Ecology (fish, birds, honeybees, other beneficial insects)	(F*R)+(D*(S+P)/2*3)+(Z*P*3) + (B*P*5)	F = fish toxicity (1-2-3) R = surface loss potential (1-3-5) D = bird toxicity (1-3-5) S = soil half-life (1-3-5) P = plant-surface residue half-life (1-3-5) Z = bee toxicity (1-3-5) B = beneficial arthropod toxicity (1-3-5)
Total	(Farmworker + Consumer + Ecology)/3 = {[C(DT*5)+(DT*P)] + [(C*(S+P)/2*SY)+(L)] + [(F*R)+(D*(S+P)/2*3)+(Z*P*3)+(B*P*5)]}/3	
Field-use rating (EI/A)	EIQ * %active ingredient * rate (lbs/A)	

Reference: Kovach et al. (1992)

Table 3. Comparison of the calculated environmental impact of herbicide regimes on GM soybean in South Carolina [1]

Herbicide		Rate lbs/A	Environmental impact							
Brand	Ingredient		Farm worker		Consumer		Ecology		Total	
			EIQ	EI/A	EIQ	EI/A	EIQ	EI/A	EIQ	EI/A
Glyphosate										
Roundup	glyphosate	0,95	16,0	15,20	7,0	6,65	74,3	70,59	32,4	30,78
Alternative herbicide mix										
Classic	chlorimuron	0,01	13,3	0,13	10,0	0,10	69,9	0,70	31,1	0,31
First Rate	cloransulam[2]	0,016	21,2	0,34	7,8	0,12	57,2	0,92	28,7	0,46
Assure-II	quizalofop	0,1	17,6	1,76	7,6	0,76	129,9	12,99	51,7	5,17
			Total	2,23	Total	0,98	Total	14,60	Total	5,94

EIQ = environmental-impact quotient;
EI/A = field-use rating of environmental-impact quotient

[1] Calculations as applied by Kleter and Kuiper (2003), based on pesticide-use data from Gianessi et al. (2002)

[2] No EIQ known for this herbicide, average EIQ for herbicide class used

Table 4. Calculated environmental impacts of glyphosate and alternative herbicide mixes used on soybeans [1]

EI/A	Alternative mixes			Glyphosate
	High	Low	Average	
Total	59,62	5,94	40,93	30,78
- Farm worker	48,70	2,23	29,89	15,20
- Consumer	18,94	0,98	12,80	6,65
- Ecology	112,89	14,60	80,13	70,59

EI/A = field use rating of environmental-impact quotient

[1] Calculations as applied by Kleter and Kuiper (2003), based on pesticide-use data from Gianessi et al. (2002)

Table 5. Apparent EIQs derived from the calculations of environmental impact of pesticides used on GM versus non-GM crops [1]

Crop (GM trait)	GM / non GM	Apparent EIQ			
		Total	Farm worker	Consumer	Ecology
Canola (HT)	GM	32,4	16,0	7,0	74,3
	non GM	28,0	15,4	8,4	60,3
Cotton (HT)	GM	28,3	17,4	7,7	59,9
	non GM	26,1	18,1	7,9	52,4
Cotton (IR)	GM	28,0	15,3	5,8	63,0
	non GM	28,0	15,3	5,8	63,0
Maize (HT)	GM	29,4	16,9	8,0	63,4
	non GM	30,3	15,4	7,9	66,3
Maize (IR)	GM	n.a.	n.a.	n.a.	n.a.
	non GM	49,4	53,4	14,0	80,7
Soybean (HT)	GM	32,4	16,0	7,0	74,3
	non GM	26,8	19,6	8,4	52,5

EIQ = environmental-impact quotient; apparent EIQ = aggregate for pesticide mixtures or parallel pesticide treatments, weighted for the relative contribution of each pesticide according to its rate (and for HT soybeans, each US state according to its share in total acreage); HT = herbicide-tolerant; IR = insect-resistant; n.a. = not applicable.

[1] Based on calculations applied by Kleter and Kuiper (2003) to pesticide-use data from Gianessi et al. (2002)

IUPAC project on GM crops

Recently, a three-year project “Impact of transgenic-crop cultivation on the use of agrochemicals and the environment” started under the umbrella of the International Union for Pure and Applied Chemistry. The purpose of this project is to investigate the environmental effects that the altered pesticide-management practices in GM-crop cultivation have, and, in a broader context, to provide tools for risk–benefit analysis for policymakers, in order to weigh the risks inherent to GM-crop cultivation against its benefits. To this end, the initial phase of this project will focus on the collection of data on altered pesticide use on GM crops and the characterization of the associated hazards. Such data may come from the surveys discussed above. Subsequently, based upon calculations, the actual risks will be characterized, for example by the use of the environmental indicators, as discussed above. Finally, a comparison will be made with other environmental issues than pesticide use that may incur during the risk assessment of GM-crop cultivation. Finally, these data will be integrated into a risk–

benefit analysis, which should enable policymakers to make decisions about GM-crop cultivation. The international project team brings together specialists from various fields of pesticide science, e.g. ecology, chemistry and toxicology. The project is scheduled to continue through February 2005 (IUPAC 2003).

Conclusions

As discussed above, there is a need to translate the figures on altered pesticide-use practices during cultivation of GM crops (including data from surveys like those of the USDA-ERS and NCFAP) into terms of impact on the environment. To this end, environmental indicators may prove instrumental in quantifying such impacts of pesticides. In our previous work, we employed the EIQ, which has the advantage that it is generally applicable, that EIQs have been established for a great number of pesticide active ingredients, and that farm worker, consumer and ecology components have been incorporated. Whereas the outcomes enable a comparison between different pesticide regimes, these results are rather abstract and may not be amenable to comparison with other issues in agriculture.

One alternative may be to apply environmental indicators that express the impact in terms of financial costs. For damage to fish, for example, direct effects, such as market value and penalties for causing fish mortality, and indirect effects, such as the attraction of sport-fishing tourists, can be taken into consideration for estimating the financial side of the environmental impact (Pimentel and Greiner 1997). Another way would be to translate the outcomes into 'low', 'intermediate' and 'high' risk levels, as is done for water contamination in the US State of Michigan by the web version of the NAPRA (National Agricultural Pesticide Risk Analysis) model (MSU 2003). NAPRA helps to predict the levels of pesticides that will be emitted from farm fields into ground and surface water following pesticide applications, and compares these values with safety threshold values, taking into account the local climatic conditions over the last 50 years and geographical characteristics (NWCC 2003). In addition, the Organisation for Economic Co-operation and Development (OECD) is developing environmental indicators for agriculture, including pesticide use, which may facilitate the holistic approach of assessing GM technology without focus on a specific issue. Three publications about these indicators have already been released, whereas the fourth volume, which describes how to incorporate the agro-environmental indicators into policy decisions, is forthcoming (OECD 1997; 1999; 2001). For example, three indicators for pesticide-use risk to aquatic systems have been developed and will be tested on national data from member states for further refinement, while additional indicators for human and terrestrial organisms will also be developed (OECD 2001).

Similar to 'health economics' for new medicines in their pre-market approval process, futuristic tools can be envisioned that enable the (mandatory) assessment of environmental and economic benefits of an agricultural technology prior to their market introduction.

Acknowledgements

The authors gratefully acknowledge contributions from the IUPAC project team on the impact of transgenic crops on pesticide use, as well as financial support from the International Union for Pure and Applied Chemistry and the Netherlands Ministry of Agriculture, Nature Management and Food Quality Research Programme 390.

References

Gianessi, L.P., Silvers, C.S., Sankula, S., et al., 2002. *Plant biotechnology: current and potential impact for improving pest management in U.S. agriculture: an analysis of 40 case studies.* NCFAP, Washington. [http://www.ncfap.org/40CaseStudies.htm]

IUPAC, 2003. Impact of transgenic crops on the use of agrochemicals and the environment. *Chemistry International,* 25 (3), 16-17. [http://www.iupac.org/publications/ci/2003/2503/pp5_2001-024-2-600.html]

James, C., 2001. *Global review of commercialized transgenic crops: 2000.* ISAAA, Ithaca. ISAAA Briefs no. 23. [http://www.isaaa.org/kc/Publications/pdfs/isaaabriefs/Briefs%2023.pdf]

James, C., 2002. *Preview: global status of commercialized transgenic crops*. ISAAA, Ithaca. ISAAA Briefs no. 27. [http://www.isaaa.org/kc/Publications/pdfs/isaaabriefs/Briefs%2027.pdf]

Kleter, G.A. and Kuiper, H.A., 2003. Environmental fate and impact considerations related to the use of transgenic crops. *In:* Voss, G. and Ramos, G. eds. *Chemistry of crop protection: progress and prospects in science and regulation: 10th IUPAC international congress on the chemistry of crop protection*. Wiley-VCH Verlag, Weinheim, 304-321.

Kovach, J., Petzoldt, C., Degni, J., et al., 1992. *A method to measure the environmental impact of pesticides*. New York State IPM Program, Cornell University. New York Food and Life Sciences Bulletin no. 139. [http://www.nysipm.cornell.edu/publications/EIQ.html]

Levitan, L., Merwin, I. and Kovach, J., 1995. Assessing the relative environmental impacts of agricultural pesticides: the quest for a holistic method. *Agriculture, Ecosystems and Environment,* 55 (3), 153-168.

MSU, 2003. *NAPRA in Michigan*. Michigan State University, Institute of Water Research, East Lansing. [http://www.iwr.msu.edu/~ouyangda/napra/]

NASS, 2003a. *Agricultural chemical usage (PCU-BB)*. National Agricultural Statistics Service, United States Department of Agriculture, Washington. [http://usda.mannlib.cornell.edu/reports/nassr/other/pcu-bb/]

NASS, 2003b. *Crop production acreage supplement (PCP-BB)*. National Agricultural Statistics Service, United States Department of Agriculture, Washington. [http://jan.mannlib.cornell.edu/reports/nassr/field/pcp-bba/acrg0603.pdf]

NWCC, 2003. *Pest management: NAPRA (National Agricultural Pesticide Risk Analysis)*. National Water and Climate Center, Natural Resources Conservation Service, United States Department of Agriculture, Portland. [http://www.wcc.nrcs.usda.gov/pestmgt/napra.html]

OECD, 1997. *Environmental indicators for agriculture. Vol. 1. Concepts and framework*. Organisation for Economic Cooperation and Development, Paris. [http://www1.oecd.org/publications/e-book/5199071E.PDF]

OECD, 1999. *Environmental indicators for agriculture. Vol. 2. Issues and design: "The York Workshop"*. Organisation for Economic Cooperation and Development, Paris. OECD Proceedings. [http://www1.oecd.org/publications/e-book/5199051E.PDF]

OECD, 2001. *Environmental indicators for agriculture. Vol. 3. Methods and results.* Organisation for Economic Cooperation and Development, Paris. [http://www1.oecd.org/publications/e-book/5101011E.PDF]

Pimentel, D. and Greiner, A., 1997. Environmental and socio-economic costs of pesticide use. *In:* Pimentel, D. ed. *Techniques for reducing pesticide use: economic and environmental benefits*. Wiley, Chichester, 51-78.

Reus, J. and Middleton, D., 1999. *Comparing environmental risk indicators for pesticides: results of the European CAPER project*. Centre for Agriculture and Environment, Utrecht. CLM no. 426.

Serecon Management Consulting and Koch Paul Associates, 2001. *An agronomic and economic assessment of transgenic canola.* Canola Council of Canada, Winnipeg. [http://www.canola-council.org/production/gmo_main.html]

Strandberg, B. and Bruus Pedersen, M., 2002. *Biodiversity in glyphosate tolerant fodder beet fields: timing of herbicide application.* National Environmental Research Institute, Silkeborg. NERI Technical Report no. 410. [http://www.dmu.dk/1_viden/2_Publikationer/3_fagrapporter/rapporter/FR410.pdf]

Wauchope, R.D., Estes, T.L., Allen, R., et al., 2002. Predicted impact of transgenic, herbicide-tolerant corn on drinking water quality in vulnerable watersheds of the mid-western USA. *Pest Management Science,* 58 (2), 146-160.

3b

Comment on Kleter and Kuiper: Environmental fate and impact considerations related to the use of transgenic crops

Terrance M. Hurley[#]

Proponents of pest resistant (pt) and herbicide tolerant (ht) transgenic crops argue that a key benefit to deployment is the potential to reduce the burden of pesticides on human health and the environment. However, evidence available to validate this argument is often mixed, which has led to heated debate.

A key problem with determining the impact of pt and ht crops on pesticide use, human health and the environment is that these crops do not eliminate pesticides. They result in a substitution from one pesticide to another. This poses a problem because kilogram for kilogram not all pesticides are equally hazardous. Most studies exploring the issue report how pt and ht crops have changed the kilograms of pesticide active ingredient released into the environment, which ignores the fact that all pesticides are not equally hazardous. To the extent that pt or ht crops result in the use of more active ingredient of a less hazardous pesticide, the results of this type of analysis can be misleading. Kleter and Kuiper rectify this shortcoming by using the Environmental Impact Quotient (EIQ, Kovach et al. 1992) to weight the kilograms of a pesticide used by measures of its hazard to human health and the environment. While their methodology represents an improvement over previous efforts, their analysis can still be criticized as biased in favour of transgenic crops. The bias comes from their treatment of *Bt* crops. Implicitly, their analysis assumes the EIQ for Bt crops is 0, which indicates no risk to human health or the environment. Kovach et al. (1992) does not report an EIQ for the toxins expressed by *Bt* crops. However, it does report an EIQ for Dipel, which is a spray formulation of toxins similar to those in *Bt* crops. While the *Bt* toxins are generally rated as less hazardous than most alternatives, they are not hazard-free. To avoid claims of bias, Kleter and Kuiper should explicitly include toxins present in *Bt* crops in accounting the effect of *Bt* crops on pesticide use and hazard.

Another issue not addressed by Kleter and Kuiper is the long term effect of pt and ht crops on the pest complex as a whole. Most pt and ht crops offer superior control of the targeted pests when compared to alternatives. This can result in less frequent and severe infestations over time, reducing the need for pesticides altogether. Alternatively, the use of pt and ht crops could result in a new pest complex that is more difficult to control without significant increases in pesticide use and hazard. How pt and ht crops ultimately affect the burden of pesticides on human health and the environment can only be known by examining changes in the pest complex and how these changes influence pesticide use over time. Still, Kleter and Kuiper add another useful piece to our incomplete knowledge of the benefits and costs of transgenic crops.

[#] Department of Applied Economics, Room 249C Classroom Office Building, 1994 Buford Avenue, University of Minnesota, St. Paul, MN 55108-6040. E-Mail: thurley@apec.umn.edu

J. H. H. Wesseler (ed.), Environmental Costs and Benefits of Transgenic Crops, 45–46.

References

Kovach, J., Petzoldt, C., Degni, J., et al., 1992. *A method to measure the environmental impact of pesticides*. New York State IPM Program, Cornell University. New York Food and Life Sciences Bulletin no. 139. [http://www.nysipm.cornell.edu/publications/EIQ.html]

4a

Biological limits on agricultural intensification: an example from resistance management[1]

Ramanan Laxminarayan[#] *and David Simpson*[#]

Abstract

Agricultural intensification could reduce pressures on natural habitats, but biological constraints may mitigate the long-term benefits of improved agricultural technologies. We consider one such constraint: that imposed by resistance to pesticides. The application of pesticides places selective evolutionary pressure on pest populations. Those organisms that survive show resistance. Resistance can be managed by planting 'refuge areas' in which susceptible pests breed. We use a simple model to characterize the optimal refuge strategy when a social planner values both agricultural output and natural habitat. We also examine land use consequences. The amount of land devoted to agriculture is an increasing function of the discount rate. A related finding is that more land would be devoted to agriculture when pest resistance must be managed than would be with a hypothetical 'neutral' technology affording the same yield-per-hectare as in the steady state, but not requiring the management of any biological stocks.

Keywords: pest resistance; biotechnology; land use; pesticides; sustainability; optimization

Introduction

Progress in agricultural technology has resulted in spectacular increases in yields over the last century. Yields of some major food crops in the US have tripled (USDA 1936; 1998), while in many developing countries, the 'green revolution' has transformed agriculture through the use of inputs such as pesticides, herbicides, fertilizers and hybrid seeds. Biotechnology now represents the cutting edge of efforts to increase agricultural yields even more.

Agricultural intensification may be the only way of mitigating the threat posed by the increasing food needs of growing and more prosperous human populations to the natural habitats on which much of the world's biological diversity depends. Many authors have argued that growing more food on the same area of land can reduce the pressure on natural habitats (Southgate 1997; Pagiola et al. 1998; Leisinger 1999). However, others have pointed out that while the intensification of agriculture may increase short-term yields, long-term prospects may be compromised by the absence of sustainable management practices (Perrings and Walker 1995; Pimentel et al. 1995; Krautkraemer 1994; Naylor and Ehrlich 1997; Albers and Goldbach 2000). New agricultural technologies may deplete soils, poison surrounding areas and organisms, or, as in the case we investigate here, induce genetic resistance among pests.

[#] Resources for the Future, Washington, DC, USA. E-mail: ramanan@rff.org

J. H. H. Wesseler (ed.), Environmental Costs and Benefits of Transgenic Crops, 47–60.

Sustainable long-term outcomes can only be achieved if deterioration and regeneration are balanced.

It is not always easy to predict how that balance will be struck, however. In this paper we construct a model of land use choice when pest resistance to pesticides must be managed. The application of any pesticide will exert evolutionary pressure in favor of organisms resistant to its toxin. An effective pesticide is one that kills the great majority of the pests it targets. In almost any large population, though, some organisms will be blessed with a fortuitous combination of genetic attributes that enable them to survive the effect of pesticides. Subsequent reproduction will then result in a greater frequency of genetically resistant pests and, consequently, reduced effectiveness of the pesticide.

In this paper we consider one option for managing resistance, the maintenance of 'refuge areas'. A refuge area is a portion of agricultural land planted with the same crop the pests attack, but not treated with a pesticide. Refuge areas promote the reproduction of pests that remain susceptible to the pesticide to which others evolve resistance. Susceptible and resistant pests interbreed, and the proportion of the latter in the population is limited.

The details of the model we develop are laid out below, but briefly, we consider a situation in which a social planner cares about two things: the production of food and the preservation of natural habitat. We abstract from other components of social welfare as well as from many real-world aspects of agricultural production. We suppose that land is the *only* costly input employed in growing a single crop[2]. We also suppose – again somewhat unrealistically, but in the interest of presenting clear results – that the conversion of land from natural habitat to agricultural use and back is costless.

Three interesting insights emerge from our simple model. The first is simply that we can compare and contrast our results with those arising from models that describe the management of other renewable resources over time, such as models of fisheries. We find in our model that maintaining 'maximum sustainable resistance' analogous to 'maximum sustainable yield' in fisheries models is typically not optimal. We also find that it can be optimal to 'exhaust' pesticide effectiveness when discount rates are high enough, a result that is also analogous to findings in the fisheries literature (Clark 1990).

The second insight concerns the allocation of land. We find that higher discount rates result in less environmentally friendly outcomes: more land is devoted to agriculture, at the expense of habitat retained for biodiversity. However, the underlying explanation is different here from that in other biological resource models. In models of fisheries, higher discount rates imply that the stock of fish – the biological resource of interest in that context – is deemed less valuable, and thus lower stocks will be maintained. Higher discount rates also motivate less conservation in our model, but in our context habitat supporting biological diversity is an argument of the period-by-period objective function (the model is similar to Hartman (1976), in this respect).

The final insight concerns a comparison between steady state with resistance management and a hypothetical alternative. In the steady state of the model, yield per hectare planted is, of course, invariant. We might regard the combination of pesticide use and resistance management as an agricultural 'technology' that affords a certain yield per hectare. In steady state, *more* land would be devoted to agriculture using this 'technology' than would be under a hypothetical one that afforded the same yield per hectare but did not involve the management of any biological stock.

This may seem counterintuitive. One consequence of devoting more land to agriculture is that a larger population of pests will be supported. The hypothetical alternative abstracts from such concerns. It might seem, then, that agricultural land use would be lower when there is a shadow price attached to its long-term biological consequences. Here, the biological conditions are crucial. If the pest-management regime is effective, the pest population is constrained by the efficacy of the pesticide rather than the availability of the crop on which it feeds. Surviving pests can be satiated in the short term. The hypothetical technology affords the same constant yield-per-hectare as in the steady-state of the pest-management regime. In the latter, however, the short-term marginal product of agricultural land is greater than the long-term average product. For the functional forms we have employed, at least, this 'marginal product' effect dominates the 'shadow price' effect, leading to our result.

The result is subtle, so we should clarify what we are *not* saying. The model is, by construction, one in which social welfare is maximized. Thus, we are not saying that an externality generates sub-optimal performance. We are also not saying that the improvements afforded by the intelligent combination of technology and resistance management are a bad thing. The point is simply that enthusiasm for these improvements should be tempered. Biotechnological improvements could improve welfare and may well be land-saving. However, the amount of land saved may be less than one might initially suppose.

In the section that follows we discuss resistance-management strategies in somewhat more detail. We introduce our model in the third section. We derive its steady state and summarize its implications in the fourth section. The fifth section summarizes and concludes.

Resistance management

We have chosen resistance management as a particularly interesting instance of a biological constraint on productivity, but have abstracted from a number of real-world considerations in order to achieve tractable results. In short, our intention is more illustrative than descriptive. Having offered that caveat, however, it may be useful to describe in somewhat more detail the principles underlying resistance management, mention some of the different ways it may be accomplished, and discuss some policy developments.

Resistance can be managed by maintaining refuge areas. It may seem strange to set aside an area of crops for the express purpose of feeding the pests one is trying to eliminate. The argument for doing so is that a population of 'susceptible' organisms from the refuge areas will interbreed with organisms that are genetically resistant to the pesticide. Resistance is often a recessive genetic trait, meaning that it will only occur in the offspring of parents that are both resistant[3]. Refuges assure that the proportion of resistant individuals in the population will remain small, and thus that it will be unlikely that two resistant individuals will mate.

A crucial consideration is that resistance typically comes with an evolutionary 'fitness cost' (Anderson and May 1991). In the absence of the pesticide, mortality is higher, or reproductive success lower, among resistant than among susceptible pest organisms. There are numerous examples of the reduced fitness of resistant strains to natural or synthetic insecticides (Ferrari and Georghiou 1981; Georghiou 1981; Beeman and Nanis 1986; Groeters et al. 1993; Alyokhin and Ferro 1999). Thus refuge areas may renew a population's susceptibility to pesticides.

Earlier studies have considered the economics of resistance management. Hueth and Regev (1974) consider the timing of pesticide applications during the growing season and its effect on resistance (see also Regev, Shalit and Gutierrez 1983). Laxminarayan and Brown (2001) and Goeschl and Swanson (2000) consider analogous issues in the management of antibiotic resistance.

Recent developments in biotechnology have spurred a renewed interest in resistance management. A gene from the bacterium *Bacillus thuringiensis* (frequently abbreviated as *Bt*) has been inserted in crops such as cotton, tobacco and corn. This gene codes for the production of a protein that is highly toxic to many insect pests. In 1990, no genetically modified organisms (GMOs) were under commercial cultivation in the United States. By 1999, nearly 100 million acres – close to a third of all land under cultivation in the US – were planted with GMOs.

In Canada, the regulation of *Bt* corn and *Bt* potato is carried out under the Seeds Act and Part V of the Seeds Regulations administered by the Canadian Food Inspection Agency (CFIA). As of 1998, the CFIA has mandated an industry-wide standard for pest-resistance management that requires growers to plant a minimum of 20% non-*Bt* corn not sprayed with insecticides on their planted acreage each year. These regulations also require that non-*Bt* corn be planted within ¼ mile of the farthest *Bt* corn in a field to provide a refuge where *Bt*-susceptible pests may exist[4]. Similar resistance-management plans have been mandated for *Bt* potato as well. Several economic analyses have now been conducted of these refuge policies (Hurley, Babcock and Hellmich 1997; Hyde et al. 1999; Livingston, Carlson and Fackler 2000).

Refuge areas are not the only option for resistance management. Pesticide applications can be timed in such a way as to allow susceptible individuals to interbreed with resistant ones (Hueth and Regev 1974; Regev, Shalit and Gutierrez 1983). Chemical and/or crop rotation may accomplish similar ends. The intensity of pesticide use may also influence the development of resistance. Invention of new pesticides may also be an option.

There are reasons for which refuge areas may be the preferred option, however. With advances in biotechnology, pesticides are increasingly bred *into* as opposed to applied *onto* crop plants, obviating timing and, to some extent, dosage and rotation, as management strategies. It may not be reasonable to suppose that better pesticides can always be developed. Pests that have developed resistance to existing pesticides may display 'cross-resistance' to newly developed toxins.

The effects of refuges may be analogous to other resistance-management options. Refuges call for sacrificing some portion of current production in order to maintain long-term productivity. Rotations among crops or using lower pesticide dosages also represent strategies for achieving a similar trade-off. Thus our results may well constitute an allegory for similar findings in somewhat broader contexts.

The model

We develop a simple model of the evolution of pest resistance. The setting is a large area in which a single crop is planted. The pest population is assumed to be local; both in-migration and out-migration are ruled out. Other conditions implicit in deriving the Hardy–Weinberg principle, such as random mating between resistant and susceptible pests, negligible mutation, non-overlapping pest generations and sexual reproduction of pests, are all assumed to hold[5]. These assumptions imply a high degree of mobility among pests. Each surviving organism is assumed to be equally likely to mate with every surviving organism of the opposite sex[6].

The pest population is denoted by D. The proportion of susceptible pests in the population is denoted by a fraction w. Susceptible pests are those that have not developed resistance to the toxin. Put in another way, w may be thought of as the stock of effectiveness of the pesticide; it is the proportion of the pest population to which the toxin is lethal.

The pest population is assumed to grow logistically with an intrinsic growth rate of g and a carrying capacity of K per unit of land planted in the crop. Total land is assumed to be fixed and is normalized to 1. The fraction of total available land area devoted to agriculture is denoted by Q. The total number of new pest organisms hatched (presuming they are the offspring of egg-bearing insects) in every period, then, is $gD(1 - D/KQ)$. From this gross increase, we must subtract mortality among both susceptible and non-susceptible pests.

A refuge strategy calls for planting a fraction q of the total land devoted to agriculture, Q, in a crop to which pesticide is applied (or in the case of GMOs, implanted). Hence, a fraction $1 - q$ of agricultural land is not treated with pesticide. Recall that a fraction w of all pests is susceptible. We suppose that all susceptible pests that feed on crops treated with pesticide die after exposure to the toxin. A fraction r of non-susceptible pests die, regardless of whether they are exposed to the pesticide[7]. The analysis can be extended to suppose that some susceptible pests survive exposure to the pesticide, but no significant generality is gained as a result.

We assume that pests distribute themselves evenly over the area planted with crops. A fraction q of the pest population D will feed on the area treated with pesticide. Of these, the fraction w that are susceptible will die. A fraction r of the fraction $1 - w$ of resistant pests will also die. Note that r may be interpreted as the 'excess mortality' among resistant pests relative to susceptible pests, with the 'baseline' mortality of susceptible pests in the absence of the pesticide subsumed in the parameters of the logistic function.

Combining our assumptions, the growth of the pest population can be modelled as

$$\text{(1)} \qquad \dot{D} = gD\left(1 - \frac{D}{KQ}\right) - wqD - (1-w)rD .$$

It can be shown that the proportion of resistant pests in the population also follows a logistic equation, with the growth parameter equal to the difference in relative mortality rates between genotypes (Bonhoeffer, Lipsitch and Levin 1997). Thus

$$\text{(2)} \qquad \dot{w} = (q-r)w(w-1).$$

Because $w < 1$, the larger the fraction of crop land treated with pesticide, the greater the decline in the effectiveness of the pesticide.

Optimal refuge areas and steady-state results

We now characterize the optimal refuge strategy. The proportion of agricultural land set aside as refuge area in each period determines the crop yield net of losses to pests in that period as well as the effectiveness of the pesticide in succeeding periods. There is, then, an intertemporal trade-off between increasing refuge size (and consequently losing more agricultural yield to pests today) and more rapidly eroding pest susceptibility to the toxin.

Suppose that each surviving pest eats an amount α. Normalize gross output per unit area planted to 1. If a fraction q of the area Q devoted to agriculture is treated with pesticide, then gross production in this area will be qQ. A fraction q of the pest population D will feed on the area exposed to the toxin. Of these, the fraction $1 - w$ that are resistant will survive. Each of these surviving organisms will consume α units. Thus, the net yield from the area exposed to the pesticide is given by $qQ - (1-w)qD\alpha$. Similarly, the gross yield from the area without toxin exposure is $(1 - q)Q$. A fraction $1 - q$ of pests will feed on the unexposed area, where they will consume $(1 - q)D\alpha$ units of the crop[8]. Net yield in the unexposed area is, then, $(1-q)Q - (1-q)D\alpha$. Net yield from agriculture, Y, is given by the sum of net yields from the area treated with pesticide and the refuge area:

$$(3) \qquad Y(q, Q, D, w) = qQ-(1-w)qD\alpha+(1-q)Q-(1-q)D\alpha=Q-(1-wq)D\alpha .$$

Expression (3) postulates that net yield has an 'additive' form. Net yield is equal to gross yield, Q, less the total amount of the crop consumed by pests. The latter quantity depends on the number of pests, *but not on the amount of food available to them.* This is both a special assumption and a crucial one for the results that follow. In ecological terms, it means that the pest population is constrained by the analogue to 'predation' imposed by the pesticide (for a discussion of predator–prey relationships, their implications for population constraints, and circumstances under which populations are regulated 'from above' by predation as opposed to 'from below' by resources Estes, Crooks and Holt 2001). Each *surviving* pest eats until it is satiated. Put in another way, our 'additive' assumption is consistent with a successful pest-control regime. The optimal management programme never adopts the trivial solution of 'managing' pest numbers by simply allowing them to expand until they are limited by the food planted for their consumption (although if pest numbers were limited by natural predators, our 'additive' assumption could continue to hold even if pest populations were not managed by pesticides).

Expression (3) is rarely *literally* true. To the extent that there is always some competition among pests, the amount they consume may be a function not only of their numbers, but also of the amount of food available to them. In situations in which pest populations are controlled at numbers significantly below the carrying capacity of their environment, though, expression (3) is a valid and reasonably accurate first approximation.

It is also worth emphasizing that expression (3) describes the net yield from agriculture at a point in time *during which the pest population is given.* Net yield is equal to gross yield less the amount the current generation of pests consumes. A choice to expand the area planted in crops will, of course, result in a larger pest population. The effects of population growth will not begin to be felt until later, however[9].

For the purposes of our very simple example, we suppose that social welfare depends on two goods: the net yield from agriculture, Y from equation (3), and the total quantity of land conserved as natural habitat, $1 - Q$. To facilitate the derivation of tractable results, we suppose that preferences are of the Cobb–Douglas form[10] $U(Y, 1 - Q) = Y^{\beta}(1 - Q)^{1-\beta}$. Let the discount rate be ρ. Then, a social planner would choose q and Q so as to maximize

$$(4) \qquad \int_0^{\infty} (Q - [1 - wq]D\alpha)^{\beta} (1 - Q)^{1-\beta} e^{-\rho t} dt,$$

subject to equations (1) and (2) for the evolution of pest populations and resistance.

The current-value Hamiltonian for our problem is

$$(5) \qquad \begin{aligned} H = (Q - [1 - wq]D\alpha)^{\beta}(1 - Q)^{1-\beta} \; + \; \lambda_1 D\left[g\left(1 - \frac{D}{KQ}\right) - wq - r(1-w)\right] \\ + \; \lambda_2(q - r)w(w-1) \end{aligned},$$

where λ_1 and λ_2 are the co-state variables associated with the stock of pests, D, and the proportion of susceptible pests, w, respectively. Necessary conditions for an optimum are given by equations (5.1) through (5.4)[11]:

$$(5.1) \quad \beta\frac{U}{Y}wD\alpha - \lambda_1 wD - \lambda_2 w(1-w) \begin{pmatrix} < \\ = \\ > \end{pmatrix} 0 \; as \; q \begin{pmatrix} = 0 \\ \in (0,1) \\ = 1 \end{pmatrix};$$

$$(5.2) \quad \beta\frac{U}{Y} - (1-\beta)\frac{U}{1-Q} + \lambda_1 g\frac{D^2}{KQ^2} = 0;$$

$$(5.3) \quad \rho\lambda_1 - \dot{\lambda}_1 = \beta\frac{U}{Y}(wq-1)\alpha + \lambda_1\left[g\left(1 - \frac{2D}{KQ}\right) - qw - r(1-w)\right];$$

$$(5.4) \quad \rho\lambda_2 - \dot{\lambda}_2 = \beta\frac{U}{Y}qD\alpha + \lambda_1(r - q)D + \lambda_2(r - q)(1-2w).$$

Let us consider results in a steady state. We will denote steady-state variable values with a superscript '*SS*'. It is obvious from equation (2) that, if (5.1) holds in the steady state (i.e., if $0 < q < 1$), then $q^{SS} = r$. If r were zero, there would be no fitness cost of resistance, resistant organisms must eventually come to dominate the population, and refuge areas would serve no purpose in the long run. In such circumstances the optimal refuge strategy would involve the management of an inevitably exhaustible resource of pesticide susceptibility.

If $D \neq 0$, then from equations (1), (2), (5.3) and (5.4), we have

$$(6) \qquad D^{SS} = \frac{g - r}{g}KQ^{SS},$$

$$(7.1) \quad \lambda_1^{SS} = \beta\frac{U}{Y}\frac{(w^{SS}r-1)\alpha}{\rho + g - r},$$

and

$$(7.2) \quad \lambda_2^{SS} = \beta\frac{U}{Y}\frac{rD^{SS}\alpha}{\rho}.$$

Equation (6) requires that $g > r$ if the pest population is to be positive in steady state. If this were not the case, then it would be optimal to eradicate the pest population.

In equation (7.1), we note that $w^{SS}r - 1$ is non-positive because both w^{SS} and r are fractions. It follows that $\lambda_1^{SS} < 0$; λ_1^{SS} represents the shadow price of a 'bad', the pest population. Similarly, from equation (7.2), $\lambda_2^{SS} > 0$, because it is the shadow price of a 'good' stock, susceptibility to the pesticide. Substituting for λ_1^{SS} and λ_2^{SS} from equations (7.1) and (7.2) and for D^{SS} from equation (6), w^{SS} can be derived from equation (5.1):

$$(8) \qquad w^{SS} = \frac{(r-\rho)(\rho+g-r)-\rho}{r(g-r)}.$$

Our results echo findings from other renewable-resource contexts. Current thinking in the management of genetically modified crops, at least, seems to be that refuge areas should be established to manage resistance (EPA 1998). Equation (8) demonstrates that such a strategy is not always optimal, however. If the discount rate, ρ, is large enough – a sufficient condition is that it be greater than the fitness cost of resistance, r – it is optimal to exhaust the stock of effectiveness at steady state[12]. This conclusion is analogous to a finding from the fisheries literature: it may be 'optimal' (ignoring possible ethical and ecological considerations) to harvest a species to extinction if its growth rate is less than the discount rate (Clark 1990).

In the limit as ρ approaches zero, w^{SS} approaches one. With a vanishing discount rate, resistant pests would be eradicated. So long as there remain any resistant alleles in the population, welfare in the indefinitely long run would be increased by eradicating those alleles[13]. This would not be an optimal strategy under a positive discount rate, of course. The reasoning recalls Tjalling Koopmans's (1960) argument that discount rates *should* be positive: it is unreasonable to sacrifice welfare in all foreseeable periods in anticipation of some date in the indefinite future at which things would be better (Koopmans 1960).

Next, consider the steady-state level of agricultural land use. To reduce clutter, we will suppress superscripts, but all variables are assumed to take their steady-state values in the expressions leading up to and including (13) below. This analysis may be conducted most easily by deriving an implicit expression for land in agriculture, Q. To do this, we start from expression (5.2). Recall that its three terms equate the marginal utility afforded by more immediate consumption to the marginal utility lost by habitat reduction and the shadow price of the increment in the pest population induced by devoting more land to agriculture. Restate expression (5.2) by adding and subtracting $\beta U/Q$,

$$\beta\frac{U}{Q} + \beta\frac{U}{Y} - \beta\frac{U}{Q} - (1-\beta)\frac{U}{1-Q} + \lambda_1 g\frac{D^2}{KQ^2} = 0,$$

and rearrange it as:

$$(9) \qquad \frac{Q-\beta}{1-Q}U = \beta\frac{(Q-Y)U}{Y} + \lambda_1 g\frac{D^2}{KQ}.$$

In a somewhat more heuristic notation, we could rewrite (9) as

$$(9.1) \qquad \frac{Q-\beta}{1-\beta}U_H = (Q-Y)U_Y + \lambda_1 g\frac{D^2}{KQ},$$

where U_H is the marginal utility afforded by natural habitat and U_Y the marginal utility afforded by agricultural consumption. The final term in (9.1) is the shadow price of the larger pest population induced by more land in agriculture times land used in agriculture. Total crop losses to pests are gross production, Q, less net production, Y. The first term on the right-hand side of (9.1) is, then, the value of agricultural production lost to pests, calculated at its marginal utility shadow price.

Our results below concerning the discount rate and land devoted to agriculture, Q, relate the relative importance of present and future losses to pests. Heuristically, as the discount rate declines, the two terms on the right-hand side of (9.1) will converge. In the limit as the discount rate approaches zero, the optimal policy calls for exactly offsetting the cost of present crop losses with that of future losses. For higher discount rates, present losses will be weighted more heavily than future[14]. The consequence is that more land will be devoted to agricultural production.

Return now to expression (9). Note that

$$(10) \quad Q - Y = (1 - wr)\alpha K \frac{(g-r)}{g} Q,$$

and

$$(11) \quad Y = \left[1 - (1 - wr)\alpha K \frac{(g-r)}{g}\right] Q.$$

From expression (8),

$$(12) \quad 1 - wr = \frac{(1 + \rho - r)(\rho + g - r)}{g - r}.$$

Using expressions (6) and (7.1) above to substitute for λ_1 and D, we can restate (9) as

$$(13) \quad \frac{Q-\beta}{1-Q} = \frac{\beta\rho(1 + \rho - r)\alpha K}{[g - (1 + \rho - r)(\rho + g - r)\alpha K]}.$$

Using (11) and (12), the denominator of (13) is positive, since $Y > 0$. Thus, the right-hand side is positive, and $Q > \beta$.

We could differentiate (9) with respect to ρ and solve for $\partial Q/\partial \rho$, but it is clear on inspection that its left-hand side is increasing in Q and its right-hand side is increasing in ρ. Thus $\partial Q/\partial \rho > 0$. The greater is the discount rate, the greater is the quantity of land devoted to agriculture in the steady state – if indeed a non-trivial steady state obtains.

Expression (13) has an interesting interpretation. Suppose that production took place under a hypothetical technology such that net yield, Y, were a constant fraction θ of total area planted regardless of pests and other factors, Q: $Y = \theta Q$. If we set the fraction θ equal to $1 - (1 - wr)\alpha K \frac{(g-r)}{g}$, the hypothetical technology would provide the same net yield per hectare as does the optimally managed pesticide programme (see expression [11]).

It is easily shown, though, that if one were to maximize the Cobb-Douglas objective function

$$(14) \quad U = (\theta Q)^{\beta} (1-Q)^{1-\beta},$$

the solution would be to set $Q = \beta$. Heuristically, expression (14) is formally identical to maximizing utility from food and habitat subject to the fixed 'budget constraint' that agricultural land plus habitat equals 1. Using land as a *numeraire*, the 'price' of food is the inverse of yield-per-hectare, $1/\theta$. As is well known, the solution to such a problem is to set the share of expenditure on food equal to its exponent in the utility function, β. But the share of expenditure when land is the *numeraire* is simply the share of land devoted to agriculture, Q.

We have just shown, then, that more land is used in the steady state of our model than would be used were a technology available that provided the same yield-per-hectare as does the optimally managed pesticide programme, but that did not involve the management of pest populations.

This may seem paradoxical, as pest control introduces an intertemporal consideration. More land in agriculture now implies more pests in the future. One might, then, suspect that a shadow price associated with the current use of land would induce the decision-maker to use less, rather than more, land in agriculture. If Y were equal to θQ, the analogue to expression (5.2) above would be

$$\beta\theta\frac{U}{\theta Q} - (1-\beta)\frac{U}{1-Q} + \lambda_1 g\frac{D^2}{KQ^2} = 0;$$

note that the θs cancel from the first term. The analogue to our expression (9) would then be

$$(15)\quad (Q-\beta)\frac{U}{Q(1-Q)} = \lambda_1 g\frac{D^2}{KQ^2}.$$

Since λ_1 denotes the shadow price of a 'bad' – the pest population – one would reach the opposite of the conclusion implied by (13). In (15), $Q < \beta$.

The reason for this difference is to be found in expression (3) above, in which we presume net yield takes an 'additive' form. This implies that the marginal product of land in agriculture is always one, while the alternative 'multiplicative' specification $Y = \theta Q$ would imply that the marginal product of land is $\theta < 1$. Since the opportunity cost of devoting land to habitat rather than agriculture is always higher under the 'additive' than the 'multiplicative' specification, less land is devoted to habitat.

The assumption that pests can be satiated in the short run is crucial to our results, but for the reasons we have discussed above, it is reasonable. Multiplicative forms have been suggested in the agricultural economics literature (see e.g. Lichtenberg and Zilberman 1986). However, our specification of an additive relationship *in the short run* in expression (3) is not inconsistent with a multiplicative *long-run* relationship in expression (11). In fact, the message of Lichtenberg and Zilberman is that one must model biological relationships carefully, considering and distinguishing short- and long-term effects.

A supplementary question that is easily explored using our current model relates to the impact of pest-resistance management practices on habitat conservation. Foregoing resistance management results in greater losses to pests. Consequently, agricultural output is smaller in this situation and the marginal utility of food is greater than if we were to manage for pest resistance. Needless to add, foregoing resistance management is not an optimal solution from a societal standpoint, but as is demonstrated below, it does have the effect of helping conserve habitat.

The steady-state allocation of land to agriculture can be calculated by solving the optimal control problem where q is invariant and set equal to one. The expression below for this problem that is analogous to equation (13) is

$$(16)\quad \frac{(Q-\beta)}{(1-Q)} = \beta\frac{\alpha(g-r)K}{g-\alpha(g-r)K}\left[\frac{\rho}{\rho+g-r}\right].$$

By contrasting this expression (16) with equation (13), we can show that the amount of land devoted to agriculture is greater than when we choose to manage for resistance[15].

Cobb-Douglas preferences are a conspicuous and restrictive assumption in our modelling. We have demonstrated in an earlier version of this work (Laxminarayan and Simpson 2000) that the same basic results obtain under a constant elasticity of substitution specification, of which the Cobb-Douglas is a special case. It seems reasonable to suppose that they generalize at least somewhat.

Conventional wisdom holds that demand for food is inelastic, suggesting a low elasticity of substitution at low levels of consumption. Moreover, environmental amenities are likely to be luxury goods. Thus a homothetic representation of preferences is probably unrealistic, and it is not immediately clear how alternatives would affect specific results. It is, then, not constructive to argue too forcefully or concretely on the basis of so simple a model. There is, however, a general principle at work. The fact that short-term yields can be increased to the detriment of long-term prospects implies not only that short-term results may not be sustainable, but also that steady-state prospects may be less optimistic than they might first appear.

Conclusion

We have employed a simple and analytically tractable model to demonstrate some implications of pesticide use and resistance for land use. It is probably unnecessary to caution readers that the model is intended to be illustrative rather than realistic. Despite these concessions to practicality, we believe that the model develops some useful insights.

Perhaps the most important of these is that biological constraints limit the intensification of agriculture in ways that are not always transparent. Whereas other examples of this phenomenon have been developed, we find the case of pesticide resistance particularly interesting in that the interactions are complicated and the implications subtle. Our results do not hinge on externalities, nor do they suggest that technological advances are not desirable. They only point to the wisdom of measured expectations concerning the potential of agricultural intensification to solve the problem of habitat and biodiversity loss.

References

Albers, H.J. and Goldbach, M.J., 2000. Irreversible ecosystem change, species competition, and shifting cultivation. *Resource and Energy Economics,* 22 (3), 261-280.

Alyokhin, A.V. and Ferro, D.N., 1999. Relative fitness of Colorado potato beetle (Coleoptera: Chrysomelidae) resistant and susceptible to the *Bacillus thuringiensis* Cry3A toxin. *Journal of Economic Entomology,* 92 (3), 510-515.

Anderson, R.M. and May, R.M., 1991. *Infectious diseases of humans: dynamics and control*. Oxford University Press, Oxford.

Beeman, R.W. and Nanis, S.M., 1986. Malathion resistance alleles and their fitness in the red flour beetle (Coleoptera: Tenebrionidae). *Journal of Economic Entomology,* 79 (3), 580-587.

Bonhoeffer, S., Lipsitch, M. and Levin, B.R., 1997. Evaluating treatment protocols to prevent antibiotic resistance. *Proceedings of the National Academy of Sciences of the United States of America,* 94 (22), 12106-12111.

Clark, C.W., 1990. *Mathematical bioeconomics: the optimal management of renewable resources*. 2nd edn. Wiley, New York.

EPA, 1998. *The Environmental Protection Agency's white paper on Bt plant-pesticide resistance management*. US Environmental Protection Agency, Washington.

Estes, J., Crooks, K. and Holt, R., 2001. Ecological role of predators. *In:* Levin, S. ed. *Encyclopedia of biodiversity*. Academic Press, San Diego, 857-878.

Ferrari, J.A. and Georghiou, G.P., 1981. Effects on insecticidal selection and treatment on reproductive potential of resistant, susceptible, and heterozygous strains of the southern house mosquito. *Journal of Economic Entomology,* 74 (3), 323-327.

Georghiou, G.P., 1981. Implications of potential resistance in biopesticides. *In:* Robert, D.W. and Granados, R.R. eds. *Biotechnology, biological pesticides and novel plant-host resistance for insect pest management.* Boyce Thompson Institute for Plant Research, Ithaca.

Goeschl, T. and Swanson, T., 2000. Lost horizons: the interaction of IPR systems and resistance management. *In:* Aledort, J., Laxminarayan, R., Howard, D., et al. eds. *International workshop on antibiotic resistance: global policies and options*. Harvard University, Cambridge. [http://www2.cid.harvard.edu/cidabx/swanson.pdf]

Groeters, F.R., Tabashnik, B.E., Finson, N., et al., 1993. Effects of resistance to *Bacillus thuringiensis* on the mating success of the diamondback moth (Lepidoptera: Plutellidae). *Journal of Economic Entomology,* 86, 1035-1039.

Hartman, R., 1976. The harvesting decision when a standing forest has value. *Economic Inquiry,* 14 (1), 52-58.

Hueth, D. and Regev, U., 1974. Optimal agricultural pest management with increasing pest resistance. *American Journal of Agricultural Economics,* 56 (3), 543-553.

Hurley, T.M., Babcock, B.A. and Hellmich, R.L., 1997. *Biotechnology and pest resistance: an economic assessment of refuges*. Center for Agricultural and Rural Development, Iowa State University.

Hyde, J., Martin, M.A., Preckel, P.V., et al., 1999. The economics of refuge design for Bt corn. *In: American Agricultural Economics Association Annual Meeting.* AAEA, Nashville.

Kamien, M.I. and Schwartz, N.L., 1991. *Dynamic optimization: the calculus of variations and optimal control in economics and management.* 2nd edn. North-Holland, Amsterdam. Advanced Textbooks in Economics no. 31.

Koopmans, T.C., 1960. Stationary ordinal utility and impatience. *Econometrica,* 28 (2), 287-309.

Krautkraemer, J.A., 1994. Population growth, soil fertility, and agricultural intensification. *Journal of Development Economics,* 44 (2), 403-428.

Laxminarayan, R. and Brown, G.M., 2001. Economics of antibiotic resistance: a theory of optimal use. *Journal of Environmental Economics and Management,* 42 (2), 183-206.

Laxminarayan, R. and Simpson, R.D., 2000. *Biological limits on agricultural intensification: an example from resistance management.* Resources for the Future, Washington. Resources for the Future Discussion Paper no. 00-43.

Laxminarayan, R. and Simpson, R.D., 2002. Refuge strategies for managing pest resistance in transgenic agriculture. *Environmental and Resource Economics,* 22 (4), 521-536.

Leisinger, K.M., 1999. *Disentangling risk issues: biotechnology for developing country agriculture, problems and opportunities*. IFPRI, Washington.

Lichtenberg, E. and Zilberman, D., 1986. The econometrics of damage control: why specification matters. *American Journal of Agricultural Economics,* 68 (2), 261-273.

Livingston, M.J., Carlson, G.A. and Fackler, P.L., 2000. Bt cotton refuge policy. *In: American Agricultural Economics Association meetings, Tampa Bay, July 30-August 2, 2000.* AAEA, Ames.

Naylor, R.L. and Ehrlich, P.R., 1997. Natural pest control services and agriculture. *In:* Daily, G.C. ed. *Nature's services: societal dependence on natural ecosystems.* Island Press, Washington, 151-174.

Pagiola, S., Kellenberg, J., Vidaeus, L., et al., 1998. Mainstreaming biodiversity in agricultural development. *Finance and Development,* 35 (1), 38-41.

Perrings, C.A. and Walker, B.H., 1995. Biodiversity loss and the economics of discontinuous land change in semi-arid rangelands. *In:* Perrings, C.A., Mäler, K.G. and Folke, C. eds. *Biodiversity loss: economic and ecological issues.* Cambridge University Press, Cambridge.

Pimentel, D., Harvey, C., Resosudarmo, P., et al., 1995. Environmental and economic costs of soil erosion and conservation benefits. *Science,* 267 (5201), 1117-1123.

Regev, U., Shalit, H. and Gutierrez, A.P., 1983. On the optimal allocation of pesticides with increasing resistance: the case of alfalfa weevil. *Journal of Environmental Economics and Management,* 10 (1), 86-100.

Secchi, S. and Babcock, B.A., 2003. Pest mobility, market share and the efficacy of refuge requirements for resistance management. *In:* Laxminarayan, R. ed. *Battling resistance to antibiotics and pesticides: an economic approach.* Resources for the Future, Washington, 94-112.

Southgate, D., 1997. Alternatives to the regulatory approach to biodiverse habitat conservation. *Environment and Development Economics,* 2, 106-110.

USDA, 1936. *Agricultural statistics.* US Department of Agriculture, Washington.

USDA, 1998. *Agricultural statistics.* US Department of Agriculture, Washington. [http://www.usda.gov/nass/pubs/agr98/acro98.htm]

---

[1] Address correspondence to ramanan@rff.org. We are grateful to seminar participants at Resources for the Future; University of Minnesota; Delhi School of Economics; American Economic Association Annual Meetings (New Orleans) and the IPGRI Workshop on Biotechnology, Environmental Policy and Agriculture in Rome for useful comments. Finally, we are grateful to Sarah Cline for research assistance. Any remaining errors are our responsibility.

[2] This may not be as unrealistic as it seems. When a toxin is expressed in a genetically modified plant, we might suppose that the social planner is deciding how much use to make of a resource that can be acquired at low marginal cost: genetically modified seed. More generally, expenditures on pesticides may be modest compared to other expenses.

[3] 'Heterozygous' individuals (those having one 'resistant' and one 'susceptible' allele) may show some resistance, but typically not as much as those with two resistant alleles.

[4] Similar regulations have been enacted in the United States by the Environmental Protection Agency (EPA), which has mandated that refuge areas be grown in conjunction with certain transgenic crops, such as *Bt* cotton and corn (EPA 1998). In fact, EPA regulations regarding GMOs are the first from any agency in the US that treat pest susceptibility as a public good (Livingston, Carlson and Fackler 2000), even though resistance issues arose with more traditional pesticides as well.

[5] A *genotype* is a particular genetic configuration. An *allele* is any one of the two or more forms that may compose a gene; for example, alleles for blue or brown eyes are common in many human populations. The Hardy–Weinberg principle of quantitative genetics holds that, for a population satisfying the assumptions we have stated and in which expected mortality is the same across different genotypes, the expected proportion of alleles and of genotypes remains constant from generation to generation.

[6] Assumptions concerning pest are important in forming refuge-regulation policy (Secchi and Babcock 2002)

[7] The continued survival of 'unfit' genes in the absence of pesticide use is problematic but might be explained by rare mutations or the infrequent occurrence of random events that temporarily favor otherwise 'unfit' organisms.

[8] Implicit in our assumptions are the notions that susceptible pests die so quickly on exposure to the pesticide as to consume only a negligible amount of the crop before expiring and that mortality to other causes occurs after consumption. Alternatively, we could have supposed that pests eat a constant amount of the crop at each instant they are alive and treated lifespan as a random variable drawn from different distributions depending on whether the pest is 'resistant' or 'susceptible' to the pesticide and whether it feeds in a 'treated' or a 'refuge' area. We could then replace losses to pests with expected losses. This does not lead to qualitatively different results, however, so we have maintained the simplifying assumptions. Both assumptions could be relaxed, but neither is important for establishing our general results.

[9] The assertion that land devoted to agriculture affects population growth rather than the amount consumed by any individual pest might be best justified by noting that a pest's mortality risk increases with the effort required to discover an area in which it can consume the crop without competing with another pest to do so.

[10] Our basic results also obtain under a CES utility function of the form $U = \left(\beta Y^{\eta} + (1-\beta)(1-Q)^{\eta}\right)^{1/\eta}$; see Laxminarayan and Simpson (2000).

[11] It can also be shown that these conditions are also sufficient for an optimal solution. A rather laborious process shows that the problem satisfies Arrow's conditions for an optimum (Kamien and Schwartz 1991, p. 222).

[12] Laxminarayan and Simpson (2002) also show that it may not be optimal to establish refuges when the proportion of resistant pests is very low. Resistance does not yet constitute a sufficient threat as to justify foregoing yield.

[13] This extreme result is an artifact of a specification of the evolution of resistance (see equation (2)) in which resistant alleles can be reduced to arbitrarily small proportions without ever being entirely eliminated. The fundamentally discrete character of alleles and the possibilities of mutation preclude interpreting the model literally. It is, nonetheless, a useful way of thinking about these issues.

[14] This interpretation of losses as 'totals' evaluated at prices determined by marginal utility or shadow price would appear to be an artifact of the Cobb-Douglas specification, and hence not necessarily generalizable.

[15] This holds true when $(1+\rho-r)(\rho+g-r) > g-r$, a condition that is satisfied for all relevant parameter values where the optimal strategy is not to exhaust pesticide effectiveness anyway (see equation 8 of the paper in this regard.)

4b

Comment on Laxminarayan and Simpson: Biological limits on agricultural intensification: an example from resistance management

Claudio Soregaroli[#]

This article is centred on evaluating the effects of an innovation in the agricultural sector, such as the biotech wave, which is increasing crop yields but, at the same time, requires sustainable management practices. In particular, it highlights the role of 'refuge areas' and the environmental effects of the likely long-term effect of pest resistance to the newly introduced tolerant crops. Results provide useful insight into the factors influencing the optimal resistance-management strategy and show the implication on the distribution of land between agricultural and natural areas in the steady state.

As the authors underline, the results of the study should be carefully interpreted: the developed model is relatively simple and some of the assumptions could be restrictive. Hence, the questions arising to the reader are related to the possible generalization of results. What are the implications of relaxing some of the assumptions? How could the model be extended and made more realistic without losing in tractability? Which of the presented results would reasonably hold under different modelling frameworks?

The answers to the above questions are certainly not simple and the effort made by the authors to qualify the different assumptions and evaluate alternative possibilities in some cases should be underlined; in particular, the focus on the specification of the 'additive' form of the yield function, of the pest population dynamics and of the utility function of the social planner.

However, other assumptions seem important. In particular, the model assumes a "high degree of mobility among pests" that implies a random mating between resistant and susceptible pests. These features are typical of pests such as insects, but are far removed from the characteristics of weeds. If the scope of the work is to cover the evaluation of the effects of biotech innovation in general, then weed control and their resistance management should also be introduced into the framework. This is probably not a trivial task given the peculiarities of weed populations: the assumption of 'proportional' yield function seems more realistic for weeds than the 'additive' form (Mitchell 2001), population dynamics are different and the management of refuge areas is probably not effective for controlling weed resistance as compared, for example, to crop rotation. All of the above points would substantially change the set-up of the presented model with likely important implications on its results. Unfortunately, there is a substantial lack of information on weed-resistance management as compared to that of insects. Few studies in the literature have focused on the topic: some examples, even if from the sole farmer's perspective, are Schmidt and Pannell (1996) and Gorddard, Pannell and Hertzler (1995).

[#] Università Cattolica del Sacro Cuore, Cremona, Italy

J. H. H. Wesseler (ed.), Environmental Costs and Benefits of Transgenic Crops, 61–62.

The inclusion of weed-resistance management would be an interesting direction for further research that could help in providing a more complete view of the global effect of the biotech innovation.

References

Gorddard, R.J., Pannell, D.J. and Hertzler, G., 1995. An optimal control model for integrated weed management under herbicide resistance. *Australian Journal of Agricultural Economics,* 39 (1), 71-87.

Mitchell, P.D., 2001. Additive versus proportional pest damage functions: why ecology matters. *In: AAEA Annual Meeting, August 5-8, 2001, Chicago, Illinois*. American Agricultural Economics Association.

Schmidt, C.P. and Pannell, D.J., 1996. Economic issues in management of herbicide-resistant weeds. *Review of Marketing and Agricultural Economics,* 64 (3), 301-308.

5a

Stability of pathogen-derived *Potato virus Y* resistance in potato under field conditions and some aspects of their ecological impact

Jörg Schubert[#], Jaroslav Matoušek[##] and Patrick Supp[#]

Abstract

The results of three years of field experiments with transgenic potato clones resistant to *Potato virus Y* (PVY) are presented. The plants were transformed either with a truncated NIb gene of PVY^N fused to the Enhanced Blue Fluorescent Protein gene or the coat-protein gene of a PVY^N strain. It was demonstrated that their resistance to PVY can be overcome by several isolates of this virus. The spectrum of PVY strains infecting transgenic plants was different from that of control plants. On resistant clones the virulent strain PVY^NW prevailed while on control plants as well as on susceptible transgenic plants PVY^N was the dominating strain. Susceptibility or resistance of the transgenic plants to several other viruses was altered too. Some of the clones were more attractive to aphids though the reproduction rate of the aphids on transgenic plants was not altered. Recombination was not observed between transgenic and viral RNAs but between RNAs of invading PVY isolates. We cannot rule out the possibility that the recombination rate between viral RNAs is enhanced in transgenic plants. The paper discusses why so far little use has been made of transgenic approaches for the induction of virus resistance.

Keywords: *Potato virus Y*; potato; transgenic; aphids; pathogen-derived resistance

Introduction

From the beginning of agriculture farmers have been faced with the problem of crop diseases. It is only since the last 200 years that we have known that diseases are mainly caused by pathogens, and the history of plant viruses is even younger. In 1892 D.J. Iwanowski, working on a disease of tobacco, described the first plant virus, *Tobacco mosaic virus* (TMV). From about half way through the last century plant virology developed rapidly. Cloning and sequencing techniques introduced in the early 70s enabled the understanding of the molecular organization of viruses and founded the basis for the genetic improvement of crops.

The history of the potato as an important staple food in many parts of the world is also relatively short. It originated from the Andes of South America and the lowlands of Southern Chile and was introduced into Europe at the end of the 16th century, at

[#] Institute of Resistance Research and Pathogen Diagnostics, Federal Centre for Breeding Research on Cultivated Plants, Theodor-Roemer-Weg 4, 06449 Aschersleben, Germany; J.Schubert@bafz.de; P.Supp@bafz.de

[##] Institute for Plant Molecular Biology, Czech Academy of Sciences, Branišovska 31, 37005 Česke Budejovice, Czech Republic; J.Mat@caz.umbr.cz

J. H. H. Wesseler (ed.), Environmental Costs and Benefits of Transgenic Crops, 63–78.

that time being a short-day crop. Later forms were selected which were adopted for the long 16-18-hour day. They were named *Solanum tuberosum* subsp. *tuberosum* (Bradshaw and Mackay 1994; Ross 1986). Potatoes suffer from a large range of diseases, with viruses playing an important role. In Europe, first reports of epidemics caused by viruses date from 1775. At that time the disease was called 'degeneration' or 'senility' of potato (Schmelzer 1974). Nowadays, natural resistance alleles are known against most of the potato viruses, which confer a high level of resistance (Solomon-Blackburn and Barker 2001; Ross 1986). Consequently, the question arises why we are still faced with the virus problem in potato. In a modern cultivar more than 50 traits have to be combined. Most of them are inherited by polygenes and only a small number of pedigrees of a crossing experiment express the desired features. The fact that the potato is a tetraploid crop increases the difficulty of combining specific traits. Therefore, any new variety is always a compromise between traits that are on a desired level and those which are, unfortunately, not. As quality characteristics have been considered to be more important than resistance to viruses, the compromise looks such that breeding clones with better quality traits but lower resistance levels are preferred to those with a high level of virus resistance but unsatisfactory quality, as viruses can be controlled by eliminating their vectors. The situation may change with the appearance of new more aggressive and harmful isolates of *Potato virus Y* (PVY).

In 1986, the first report appeared on induction of resistance to a virus by means of genetic methods (Beachy, Harrison and Wilson 1999): introducing the coat-protein (CP) gene of TMV into tobacco rendered the plants resistant to infection by the virus. This principle of induction of resistance was named pathogen-derived resistance (PDR). Meanwhile, it was demonstrated for almost any of the important crops and harmful viruses infecting them, that expression of a viral gene or part of it may confer resistance to the corresponding virus. Later on it was demonstrated that expression of the mRNA is often sufficient to induce resistance. There are different models explaining mechanisms underlying the process of resistance induction. Nowadays, in most cases it is explained by post-transcriptional gene silencing (PTGS). In this case only low levels of transgenic RNA are detected in the host plant. Upon infection with a virus, which contains a RNA molecule that is highly homologous to the transgenic sequence (homology-dependent resistance), a mechanism is activated in the host leading to the degradation of both the transcript of the transgene and the viral RNA.

Though nearly 20 years have passed since the first report of PDR, the principle is not yet widely used in practical agriculture (Kawchuk and Prufer 1999). There are several reasons for this. One of the most important is a broad public concern on – never demonstrated – adverse environmental effects of the transgenes. Another one is the complicated patent situation, which makes it almost impossible for small breeding companies to market corresponding cultivars. Approved cultivars with PDR originate mainly from the USA, the world leader in growing genetically improved crops. These are the papaya cultivars 'Rainbow' and 'SunUp' with CP-gene-based resistance to *Papaya ringspot virus* (Gonsalves 1998), the squash cultivars Liberator III, Destiny III and Prelude II with combined resistance to *Zucchini yellow mosaic virus*, *Watermelon mosaic virus 2* and *Cucumber mosaic virus* (CMV) (Tricoli et al. 1995; Lin et al. 2003) and the potato cultivars NewLeaf®, Shepody and Russet Burbank with either *Potato leafroll virus* (PLRV) or PVY resistance (Lawson et al. 2001; Duncan et al. 2002). Except for the NewLeaf® potato cultivars it was demonstrated that some virus isolates exist which can overcome transgenic resistance (Flasinski et al. 2002). Unfortunately, in most publications dealing with PDR to viruses, resistance was tested only with a small number of virus isolates; this does not reflect the natural conditions where the transgenic plants are faced with a large number of different virus isolates. Data on long-term stability of PDR are usually missing. The fact that only limited data exist on the ecological impact of transgenic plants with virus resistance is also an unsatisfactory situation.

Using transgenic potato plants engineered for resistance to PVY we tried to get some answers to open questions such as the stability of resistance, the influence of transgenic plants on aphids settling on them and changes in virus populations occurring in plots with transgenic potatoes. This paper reports the results of field experiments from 2000 until 2002.

PVY is a highly variable virus. Depending on its biological features one can distinguish several strains (Valkonen 1994; Kerlan et al. 1999). PVY^{O} and PVY^{N} have been the most common strains for Europe. The first, also known as the ordinary strain, is characterized by causing heavy leaf symptoms on potatoes while the second, the necrotic strain, does not. Necrotic means that it causes vein necrosis on *Nicotiana tabacum* L. Both strains can also be differentiated by monoclonal antibodies (MAb). In the early 80s of the last century two new strains appeared: PVY^{NTN} (Beczner et al. 1984) and $PVY^{N}W$ (Chrzanowska 1987). The first is the so-called tuber necrosis strain, the second the Wilga strain. $PVY^{N}W$ is a highly virulent biological N-, but serological O-strain. This means that it will react with O-strain specific MAb and cause vein necrosis on *N. tabacum*.

Material and methods

Transgenic potato plants – generation and characteristics

We have reported on the generation of transgenic potato plants with PVY resistance used for field experiments elsewhere (Schubert et al. 2000). Two types of constructs were used for *Agrobacterium tumefaciens*-mediated transformation. In one case the CP gene of strain PVY^{N} CH605 (P. Gugerli, Switzerland), including the 3'-NTR, driven by the CaMV 35S-promoter and terminated by the CaMV 35S-polyadenylation signal, was used (plants provided by G. Barchend, BAZ, Aschersleben). In the other case a truncated NIb gene (lacking at the 3'-end 400 nt) fused in frame to the N-terminus of the Enhanced Blue Fluorescent Protein gene (EBFP, Clontech) was used. The fused genes were also regulated by the CaMV 35S-promoter/terminator sequences. In both cases the cassettes were cloned into a binary plasmid (for NIb-EBFP: pGPTV-Kan, (Becker et al. 1992); for CP: the same plasmid but lacking the NPTII gene). While the CP gene was transferred separately from the plant selection marker NPTII, using a double transformation method including pBIN19 as donor of the NPTII gene (Bevan 1984), in case of NIb-EBFP the selection marker was transferred together with the viral gene on the same binary plasmid. For transformation the variety 'Linda' as well as the dihaploid line DH59 (BAZ, Groß Lüsewitz) were used. Plants were tested for virus resistance by manual inoculation with strain PVY^{NTN}-Hessen (DSMZ, Braunschweig, NIb construct) or PVY^{N}-CH605 (CP construct). Among more than two hundred tested clones 5 revealed a high level of resistance: plants did not get infected systemically when tested 35 days after infection (dai) by DAS-ELISA.

Selected resistant genotypes revealed different mechanisms of resistance when tested with the isolates PVY^{N}-CH605 as well as PVY^{NTN}-Hessen. Linda Nb58 showed extreme resistance to infection: no virus was detected in inoculated and following leaves. Lines DH59 Nb93, Nb146 and Nb156 as well as CP102 revealed recovery from infection: the virus was detected in the inoculated leaves but did not spread. Clones DH59 CP39 and CP41 showed partial resistance (most plants remained free of virus after inoculation). Clones DH59 Nb36, Nb51, Nb80 and Nb88 did not reveal a markedly improved resistance to PNY^{NTN}-Hessen. They were used as susceptible transgenic controls. Linda Nb58 was also extremely resistant to *Potato virus A* (PVA) isolate B11 (BAZ, Aschersleben).

No correlation was found between level of resistance and number of gene copies of the transgene nor their expression level. Apparently, the mechanism of resistance was not based on PTGS as the level of transgenic RNA was equal in resistant and susceptible transgenic plants and did not change after inoculation with PVY (Schubert et al. 2004).

Field experiments for investigation of stability of PVY resistance and changes in virus population in transgenic plants

Field experiments were performed at a location near Aschersleben, Saxony-Anhalt, Germany. This location is characterized by a high incidence of aphid-borne potato viruses.

Tubers from the PVY-resistant as well as susceptible transgenic potato clones were multiplied up under insect-proof glasshouse conditions starting from *in vitro* material. Samples from them were checked for any contamination with PVY, *Potato virus S* (PVS), *Potato virus M* (PVM), PVA and PLRV by DAS-ELISA. Non-transformed tubers of clone DH59 and cvs. Linda, Bettina, Arosa and/or Ute were used as controls. Planting was done at the end of May to ensure successful infection by naturally occurring aphids. Plots consisted of 15 tubers in three replicates with a randomized block design. In the first two years each plot was flanked on two opposite sides by one row of plants of cv. Hansa infected either with PVY^{N}-CH605 (year 2000) or PVY^{NTN}-Hessen (year 2001), which served as additional sources of infection. Two fields of the same design but with different controls were planted. In case of design B the medium resistant clones DH59 CP39 and CP41 were included while they were omitted in plots with design A. Insecticides were not applied other than *Bacillus thuringiensis* preparations against Colorado potato beetle (Novodor FC, Agrinova). Tubers were harvested at the end of August/beginning of September. Three medium-sized tubers per plant were collected and stored in a cold room at 4°C for testing for secondary virus infection.

Field experiments on the influence of transgenic plants on aphid populations

The cultivars/clones Bettina, Ute, Linda, Hansa, DH59 Nb146, DH59 Nb156, Linda Nb58 and DH59 CP102 were used for this investigation. Plots were each planted with 15 tubers in three replicates with a randomised block design.

Aphids were collected in the year 2001 twice and in 2002 once per week from 5 randomly chosen leaves of each plant of the test field. The ratio adult/larvae was determined. It served as an indicator for the reproduction rate of aphids.

Detection of virus infection

Plants from the field were tested in duplicates for primary infection with PVY, PVA, PVS, PVM and PLRV by means of DAS-ELISA (PVY: MAb cocktail, Bioreba; PVA, PVS, PVM and PLRV: polyclonal IgG/alkaline-phosphatase conjugate; PVY^{N}: specific monoclonal antibody; all provided by F. Rabenstein, BAZ, Aschersleben) at the end of July, beginning of August. For this purpose mixed samples from top leaves of each plant were collected and analysed in duplicates.

Plants were scored as infected if OD_{405nm} values exceeded the mean value for healthy control plants plus three times the corresponding standard deviation (threshold value). Standard deviation was calculated from 10 healthy leaf samples. Sprout testing of tubers was carried out in a glass house from January to March. Two tubers per plant were analysed independently. Mixed samples from top leaves of the sprouts were collected and tested in duplicates.

For serotyping of PVY isolates a MAb-based ELISA kit was used (Adgen). For analysis of the presence of $PVY^{N}W$ leaf-sap extracts from plants negatively reacting with the PVY^{N} specific MAbs were obtained and used to inoculate *N. tabacum* plants. Symptoms were scored 21 dai. Veinal necrosis and mosaic symptoms were indicative for the presence of $PVY^{N}W$ as isolates of PVY^{O} induce only mosaic symptoms.

Investigation of recombination events between viral and transgenic RNAs

Leaf samples were analysed for recombinants from sprouts originating from different PVY-infected tubers. When starting the experiments it was not clear whether some resistance-breaking PVY isolates would appear. Thus, in contrast to the experiments described above, for these investigations the susceptible transgenic potato clones DH59 Nb36, Nb51, Nb80 and Nb88 had been included, which revealed the expression of recombinant viral RNA providing the possibility of recombination. Approximately 25 PVY-infected plants were analysed from each transgenic line per year. Immunocapture reverse-transcription PCR (IC-RT-PCR) was used to ensure that viral and not plant recombinant RNA was evaluated. Two hundred μl of goat PVY antiserum (IgG-fraction, 100 ng protein) was used to coat a PCR tube. The same volume of PBS-extracted leaf sap was added to the tube and incubated overnight at 4°C. After washing the tube with PBS and distilled water, a reverse transcription reaction was performed in a final volume of 25 μl (Superscript II, Gibco) with an appropriate 3'-end primer. For the subsequent PCR reaction (Triple master polymerase, Eppendorf), 5 μl of the RT-mix was used in 50 μl final volume with appropriate 5'- and 3'-end primers:

CP: 3'-end primer - 5'-TACAGCCACTGCTATGACAGAATC-3',
5'-end primer - 5'-GCCAACTGTGATGAATGGGCTTATG-3'.
NIb: 3'-end primer - 5'-CCAATTYTCAGGTARACGCCGAAGC-3',
5'-end primer - 5'-TTCTTCAGGCCTTTGATGGATGC-3'.

Supposing that recombination between transgenic and viral RNA is a rare event under the lack of a strong selection pressure it was necessary to pre-select different sequence variants of viral RNAs to reduce the number of clones that had to be sequenced. For pre-selection the Decode Universal Mutation Detection System (BioRad) with an 8% polyacrylamide gel was used at 58°C to do a Constant Denaturing Gel Electrophoresis (CDGE). Five hundred ng of each PCR product, including probes from genomic DNA, were loaded per lane and bands that represent sequence variants of the viral RNA were visualized with ethidium bromide. DNA with altered melting properties and hence different electrophoretic mobility was isolated from the gel matrix, reamplified by PCR, ligated into pGEM-T (Promega), transformed into *Escherichia coli* XL-1 cells (Stratagene) and sequenced. Sequence alignments were performed with DNASIS software (Hitachi).

Experiments on stability of resistance to isolates of different strains of PVY

When it became obvious that resistance of transgenic plants can be overcome by certain PVY isolates, experiments were set up in a growth chamber (22°C, 16h/8h light/dark) to test whether there is a correlation between PVY strains and their ability to overcome resistance. They were done twice. In each experiment three plants of the clones DH59 Nb93, DH59 Nb146, DH59 Nb156 and Linda Nb58 were included. Isolates PVY^{O}-Adgen, PVY^{C}-Adgen and PVY^{N}-Adgen were supplied by Adgen Inc., PVY^{O}-BBA by the DSMZ (Braunschweig, Germany) and PVY^{O}-CZ by P. Dedič (IPB, Havlíčkov Brod, Czech Republic). PVY isolates 2 and 5 belonging to the strain PVY^{N}W were isolated in 2000 at Aschersleben from primarily infected transgenic plants of clone Linda Nb58. Samples from inoculated leaves were tested 10 dai and from uninoculated newly expanding leaves 28 dai.

In another experiment plants of clone Linda Nb58 were inoculated with different isolates of PVY^{NTN} under the same conditions: isolates Hessen, Igor, Lukava (both from P. Dedič), 12/94, Ditta, Gru99 (M. Chrzanowska, PBAI Mlochow, Poland) and - Langenweddingen1 (BAZ, Aschersleben). Infection was tested by DAS-ELISA 7 dai on inoculated and 35 dai on uninoculated leaves.

Results

Stability of resistance to PVY under field conditions

Results of testing for primary and secondary infection with PVY are given in Table 1. Plants with recovery type of resistance became heavily infected in summer time but recovered from virus infection after storage of tubers. In case of lower infection pressure, as in the years 2001 and 2002, these plants revealed some basal level of resistance as only a part of them became infected Extreme resistance of Linda Nb58 was overcome in each year and the plants did not recover from infection.

In all three years an interesting phenomenon was detected. In the case of DH59 Nb146, Nb156 and CP102 single plants were identified that did not recover from the infection. All were grown in the field with design B, which differed from design A only in that there clones DH59 CP39 and CP41 with partial resistance were grown. In 2003 the same was true for clone Linda Nb58.

Appearance of other potato viruses on transgenic plants

According to di Serio et al. (2002) viruses with known suppressers of gene silencing can block transgenic resistance. On the other hand some biochemical processes may have occurred in the transgenic plants rendering them susceptible to, or more resistant against, other viruses. For this reason we investigated whether the incidence of aphid-transmitted viruses was changed on transgenic plants. Results are given in Table 2. It is notable that the appearance of viruses revealed extreme variations during the three investigated years. Thus it is only possible to recognize tendencies. DH59 CP41 appears to be more susceptible to PLRV and more resistant against PVS. DH59 Nb146 seems also to be more resistant against PVS while Linda Nb58 is more susceptible.

An influence of other viruses on the appearance of PVY, acting as suppressers of the resistance reaction to PVY, was not noticed. For instance, in the case of DH59 Nb146 and Nb156, in 2001 most plants were infected with PVM but remained free from PVY. To verify field data we tested the influence of PVS, PVM and PVA on resistance of Linda Nb58 against PVY in a climate-chamber experiment (Schubert et al. 2004). No influence, either on type or level of resistance, was observed thus supporting field data.

Table 1. Incidence (%) of PVY in transgenic potato plants in three field experiments (design A)

a) 2000 growing season

Clone/cultivar	Primary infection with PVY	Secondary infection with PVY
DH59 Nb93	100#	0
DH59 Nb146	100	0
DH59 Nb156	100	0
Linda Nb58	13	42
Controls (DH59, Linda)	100	100

b) 2001 growing season

Clone/cultivar	Primary infection with PVY	Secondary infection with PVY	Secondary infection with PVY (B)
DH59 Nb93	82	7*	n.t.
DH59 Nb146	58	0	27
DH59 Nb156	27	0	7
Linda Nb58	0	30	15
Controls (DH59, Linda)	95	92	

c) 2002 growing season

Clone/cultivar	Primary infection with PVY	Secondary infection with PVY	Secondary infection with PVY (B)
DH59 Nb93	71	5	n.t.
DH59 Nb146	31	0	0
DH59 Nb156	18	0	0
Linda Nb58	2	0	33
DH59 CP102	4	0	7
Controls (DH59/ Linda)§	80/82	18/100	

For each clone 45 plants were tested.

*Only slightly above threshold value.

(B): Design B. In this experimental field, resistant transgenic plants were grown mixed with partially resistant transgenic plants (CP39 and CP41). n.t.: not tested

§ In 2002 pronounced differences existed in secondary infection between both controls.

Table 2. Incidence (%) of four other viruses in transgenic potato clones in field experiments

Clone/cultivar	PVS	PLRV	PVM	PVA
DH59 CP39	-: -: 2*	-: -:13	-: -: 4	-: -: 8
DH59 CP41	-: 0: 7	57:21:16	-:64: 0	2: 0: 0
DH59 CP102	-: -:11	-: -: 4	-: -: 0	-: -: 0
DH59 Nb146	0: 2: 2	14:13: 4	33:53: 0	7: 0: 0
DH59 Nb156	20: 7:16	27: 2:18	23:62:13	14: 2: 9
Linda Nb58	58: 3:60	44: 0: 7	51:64: 2	0:18:18
Ute	60:20: -	9:20: -	40:91: -	4: 2: -
Bettina	13:18:33	11: 9: 4	29:75: 0	11:11:24
Hansa	45:13:77	33: 4: 2	58:62: 3	4:22:42
Linda	4: 5:64	4: 7: 7	52:82: 4	4: 0:62
DH59	-: -: 7	-: -: 0	-: -: 0	-: -: 0
Arosa	-: -:21	-: -: 2	-: -: 0	-: -:33

* In year 2000 (left figure), year 2001 (middle figure) and 2002 (right figure).

-: not tested. Shaded fields: data mentioned in the text.

Resistance to different isolates of PVY

As field experiments revealed that some PVY isolates can overcome resistance of transgenic clones we tested whether this effect is strain-dependent. Data are given in Table 3. They clearly indicate that no correlation exists between the strain used for inoculation and its ability to overcome resistance. The ability to overcome resistance does not depend on differences in the sequences between transgene and invading virus (Schubert et al. 2004).

After storage of tubers harvested in 2001, ring-necrosis symptoms were observed in ca 8% of the tubers of cv. Linda. Tuber ring necrosis was never noticed on transgenic Linda Nb58. The question arose whether Nb58 is resistant against different PVY^{NTN} isolates. Manual inoculation of plants of clone Linda Nb58 with 6 PVY^{NTN} isolates of different geographical origin revealed that this clone was immune to these isolates too. PVY was detected neither on inoculated nor on non inoculated leaves (Table 3).

Table 3. Resistance reactions in four transgenic clones manually inoculated with diverse PVY isolates

PVY isolates	Transgenic clones			
	DH59 Nb93	DH59 Nb146	DH59 Nb156	Linda Nb58
PVY^{N}-CH605	S	S	R	R
PVY^{N}-Adgen	S	RE	RE	S
PVY^{NTN}-Hessen	RE	RE	RE	R
PVY^{NTN}-Igor	-	-	-	R
PVY^{NTN}-Lukava	-	-	-	R
PVY^{NTN}-12/94	-	-	-	R
PVY^{NTN}-Ditta	-	-	-	R
PVY^{NTN}-Gru99	-	-	-	R
PVY^{NTN}-LW1	-	-	-	R
PVY^{O}-Adgen	RE	RE	RE	S
PVY^{O}-BBA	R	S	R	R
PVY^{O}-CZ	S	S	R	S
PVY^{C}-Adgen	S	RE	RE	RE
PVY^{N}W-isolate 2	S	S	R	S
PVY^{N}W-isolate 5	S	R	R	S

S: susceptible (inoculated and uninoculated leaves infected), R: resistant (no virus detected in inoculated or uninoculated leaves), RE: recovery resistance (inoculated leaves infected, no virus detected in uninoculated leaves). Leaves tested 10 and 28 dai by DAS-ELISA.
LW1: isolate Langenweddingen 1; -: not tested.

In 2002 we also compared changes of the strain spectrum of PVY on transgenic and non-transgenic plants. Results are presented in Figure 1. Isolates found on control plants as well as on the resistant clone DH59 Nb93 belonged to PVY^{N}. The same strain prevailed on the susceptible clones DH59 Nb36 and Nb88. In contrast, on the susceptible clones DH59 Nb51 and Nb80 as well as on the other resistant clones mainly isolates of PVY^{O}/PVY^{N}W were detected. All isolates found on DH59 Nb146 and Nb156 belonged to PVY^{N}W too.

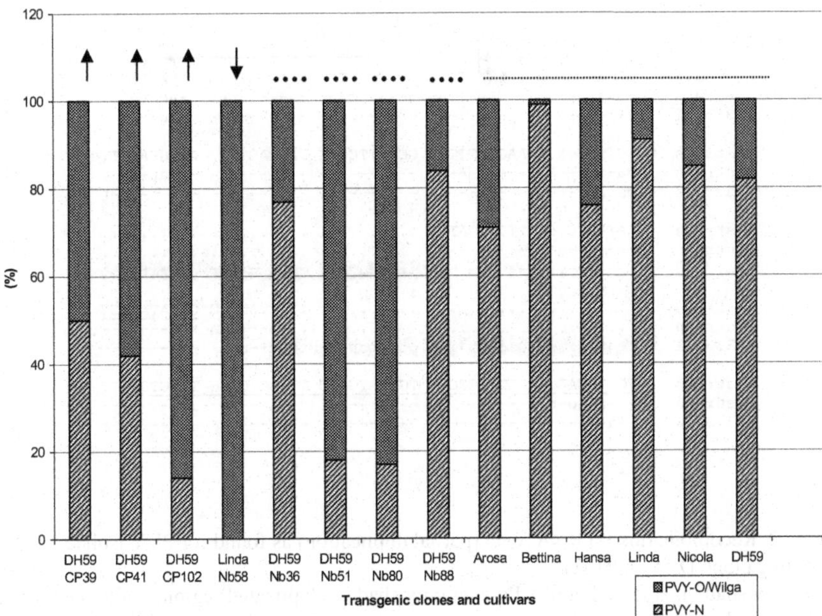

Figure 1. Incidence (%) of PVY^N and PVY^O/PVY^NW^* strains on transgenic potato plants grown in the year 2002.
Data for DH59 Nb93, Nb146 and Nb156 have been omitted as no or only single plants have been infected. ↑: clones with partial resistance and recovery type of resistance; ↓: clone with extreme resistance; ••••: susceptible clones; ....: non-transformed, susceptible cultivars/line. In case of Arosa and Bettina the dealer provided us with false cultivars

Recombination on RNA level

For transformation we used the cDNA of a PVY^N strain. In 2000, the year in which we started the field experiments, most of the PVY isolates identified belonged to PVY^N strains according to their nucleic-acid sequences. Thus, it was complicated to identify recombination between transgenic and viral RNA as differences in sequences have been only marginal. Investigating several hundred pre-selected cDNA clones no hint of a recombination was obtained. On the other hand some mixed infections took place. In this case it was not possible to recognize whether a recombination between an N- and an O-strain was the result of a recombination between viral RNAs or viral and transgenic RNAs. For this reason potatoes were planted in 2002 in a screen house and infected manually with PVY^NW isolate 5. Investigations of the RNA are in progress.

Among investigated clones that had been pre-selected by CDGE three recombinants were identified, two in experiments of the year 2000 and one in the year 2001. They represent a recombination between O- and N-strains and appeared solely on transgenic plants. The most interesting was the mutant detected in 2001 that originated from a plant of clone DH59 CP102. It was the dominating isolate in this plant. The recombination was rather complex. It was based on a partial duplication in the 3'-non-coding region as well as an exchange between sequences of a PVY^N and PVY^O strain (Figure 2). Whether or not it is competitive enough to survive under conditions possibly favouring non-recombinant isolates will be investigated.

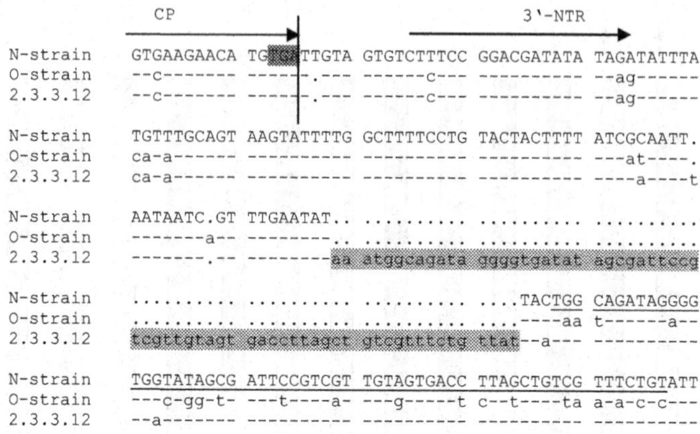

Figure 2. Recombination on RNA level (partial duplication) as found for PVY isolate 2.3.3.12 (infecting clone DH59 CP102).
Dark grey shaded: stop codon of CP gene; grey shaded: duplicated region; underlined: origin of duplicated sequence; ---: identical nucleotide; ...: missing nucleotide

Aphid populations

In 2001 we could not notice any pronounced differences in colonization with aphids of transgenic and non transgenic plants.

In 2002 the aphid population was evenly distributed over the experimental field. While in 2001 the overall number of aphids counted was ca 3,000, this number being more than 36,000 individuals in 2002. In this year differences in population density (adults plus larvae) on different clones and varieties were obvious. Data are given in Table 4.

Table 4. Number of aphids settling on transgenic potato plants in 2002

Clone/cultivar	Adults	Larvae	Overall	Ratio larva/ adults
DH59 Cp39	1,434	4,419	5,853	3.1
DH59 Cp41	1,161	3,256	4,417	2.8
DH59 Cp102	1,123	3,260	4,383	2.9
DH59 Nb146	1,143	2,784	3,927	2.4
DH59 Nb156	826	2,191	3,017	2.7
Linda Nb58	657	1,781	2,438	2.7
			Ø 4,006	**2.8**
DH59	826	2,061	2,887	2.5
Bettina	617	1,293	1,910	2.1
Linda	714	2,055	2,769	2.9
Hansa	854	1,784	2,638	2.1
Arosa	511	1,752	2,263	3.4
			Ø 2,493	**2.6**
			∑ 36,502	

On all DH59 CP clones and on the clone DH59 Nb146 the density of aphids was significantly increased as compared to DH59 Nb156, Linda Nb58 and the control plants.

In 2001 the aphid species were determined too. The prevailing species was *Macrosiphum euphorbiae* making up approximately 99% of the adult population. No differences in appearance of aphid species were found on the different transgenic clones and controls.

Discussion

During three growing seasons we investigated two main aspects of cultivating transgenic potato plants resistant to PVY, viz., stability of the resistance and biosafety aspects.

It is known that transgenic virus resistance of solanaceous hosts can be rather stable, provided that the sequence of the corresponding viral RNA is invariable. This was shown, for example, for *N. benthamiana* L. transformed with the CP gene of *Potato mop-top virus* (Reavy et al. 1997). For most other transgenic plants with PDR it is known that resistance can be overcome by some isolates of the virus if resistance is homology-dependent and based on PTGS. Ability to overcome resistance of our transgenic lines was not associated with differences in the sequences of the transgene and the invading virus. Probably, some of the virus's genes can block the resistance reaction of the host. Similar results were obtained by Maki-Valkama et al. (2000) using transgenic potato plants with the P1 gene of PVY^{O}. In this case resistance was overcome by a PVY^{N} strain. Sequence differences in the corresponding gene were not observed. Lin et al. (2003) investigated squash plants with CP-mediated resistance to CMV. When grown under field conditions plants were infected by several isolates. Again, it was demonstrated that there was no correlation between ability to overcome resistance and degree of sequence homology between the transgene and the invading virus. Several viral genes, as the HC-Pro of PVY and 2b of CMV, are responsible for suppression of the host resistance reaction (Brigneti et al. 1998; Mlotshwa et al. 2002). Virus-mediated suppression of the PTGS-based resistance reaction against one virus can influence the resistance reaction to another one. A similar phenomenon was observed for clone Linda Nb58, which became susceptible to PVA when infected with PVY^{N}W-isolate 5, able to overcome resistance. Similar results were shown by Savenkov and Valkonen (2001) for *N. benthamiana* L., resistant against PVA but susceptible to PVY. If infected with PVY the plants became susceptible to PVA. This illustrates that transgenic resistance against both viruses is possibly based on a common mechanism and that suppression of resistance is an active mechanism. In the case of potyviruses the viral proteins 6K2 and VPg play an important role in this process (Rajamaki and Valkonen 1999). One of such mechanisms blocked by the pathogen could be the process of recognition in which the recently described initiation factor 4E (eIF4E) (Ruffel et al. 2002) is involved.

Level and type of resistance seem to be highly influenced by the site of integration of the transgene. Clones containing the same construct revealed different mechanisms of resistance as well as differences in resistance to several isolates of PVY (Table 3).

The resistance of Linda Nb58 was stable against any of the tested PVY^{NTN} isolates. One can speculate that the gene(s) or part of them responsible for suppression of the resistance reaction of this potato clone is (are) lacking in all tested PVY^{NTN} isolates.

The results indicated that in many cases PDR is not stable. For this reason approaches have to be developed for its stabilization. One such possibility would be to combine different resistance mechanisms or types of resistance. In our case we started to fuse DH lines, which revealed differences in resistance to PVY isolates, e.g. DH59 CP102 with Nb156 or Nb146 with Nb156.

The presented data on isolates capable of overcoming resistance and probably originating from partially resistant plants underline that it is of utmost importance to test stability of resistance before release of a transgenic cultivar with PDR. Otherwise it could be that, due to the high selection pressure, new virus strains or quasispecies evolve in transgenic plants which can overcome the transgenic resistance and also, something that has never been excluded, natural resistance genes. In 2001 and 2002 we identified PVY isolates that were able to overcome resistance genes from *S. stoloniferum* (Ry_{sto}, present in cvs. Bettina and Ute) and *S. demissum* (breeding line from BAZ). These isolates were not competitive with contaminating isolates of PVY, which replaced them when transferred to tobacco. Thus it was not possible to characterize them. One can speculate that this was the result of growing transgenic plants with partial resistance. Of course, we cannot rule out that the identified virulent PVY isolates were already present in the virus population. Growing squash plants with CP-mediated CMV resistance, Lin et al. (2003) also identified several isolates overcoming this type of resistance, but these were already present before growing transgenes.

A possibility to avoid the process of erosion of the transgenic resistance would be a resistance-gene management: in different years cultivars with different resistance mechanisms should be grown. Unfortunately, it was not possible to realize such a concept for common resistance genes and it is highly questionable whether or not this can be done with transgenic plants.

Recombination between transgenic and viral RNAs was not detected. On the other hand several recombinants between O- and N-strains of PVY were identified solely on transgenic plants. One can speculate that replication of viruses is influenced in transgenic plants resulting in an enhanced level of recombination between viral RNAs. An indication that recombination of PVY in transgenic tobacco was greatly enhanced was also obtained by Jakab et al. (2002). In this case isolates appeared as a result of recombination between N- and O-strains that could overcome resistance. Results presented in Table 1 can be interpreted as an indication that incompletely resistant plants fulfil a bridge function for PVY between susceptible and resistant plants in that they exert a selection pressure on the virus to adapt to PDR. Once mutated and able to overcome an incomplete resistance a second mutation could enable the virus to infect plants with a stronger resistance. Probability for such double mutations, necessary to overcome strong resistance, might be too low for it to happen in highly resistant clones.

Although other authors have described how plant viruses with known suppressers of PTGS have enabled the target virus to infect transgenic plants, this phenomenon was not observed in the case of our transgenic lines. One reason could be that transgenic resistance of our lines is not based on PTGS. For practical use of plants with PDR these results imply that clones revealing protein-based resistance should be preferred for practical use as their resistance is more stable against strains differing in their sequence and this type of resistance will not be suppressed by unrelated viruses.

Another issue of the experiments was the investigation of the influence of transgenic resistance on the appearance of other than the target virus PVY as well as on its strain spectrum. Data in Table 2 illustrate that such influence exists. Unfortunately, the data are not fully conclusive, because there are sometimes large differences between the data of different years. Nevertheless, some tendencies can be recognized. While Linda Nb58 is more susceptible to PVS the DH59 clones CP41 and Nb146 seem to be more resistant against PVS. Though more resistant against PVS, DH59 CP41 is more susceptible to PLRV. These data underline that transgenic virus-resistant plants must be tested for susceptibility to other unrelated viruses before being released.

Besides influence on non-target viruses, some influence of the transgenic plants was detected on strains of PVY (Figure 1). As a rule, control plants were spontaneously infected by PVY^N. The same is true for the susceptible clones DH59 Nb36 and Nb88. In contrast to this the resistant transgenic plants were infected mainly by PVY^NW. One explanation for this may be that only more aggressive strains – and actually PVY^NW is the most aggressive among them – can infect transgenic plants. Changes in the spectrum of PVY strains had to be expected as this is a common process when growing plants with resistance genes. The question of whether the isolates infecting the (normally) resistant transgenic clones are more virulent on common potato cultivars and, if so, if this is the result of some unusual recombination remains to be answered.

In the literature no data are available describing the influence of transgenic plants on aphids. Using the ubiquitous active CaMV 35S promoter one has to expect high expression levels of the transgenic proteins in phloem tissue too. This could render transgenic plants more attractive for aphids as it possibly improves the nutritional value of the plants due to an enhanced level of proteins in the phloem. On the other hand, a reduced multiplication rate of aphids on transgenic plants would be an indicator for some intrinsic toxic compounds of these plants. Such toxic effects were not observed, as the multiplication rate remained unchanged. In 2001 the density of the aphid population was low. In that year the non-transgenic control cvs. Bettina and Ute were covered slightly more densely with aphids than other control plants and transgenic clones. Probably, data for 2002 are more reliable. In this year we detected a statistically significant denser population on all transgenic CP clones and on clone DH59 Nb146 (Figure 3). Possibly, these plants produce some compounds which make them more attractive for aphids, or the nutritional value of such plants is really enhanced despite the fact that their reproduction rate is not influenced at all. The epidemiological consequence of this feature would be that plants that are more attractive to aphids should be treated with insecticides. In addition, these plants should have high levels of resistance to other aphid-transmitted viruses because the infection pressure would grow with an increased number of aphids attracted by these plants. Consequently, before releasing a new cultivar with transgenic virus resistance, or any other transgenic feature, it is necessary to test whether this feature is influencing attractiveness of the plants to insects. This might also hold true for the Colorado potato beetle.

Summarizing the presented data we conclude that no obvious risk is expected from growing transgenic plants with PDR against viruses. However, it is necessary to investigate whether the suspected enhanced recombination frequency among virus isolates exists on transgenic plants.

The principle of PDR against plant viruses has been known for nearly 20 years. The question arises why only a limited number of commercially available cultivars with this type of resistance exist. One of the answers is provided by the presented results – the probable instability of this type of resistance in the case of highly variable viruses. Plant-breeding companies have to look at this question mainly from the viewpoint of economics: which development costs are necessary and what benefits they can expect. At least five to six years are necessary to develop a transgenic cultivar with virus resistance. The cost of license fees that would have to be paid is an unknown variable, especially, as the patent situation is confusing at the moment. Additional expenditure is necessary to solve this problem. Further investments have to be made in promotion of such cultivars. At a time of low acceptance expenditures for this must be expected to be high. In addition, the reputation of a breeding firm could suffer from announcing that it is involved in the production of transgenic plants, at least in Europe at the current time. On the other hand, the expected benefits are – under European conditions – marginal. In most cases virus vectors can be controlled by insecticides. Consequently, taking into account the comparably small sizes of European breeding companies one should not expect the appearance of crops with PDR against viruses in the near future. This situation may change rapidly, if methods are found to stabilize the resistance, problems with virus diseases appear that cannot be controlled by conventional methods, public acceptance of GMOs is improved and the patent situation becomes clearer.

Acknowledgements

P. Supp was supported by a grant from the German BMBF (BEO31/0312318), J. Matoušek was supported by grants from the German BMBF (BEO31/0312318) and the Czech Grant Agency (GACR 522/00/0227). The authors wish to thank B. Lorenz, M. Nielitz, B. Deumlich and K. Röse for their excellent technical assistance.

References

Beachy, R.N., Harrison, B.D. and Wilson, T.M.A., 1999. Coat-protein-mediated resistance to *Tobacco mosaic virus*: discovery mechanisms and exploitation. *Philosophical Transactions of the Royal Society of London. Series B. Biological Sciences,* 354 (1383), 659-664.

Becker, D., Kemper, E., Schell, J., et al., 1992. New plant binary vectors with selectable markers located proximal to the left T-DNA border. *Plant Molecular Biology,* 20 (6), 1195-1197.

Beczner, L., Horvath, J., Romhanyi, I., et al., 1984. Studies on the etiology of tuber necrotic ringspot disease in potato. *Potato Research,* 27, 339-352.

Bevan, M., 1984. Binary *Agrobacterium* vectors for plant transformation. *Nucleic Acids Research,* 12 (22), 8711-8721.

Bradshaw, J.E. and Mackay, G.R., 1994. *Potato genetics*. CAB International, Wallingford.

Brigneti, G., Voinnet, O., Li, W.X., et al., 1998. Viral pathogenicity determinants are suppressors of transgene silencing in *Nicotiana benthamiana. EMBO Journal,* 17 (22), 6739-6746.

Chrzanowska, M., 1987. Nowe izolaty wirusa Y zagrazajace ziemniakom w Polsce. *Hodowla Roslin I Nasienn,* 5-6, 8-11.

Di Serio, F., Rubino, L., Russo, M., et al., 2002. Homology-dependent virus resistance against *Cymbidium* ringspot virus is inhibited by post-transcriptional gene silencing suppressor viruses. *Journal of Plant Pathology,* 84 (2), 121-124.

Duncan, D.R., Hammond, D., Zalewski, J., et al., 2002. Field performance of transgenic potato, with resistance to Colorado potato beetle and viruses. *HortScience,* 37 (2), 275-276.

Flasinski, S., Aquino, V.M., Hautea, R.A., et al., 2002. Value of engineered virus resistance in crop plants and technology cooperation with developing countries. *In:* Evenson, R.E., Santaniello, V. and Zilberman, D. eds. *Economic and social issues in agricultural biotechnology*. CABI Publishing, Wallingford, 251-268.

Gonsalves, D., 1998. Control of *Papaya ringspot virus* in papaya: a case study. *Annual Review of Phytopathology,* 36, 415-437.

Jakab, G., Droz, E., Vaistij, F.E., et al., 2002. Durability of transgene-mediated virus resistance: high-frequency occurrence of recombinant viruses in transgenic virus-resistant plants. *In: Perspektiven der Biosicherheit: Rückblick und Ausblick auf die Biosicherheitsforschung in der Schweiz, 5. April 2002, Bern, Hotel Alfa*. Zentrum BATS, Bern. [http://www.bats.ch/bats/wissen/bern/downloads/bern4.pdf]

Kawchuk, L.M. and Prufer, D., 1999. Molecular strategies for engineering resistance to potato viruses. *Canadian Journal of Plant Pathology,* 21 (3), 231-247.

Kerlan, C., Tribodet, M., Glais, L., et al., 1999. Variability of *Potato virus Y* in potato crops in France. *Journal of Phytopathology,* 147 (11/12), 643-651.

Lawson, E.C., Weiss, J.D., Thomas, P.E., et al., 2001. NewLeaf Plus(R) Russet Burbank potatoes: replicase-mediated resistance to *Potato leafroll virus*. *Molecular Breeding,* 7 (1), 1-12.

Lin, H.X., Rubio, L., Smythe, A., et al., 2003. Genetic diversity and biological variation among California isolates of *Cucumber mosaic virus*. *Journal of General Virology,* 84 (1), 249-258.

Mäki-Valkama, T., Valkonen, J.P.T., Kreuze, J.F., et al., 2000. Transgenic resistance to PVY^{O} associated with post-transcriptional silencing of P1 transgene is overcome by PVY^{N} strains that carry highly homologous P1 sequences and recover transgene expression at infection. *Molecular Plant-Microbe Interactions,* 13 (4), 366-373.

Mlotshwa, S., Verver, J., Sithole Niang, I., et al., 2002. Transgenic plants expressing HC-Pro show enhanced virus sensitivity while silencing of the transgene results in resistance. *Virus Genes,* 25 (1), 45-57.

Rajamaki, M.L. and Valkonen, J.P.T., 1999. The 6K2 protein and the VPg of *Potato virus A* are determinants of systemic infection in *Nicandra physaloides*. *Molecular Plant Microbe Interactions,* 12 (12), 1074-1081.

Reavy, B., Sandgren, M., Barker, H., et al., 1997. A coat protein transgene from a Scottish isolate of *Potato mop-top virus* mediates strong resistance against Scandinavian isolates which have similar coat protein genes. *European Journal of Plant Pathology,* 103 (9), 829-834.

Ross, H., 1986. *Potato breeding: problems and perspectives*. Parey, Berlin. Fortschritte der Pflanzenzuechtung no. 13.

Ruffel, S., Dussault, M.H., Palloix, A., et al., 2002. A natural recessive resistance gene against *Potato virus Y* in pepper corresponds to the eukaryotic initiation factor 4E (eIF4E). *Plant Journal,* 32 (6), 1067-1075.

Savenkov, E.I. and Valkonen, J.P.T., 2001. Coat protein gene-mediated resistance to *Potato virus A* in transgenic plants is suppressed following infection with another potyvirus. *Journal of General Virology,* 82 (9), 2275-2278.

Schmelzer, K., 1974. Die Viren. *In: Urania Pflanzenreich. Bd. 1: Niedere Pflanzen.* Urania-Verlag, Leipzig, 28-29.

Schubert, J., Matoušek, J., Mattern, D., et al., 2004. Pathogen-derived resistance in potato to *Potato virus Y*: aspects of stability and biosafety under field conditions. *Virus Research,* 100 (1), 41-50.

Schubert, J., Matoušek, J., Rabenstein, F., et al., 2000. NIb-mediated resistance of potatoes to PVY infection. *Beiträge zur Züchtungsforschung,* 6 (3), 121-125.

Solomon-Blackburn, R.M. and Barker, H., 2001. A review of host major-gene resistance to *Potato viruses X, Y, A* and *V* in potato: genes, genetics and mapped locations. *Heredity,* 86 (1), 8-16.

Tricoli, D.M., Carney, K.J., Russell, P.F., et al., 1995. Field evaluation of transgenic squash containing single or multiple virus coat protein gene constructs for resistance to *Cucumber mosaic virus*, *Watermelon mosaic virus 2*, and *Zucchini yellow mosaic virus*. *Bio/Technology,* 13 (13), 1458-1465.

Valkonen, J.P.T., 1994. Natural genes and mechanisms for resistance to viruses in cultivated and wild potato species (*Solanum* spp.). *Plant Breeding,* 112 (1), 1-16.

5b

Comment on Schubert, Matoušek and Supp: Stability of pathogen-derived *Potato virus Y* resistance in potato under field conditions and some aspects of their ecological impact

R. Laxminarayan[#]

Modern methods of pest control have tended to rely on a single method, whether it be a pesticide or a gene coding for resistance embedded using modern biotechnology tools in the crop itself. An important advantage of integrated pest management (IPM) over reliance on a single tool is that the repeated application of the single method rapidly leads to pest resistance to the tool. Using a variety of methods reduces the likelihood that any single method will fail. For instance, in the case of *Bt* transgenic crops that express a toxin that kills lepidopterans, the use of multiple (stacked) genes is thought to minimize the likelihood that a single mutant pest will be able to overcome a number of different mechanisms of action simultaneously.

The importance of integrated control is an important policy conclusion that might be drawn from the paper by Schubert and colleagues. They present results from field experiments with transgenic potato clones resistant to *Potato virus Y* (PVY), and find that resistance to PVY can be overcome by some strains. Moreover, the strains infecting transgenic plants appeared to be more virulent than those affecting the non-transgenic plants. Finally, engineering potatoes for resistance to PVY alters their defences against other pests such as aphids. These results highlight two pitfalls with genetically engineered resistance against pests.

First, there may be potential problems associated with a single mode of control in that a fast reproducing virus can rapidly generate mutations that can overcome the defenses offered by the resistance gene. This can be addressed either by embedding multiple genes for resistance against viruses in the potato or by combining biotechnology with other methods of pest management to reduce the selection pressure placed on viruses.

Second, methods of pest control may have unintended ecological side-effects, which in turn have economic consequences. In the Schubert study, an example of such a side-effect is the occupation of the ecological niche of a less virulent strain of virus by a potentially more dangerous strain. Excessive use of control could lead to the development of a more virulent strain that may be harder to control. Tolerating less virulent strains as a relatively lesser evil may be a wise alternative.

[#] Resources for the Future, Washington, DC, USA. E-mail: ramanan@rff.org

J. H. H. Wesseler (ed.), Environmental Costs and Benefits of Transgenic Crops, 79.

6a

Bacillus thuringiensis resistance management: experiences from the USA

Terrance M. Hurley[#]

Abstract

The role of the US Environmental Protection Agency (EPA) in the regulation of *Bt* crops is discussed with an emphasis on resistance management. A stochastic bio-economic simulation model is presented to show how previous analyses of insect-resistance management (IRM) policy can be improved by including the effect of farmer adoption and compliance behaviour on the evolution of *Bt* resistance. An example shows that the traditional assumptions of full adoption and compliance over-estimate the risk of *Bt* resistance. However, since adoption and compliance behaviour have countervailing effects on the evolution of resistance, the result should be interpreted with caution until better information is available on farmer adoption and compliance behaviour.
Keywords: *Bt* corn; *Bt* cotton; resistance management; bio-economic; simulation; plant-incorporated protectants

Introduction

Bt crops are engineered with genetic material from the soil bacterium *Bacillus thuringiensis*. This genetic material instructs plants to produce proteins that are toxic when consumed by certain insect pests. *Bt* crops commercialized in the United States (US) include varieties of corn, cotton and potato that control agricultural pests such as the European corn borer, corn rootworm, tobacco budworm, pink bollworm and Colorado potato beatle. In 2002, 24% of nearly 78.8 million acres of corn and 35% of nearly 14.3 million acres of cotton were planted with *Bt* varieties (NASS 2002). While *Bt* corn and cotton adoption has been rapid, Monsanto removed *Bt* potatoes from the market in 2001 because consumer concerns led companies like McDonald's, Burger King, McCain's and Pringles not to buy them (Brammer, Dixon and Ambrose 2003).

The US Environmental Protection Agency (EPA) is responsible for registering pesticides for commercial use in the U.S. While the EPA does not require companies to register the genetic material in herbicide-tolerant crops like Roundup Ready® soybean, it does require companies to register the genetic material in *Bt* crops, which it refers to as plant-incorporated protectants (PIPs). The EPA requires the registration of PIPs because they enable the plant to produce a pesticide. Alternatively, herbicide-tolerant crops do not produce a pesticide. They are treated with a pesticide that is independently registered by the EPA.

[#] Department of Applied Economics, Room 249C Classroom Office Building, 1994 Buford Avenue, University of Minnesota, St. Paul, MN 55108-6040. E-Mail: thurley@apec.umn.edu

J. H. H. Wesseler (ed.), Environmental Costs and Benefits of Transgenic Crops, 81–93.

Companies registering *Bt* PIPs with the EPA are required to develop and implement an approved insect-resistance management (IRM) plan, which is contrary to requirements for other pesticides. The EPA has more requirements for *Bt* PIPs because it wants to promote the sustainable use of what it believes are reduced-risk pesticides (EPA 1998). The potential for *Bt*-resistant insects to evolve is established in the literature (e.g. Tabashnik 1994; Bauer 1995; McGaughey and Beeman 1988; Gahan, Gould and Heckel 2001; Morin et al. 2003) and poses a threat to the sustainable use of *Bt*. With effective IRM, the EPA believes it can conserve the efficacy of *Bt* in order to accomplish greater reductions in human and environmental exposure to more hazardous conventional pesticides.

EPA approved IRM plans are currently based on a high-dose refuge strategy. For a high-dose, the crop is engineered to produce enough toxins to kill all but the most resistant insects (resistant homozygotes). For refuge, farmers are required to plant some crop with a non-*Bt* variety. Refuge slows the evolution of resistance by allowing *Bt*-susceptible insects (heterozygotes and susceptible homozygotes) to thrive and mate with *Bt*-resistant ones. With a high-dose, the majority of progeny are *Bt*-susceptible. The potential for delaying resistance using a high-dose refuge strategy has been demonstrated with simulation models (e.g. Alstad and Andow 1995; Roush and Osmond 1997; Caprio 1998; Gould 1998; Onstad and Gould 1998b; Peck, Gould and Ellner 1999) and experimentally (e.g. Liu and Tabashnik 1997; Tang et al. 2001).

Debate surrounding what constitutes an acceptable IRM plan has centred around three factors: (i) refuge size, (ii) refuge configuration, and (iii) refuge treatment with non-*Bt* insecticides. The proportion of refuge plays a role in determining how fast resistance evolves because it determines the proportion of pests exposed to *Bt*. Refuge configuration, where the refuge is planted in relation to the *Bt* crop, plays a role because it influences the degree to which susceptible pests mate with resistant ones. Treating refuge with non-*Bt* insecticides may speed the evolution of resistance because fewer susceptible pests survive to mate with resistant ones. However, requiring farmers to leave part of their crop unprotected may discourage the adoption of *Bt* crops and encourage the use of more hazardous pesticides.

Early economic models of IRM (e.g. Hueth and Regev 1974; Taylor and Headley 1975; Regev, Gutierrez and Feder 1976; Regev, Shalit and Gutierrez 1983; Gorddard, Pannell and Hertzler 1995) framed the problem as a joint renewable-/nonrenewable-resource problem. It is a renewable-resource problem because pests can rapidly re-establish their populations in the absence of pesticides. It is a nonrenewable-resource problem because pest susceptibility (the converse of resistance) tends to regenerate slowly in the absence of pesticides. Another key feature of the problem is that the marginal productivity of a pesticide depends on the level of resistance. This literature shows how a farmer's long-run economic returns can improve by optimally varying pesticide application rates over time in response to pest abundance and the scarcity of pest susceptibility. Laxminarayan and Simpson (2002) extends this work to *Bt* crops relaxing the assumption that pest susceptibility is nonrenewable. This literature provides additional justification for EPA policy. Since farmers treat pests as common property (Clark and Carlson 1990), they are unlikely to manage pest resistance optimally.

Entomologists and simulation models dominate the IRM literature directly related to EPA policy. The majority of this literature focuses on characterizing how fast resistance evolves under alternative assumptions regarding IRM policy and pest and crop biology (Alstad and Andow 1995; Roush and Osmond 1997; Caprio 1998; Gould 1998; Onstad and Gould 1998b; 1998a; Peck, Gould and Ellner 1999; Caprio 2001;

Onstad et al. 2001; Andow and Ives 2002; Ives and Andow 2002). Others also explore the effect of IRM on agricultural productivity and pesticide use (Onstad and Guse 1999; Hurley, Babcock and Hellmich 2001; Hurley et al. 2002). The strength of these models is their attention to insect behaviour and the biological processes that govern resistance. A notable weakness is the lack of attention given to how farmer behaviour influences the risk of resistance. Two particularly relevant factors are *Bt* crop adoption rates and farmer compliance with IRM guidelines. Both these factors are important because they influence the proportion of pests exposed to *Bt*. One reason for a lack of attention to farmer behaviour is the lead role entomologists have played in the formulation of EPA guidelines. Another is a lack of information on farmer adoption and compliance behaviour.

The purpose of this paper is to demonstrate how ignoring adoption and compliance behaviour can result in an inaccurate assessment of the efficacy of IRM policy. To accomplish this goal, the model developed in Hurley et al. (2002) is extended to include a behavioural model of partial adoption and compliance. Adoption increases with the expected benefits of *Bt* crops, while compliance decreases with the size of refuge and expected benefits of *Bt* crops. Results for partial adoption and compliance are compared with full adoption and compliance.

Model

US EPA (1998, p. 1) expresses the agency's objectives for IRM: "pesticide resistance management is likely to benefit the American public by reducing the total pesticide burden on the environment, and by reducing the overall human and environmental exposure to pesticides". It also illuminates the important trade-offs and constraints that concern the EPA: "It is desire of the EPA that this focus on pesticide resistance management not overly burden the regulated community, jeopardize the registration of reduced risk pesticides, or exclude conventional pesticides or other control practices which can contribute to the further adoption of integrated pest management (IPM)".

Most entomological IRM models focus on quantifying the rate of resistance evolution and do not quantify the effect of IRM on the use of conventional pesticides, the burden to the regulated community or the incentive for industry to develop new reduced-risk pesticides. The models also do not consider the role of IPM. Economic models focus on optimizing the benefits of IRM to farmers, but not on the effect of IRM on environmental loadings of conventional pesticides or incentives for industry to develop new reduced-risk pesticides. IPM is seldom considered. Hurley et al. (2001; 2002) are exceptions who look at agricultural productivity and conventional pesticide use as well as resistance. Still, the models do not distinguish between the burden to the regulated community and incentives for industry to develop new reduced-risk pesticides. All of these models assume full adoption and compliance.

Following Hurley et al. (2002), consider a simplified production region with a single crop and pest. The region is divided between two crop varieties. The first, denoted by $i = 0$, is a conventional variety that also serves as refuge. The second, denoted by $i = 1$, is a *Bt* variety that is toxic when consumed by susceptible pests. Let $1.0 \geq \phi_t \geq 0.0$ be the proportion of conventional acreage planted in season t. The pest reproduces with G generations per season where g denotes the generation in season t. Let $1.0 \geq \iota_{tg}^{i} \geq 0.0$ be the proportion of crop i that receives a conventional pesticide application in season t and generation g. The model allows conventional pesticide treatments on *Bt* acreage because if *Bt* fails due to resistance, farmers may turn to conventional pesticides for supplemental control.

The number of pests per plant emerging to damage crops and reproduce is $n_{tg} \geq 0.0$. Pest populations are random due to environmental events such as storms, though not independent from past populations due to reproduction:

$$n_{tg} \sim \begin{cases} N_g\left(n^S_{tg-1}\right), & \text{for } g > 1 \\ N_g\left(n^S_{t-1G}\right), & \text{for } g = 1 \end{cases} \tag{1}$$

where n^S_{tg} is the number of pests that escape control and survive to damage crops and reproduce, and $N_g(\cdot)$ is a conditional distribution function.

The Hardy-Weinberg model characterizes resistance, assuming it is conferred by a single allele that is not sex-linked. There are two types of alleles: resistant and susceptible. The proportion of resistant alleles is $1.0 \geq r_{tg} \geq 0$. Each pest has two alleles, one from its mother and one from its father, and can be one of three genotypes: resistant homozygote – with two resistant alleles; heterozygote – with one resistant allele; or susceptible homozygote – with no resistant alleles. The Hardy-Weinberg model implies the proportion of each genotype is

$$\eta_{tg} = [r_{tg}^2,\ 2\, r_{tg}(1 - r_{tg}),\ (1 - r_{tg})^2] \tag{2}$$

corresponding to resistant homozygotes, heterozygotes, and susceptible homozygotes.

The Hardy-Weinberg model assumes no selection pressure – survival rates are the same for all genotypes. *Bt* crops select for resistant pests. Let σ_g^i be a 1×3 vector of genotypic survival rates for pests on crop i in generation g with elements corresponding to resistant homozygotes, heterozygotes and susceptible homozygotes. The survival rate of all genotypes treated with conventional pesticides is σ_g^ι. The vector of genotypic survival rates for each crop is then $\rho_{tg}^i = \sigma_g^i + \iota_{tg}^i(\sigma_g^i\sigma_g^\iota - \sigma_g^i)$, implying the number of pests that survive to damage crop i and reproduce is $n^{S}{}_{tg}^{i} = \rho_{tg}^i \cdot \eta_{tg}^i n_{tg}$. The vector of genotypic survival rates for the region is $\rho_{tg} = \rho_{tg}^1 + \phi_t\,(\rho_{tg}^0 - \rho_{tg}^1)$, implying the number of pests that survive to reproduce is $n^S_{tg} = \rho_{tg}\eta_{tg}n_{tg}$.

Since each surviving pest contributes two alleles, resistant homozygotes contribute two resistant alleles and heterozygotes contribute one resistant allele, the proportion of resistant alleles in the subsequent generation is

$$r_{tg} = \begin{cases} \dfrac{\rho_{tg-1} M \eta_{tg-1}}{\rho_{tg-1} \cdot \eta_{tg-1}}, & \text{for } g > 1 \\[2ex] \dfrac{\rho_{t-1G} M \eta_{t-1G}}{\rho_{t-1G} \cdot \eta_{t-1G}}, & \text{for } g = 1 \end{cases} \tag{3}$$

where M is the 3×3 diagonal matrix [1.0, 0.5, 0.0].

Equations (1) – (3) and the initial conditions $n_{01} = N_0$ and $r_{01} = R_0$ describe a dynamic stochastic biological system, which is controlled by the parameters for the proportion of conventional acreage, ϕ_t for t =0,..,T, and conventional pesticide use, ι_{tg}^i for t = 0,..,T, g = 1,..,G, and i = 0, 1. The performance of this system under alternative IRM plans is compared using measures of the risk of resistance, conventional pesticide use, production value to farmers and production value to industry.

The probability that the proportion of resistant alleles exceeds 0.5 within T years measures the risk of resistance to *Bt*:

$$\Theta = \Pr(r_{1T} \geq 0.5) \tag{4}$$

where the probability is defined over the random distribution of pests for t = 0,..,T and g = 1,..,G. The expected number of conventional pesticide applications measures conventional pesticide use:

$$\Gamma = E_n\left[\sum_{t=0}^{T-1}\sum_{g=1}^{G}\frac{\phi_t \iota_{tg}^0 + (1-\phi_t)\iota_{tg}^1}{T}\right] \tag{5}$$

where E_n is the expectation operator defined over the random distribution of pests for t = 0,..,T and g = 1,..,G. The expected annualized net present production value to farmers is a measure of value of the *Bt* crop to the regulated community:

$$\Pi_F = E_n\left[\frac{\sum_{t=0}^{T-1}\delta^t\left(\phi_t\pi_t^0 + (1-\phi_t)\pi_t^1\right)}{\sum_{t=0}^{T-1}\delta^t}\right] \tag{6}$$

where π_t^i is the annual production value to farmers in season t for variety i, δ = 1 / (1 + r) is the discount rate and r is the real rate of interest. The annual production value for variety i is

$$\pi_t^i = P_t^i Y_t^i\left[1 - D\left(n_{t1}^{S\,i},\dots,n_{tG}^{S\,i}\right) - FC_t^i - \sum_{g=1}^{G}\iota_{tg}^i VC_{tg}^i\right] \tag{7}$$

where Y_t^i bushels/acre and P_t^i \$/bushel are the pest-free yields and crop prices; FC_t^i \$/acre is the production cost for items such as seed, fertilizer and labour that are exclusive of the cost of a conventional pesticide application; VC_{tg}^i \$/acre is the cost of a conventional pesticide application; and $D_t^i(n^S{}_{t1}{}^i,\dots, n^S{}_{tG}{}^i)$ is the seasonal proportion of yield lost to pests. The expected annualized net present value of farmer payments to industry for the use of the *Bt* variety is another measure of the production value of *Bt* crops to the regulated community as well as a measure of incentives to develop new reduced-risk pesticides:

$$\Pi_I = E_n\left[\frac{\sum_{t=0}^{T-1}\delta^t(1-\phi_t)TF_t}{\sum_{t=0}^{T-1}\delta^t}\right] \tag{8}$$

where TF_t is technology fee paid by farmers for the right to use the *Bt* variety.

Equations (4)-(6) and (8) are conditional on values assigned to the generations of pest per season, genotypic survival rates, survival rates for conventional pesticides, number of time periods, prices, pest-free yields, production costs, technology fee, discount rate, initial pest population and initial proportion of resistance. While reasonable values are available for many parameters, others are not known for certain. Typically, this uncertainty is addressed using sensitivity analysis for reasonable variations in parameter values. However, with suitable data, this uncertainty can be captured more directly using estimated distributions for the parameters. Let E_{UP} be the expectation operator defined over the distribution of uncertain parameters. Combined with equations (1)-(3) and (7), $E_{UP}[\Theta]$, $E_{UP}[\Gamma]$, $E_{UP}[\Pi_F]$, and $E_{UP}[\Pi_I]$ can be compared for alternative IRM policies to assess how well each meets the EPA's objectives.

Analyses of EPA policy alternatives have focused on the proportion of refuge assuming full adoption and compliance. These analyses hold ϕ_t constant over time assuming it is equal to the EPA's mandated proportion of refuge, ϕ. The adoption rate of a *Bt* variety in season t will depend on the expected return of conventional and *Bt* varieties: $\alpha_t = \alpha[EF(\pi_t 0), EF(\pi_t 1)]$ where $EF(\pi_t 0)$ and $EF(\pi_t 1)$ is the expected return for the conventional and *Bt* variety in season t. Adoption of *Bt* varieties of corn and cotton is far from full even after seven years in the field (NASS 2002). Furthermore, Carrière et al. (2003) found that *Bt* cotton leads to the regional suppression of the pink bollworm, a result predicted by many simulation models. The regional suppression of pests by *Bt* crops serves to reduce the value of these crops to farmers over time and puts downward pressure on adoption. Both of these factors reduce the proportion of the pest population exposed to *Bt* and the risk of resistance.

Becker (1968) argues that compliance costs play an important role in determining compliance rates, suggesting compliance with IRM will depend on the required proportion of refuge and expected return of conventional and *Bt* varieties: $\varphi_t = \varphi[EF(\pi_t 0), EF(\pi_t 1), \phi]$. Agricultural Biotechnology Stewardship Technical Committee (2002) found that 13% of *Bt*-corn farmers surveyed in the Midwestern US did not plant at least 20% refuge as required by the EPA. In the Southern US, where the EPA requires 50% refuge and pest pressure is more severe, 23% of *Bt*-corn farmers reported not planting enough refuge. These results provide anecdotal evidence against full compliance. They also suggest compliance rates are lower for higher refuge requirements and more severe pest pressure.

Given these behavioural adoption and compliance functions, the proportion of conventional variety planted in season t can be written as

(9) $\phi_t = (1 - \alpha_t) + \alpha_t \varphi_t \phi.$

To the extent that *Bt* varieties are not fully adopted, previous analyses tend to underestimate the proportion of conventional acreage. To the extent that farmers violate refuge-size requirements, previous analyses tend to overestimate the proportion of conventional acreage.

Model implementation

Partial adoption and compliance have countervailing effects on the proportion of conventional variety planted and the evolution of resistance. Complex interactions between the biological processes governing resistance and economic incentives governing farmer and industry behaviour make the model generally intractable. Therefore, simulation provides a useful tool for understanding how partial adoption and compliance influence the efficacy of IRM policy. As an example, simulation results are constructed for European corn borer (ECB) *Bt* corn, based on parameter values that are consistent with corn production in the North-Central US.

Table 1 summarizes a variety of the parameter values used for this example. With the exception of the technology fee paid for *Bt* corn, these parameters come from Hurley et al. (2002). Other information not provide in Table 1 includes the distribution of random pests ($N_g(\cdot)$), initial frequency of resistance (R_0), heterozygote survival rate on *Bt* corn (σ_{RS}), frequency of conventional pesticide applications (ι_{tg}^{i}), *Bt*-corn adoption rates (α_t), compliance rates (φ_t), and expectations for the benefit of conventional and Bt corn ($E_F(\pi_t^0)$, $E_F(\pi_t^1)$).

The log-normal distribution of random pest populations is also taken from Hurley et al. (2002). The mean of this distribution for first and second generation ECB are $-3.52 + 1.81 n^S_{t-1\,2} - 0.39 n^S_{t-1\,2}{}^2$ and $-1.59 + 9.47 n^S_{t\,1} - 11.31 n^S_{t\,1}{}^2$ (ECB/Plant). The standard deviation for first and second generation ECB is 0.96 and 1.11 (ECB/Plant). The distributions were estimated using field data from the North-Central US and imply an intergenerational dependence in random ECB populations that can result in the type of regional suppression reported by Carrière et al. (2003). The distributions also imply that population growth is naturally limited.

The initial frequency of resistant alleles and heterozygote survival rate on *Bt* corn are key parameters that are not known for certain. Hurley et al. (2002) use Bayesian methods to estimate a joint distribution for these parameters based on field data. This distribution is used for calculating the expectation E_{UP}. The mean and standard deviation of the initial frequency of resistance is 1.1×10^{-3} and 1.1×10^{-3}. The mean and standard deviation of the heterozygote survival rate on *Bt* corn is 0.026 and 0.027. The correlation is -0.49.

Conventional pesticide applications are simulated based on an IPM economic threshold. Following Mason et al. (1996), the threshold used in the model for first and second generation ECB are $\frac{VC_{t1}^{i}}{0.055P_t^iY_t^i\sigma_1^i\eta_{t1}\left(1-\sigma_1^i\right)}$ and $\frac{VC_{t2}^{i}}{0.028P_t^iY_t^i\sigma_2^i\eta_{t2}\left(1-\sigma_2^i\right)}$. When ECB populations exceed the threshold, conventional pesticides are applied. The thresholds imply that conventional pesticides are used only when the within-season value of an application exceeds the cost.

Table 1. Simulation parameter values

Parameter	Values
Biological parameters	
Pest generation	$G = 2$
Genotypic survival rates	$\sigma_g^0 = [1.0, 1.0, 1.0]$, $\sigma_g^1 = [1.0, \sigma_{RS}, 0.0]$
Conventional pesticide survival rate	$\sigma_1^i = 0.20$, $\sigma_2^i = 0.33$
Initial pest population (pests/plant)	$N_0 = 0.12$
Economic parameters	
Planning horizon (years)	$T = 15$
Interest rate	$r = 0.04$
Price of corn ($/bushel)	$P_t^i = \$2.35$
Pest-free yield (bushels/acre)	$Y_t^i = 130$
Fixed production costs ($/acre)	$FC_t^0 = \$185.00$, $FC_t^1 = \$193.00$
Variable production costs ($/acre)	$VC_t^0 = \$14.00$, $VC_t^1 = \$14.00$
Bt-corn technology fee ($/acre)	$TF_t = \$8.00$
Yield loss (bushels/acre)	$D_t^i(n_{t1}^{S\ i}, n_{t2}^{S\ i}) = \text{Min}\{0.055\, n_{t1}^{S\ i} + 0.028\, n_{t2}^{S\ i}, 1.0\}$

There continues to be a lack of the farm-level data necessary to estimate how adoption and compliance are influenced by the expected production value of *Bt* corn, the size of refuge, and other factors. To illustrate the need for this type of information, exponential adoption and compliance functions are employed. The functions are based on the increase in production value to farmers for switching from conventional to *Bt* corn. The compliance rate also depends on the required refuge size. Specifically, $\alpha_t = e^{103.0\frac{E_F(\pi_t^1)-E_F(\pi_t^0)}{E_F(\pi_t^0)}-2.9}$ and $\varphi_t = e^{46.4\frac{E_F(\pi_t^1)-E_F(\pi_t^0)}{E_F(\pi_t^0)}\phi-2.4}$. The adoption equation assumes 5% of farmers adopt *Bt* corn even when there is no expected increase in production value due to a risk or convenience benefit. Adoption reaches 90% when *Bt* corn is expected to increase the production value by 5%. This adoption equation does not account for the typical technology-adoption cycle, so it tends to over-predict observed adoption trends. The compliance equation assumes that 13% of farmers will not plant a required 20% refuge and 23% will not plant a required 50% refuge when the expected increase in the production value is 5%. These results are roughly consistent with anecdotal evidence (Agricultural Biotechnology Stewardship Technical Committee

2002). The expectation for the production value of conventional and *Bt* corn is the five-year moving average of the past production value: $E_F(\pi_t^i) = \sum_{\tau=t-5}^{t-1} \frac{\pi_\tau^i}{5}$.

The model is implemented in C++ using algorithms in Press, Teukolsky and Vetterling (1992). Monte-Carlo techniques are used to evaluate expectations of the distribution of pest and uncertain parameters.

Results

Comparing the stylized model of full adoption and compliance ($\phi_t = \phi$) to partial adoption and compliance ($\phi_t = (1 - \alpha_t) + \alpha_t\varphi_t\phi$) illustrates the important role human behaviour plays in influencing the efficacy of IRM policy. Figure 1 shows this comparison for the probability that the proportion of resistant alleles exceeds 0.5 in 15 years (risk of resistance), expected percentage decrease in conventional pesticide use (environmental benefit), expected percentage increase in the production value to farmers (farmer benefit), and annualized production value to industry (industry benefit) for refuge requirements ranging from 0 to 50%.

The full model predicts a higher risk of resistance and larger industry benefits than the partial model. For the environmental and farmer benefit, the full model predicts lower values than the partial model for low refuge requirements and higher values for high refuge requirements.

Both models indicate that the risk of resistance falls with an increase in the refuge requirement. However, sensitivity analysis shows that the risk of resistance in the partial model can inrease with the refuge requirement when compliance rates are more sensitive to the cost of compliance. For example, the risk of resistance can be reduced to less than 5% with a refuge requirement of at least 26% in the full model and 7% in the partial model.

Both models show there are limited environmental and farmer benefits to requiring additional refuge. The full and partial models predict environmental benefits are maximized with a 23 and 9% refuge requirement, while farmer benefits are maximized with a 28 and 10% refuge requirement. What is in the interest of the environment is in the interest of farmers. Farmer and environmental benefits move similarly with changes in the refuge requirement because the more farmers can take advantage of *Bt* corn to increase their production value the less they rely on conventional pesticides. Adding human behaviour makes this result more pronounced. A disturbing result in the full model that demonstrates the weakness of the underlying assumptions is a decrease in the production value to farmers when the required refuge is small. Why would farmers ever plant *Bt* corn if it reduced their production value?

The full and partial models produce somewhat conflicting results for industry benefits. The full model predicts no industry benefit from refuge requirements (the industry benefit is maximized with no refuge requirement), while the partial model predicts limited benefits (the industry benefit is maximized with a 4% refuge requirement). The full model assumes farmers must plant *Bt* corn even if it is not in their interest, so the industry has a captive market. Maximizing the amount of *Bt* corn or minimizing the amount of refuge maximizes the industry benefit from this captive market. For the more plausible assumptions of the partial model, farmers choose not to plant *Bt* corn if it is not in their interest. Therefore, if industry wants to sell more *Bt* corn it must ensure its product remains both effective and necessary. Two factors work against the effectiveness and necessity of *Bt* corn. First, as resistance increases

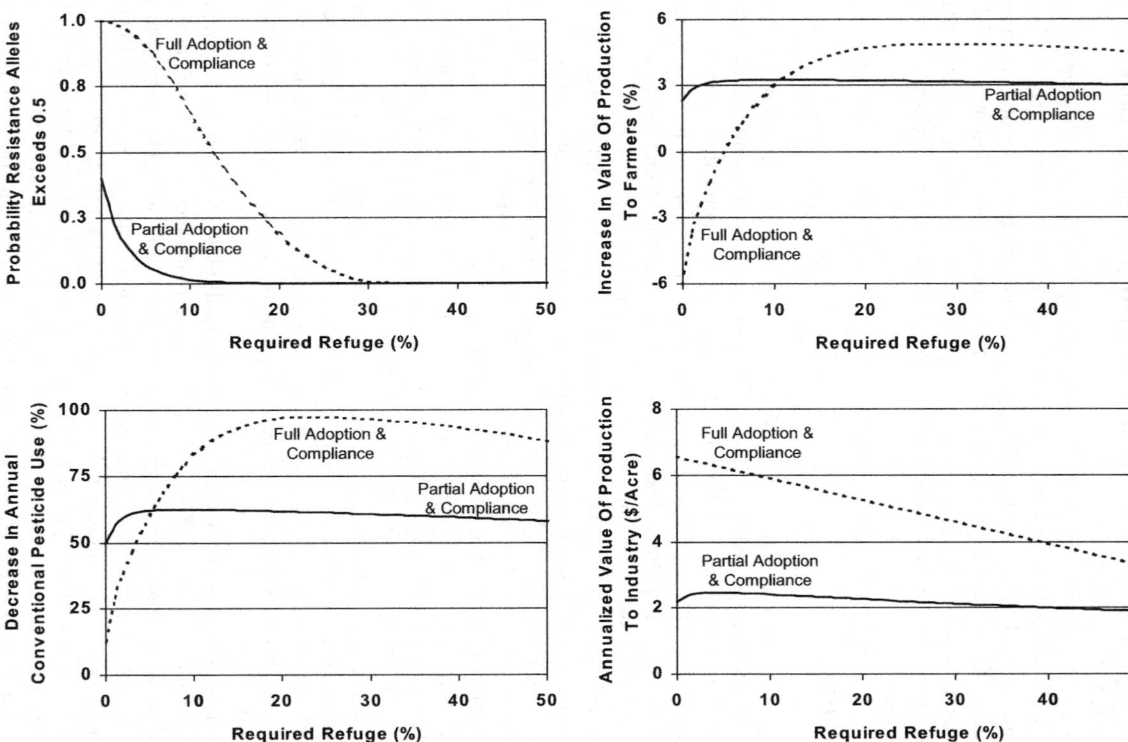

Figure 1. Risk of resistance and environmental, farmer and industry benefits for *Bt* corn by the percentage of required refuge

the product becomes less effective. Since refuge slows the evolution of resistance, it provides an important industry benefit by maintaining the efficacy of *Bt*. Second, regional suppression of ECB reduces the necessity of *Bt* corn, as well as other pesticides. Planting refuge reduces suppression and increases the long-run need for *Bt* corn to the benefit of industry. Interestingly, industry never strongly opposed having refuge requirements. Instead, they argued about how much refuge was necessary. Anecdotally, Monsanto's original voluntary IRM plan required farmers to plant 5 percent refuge for *Bt* corn and cotton.

Most of the entomological literature has relied on the risk of resistance to gauge the efficacy of IRM policy. Figure 1 shows why the risk of resistance may not be a good measure of IRM success given the EPA's stated objectives. While the benefits of increasing the required refuge on the risk of resistance appear unlimited, the benefits to the environment, farmers and industry are. The risk of resistance is only positively correlated with specific measures of EPA objectives for relatively low refuge requirements.

Conclusions

The EPA has determined that IRM for *Bt* crops is in the interest of the American public because it will reduce the use of more hazardous conventional pesticides. To develop IRM guidelines, the EPA has relied heavily on simulation models due to the novelty of *Bt* PIPs. Entomologists have taken a lead role in the developing IRM models to inform policy. These models focus on the insect behaviour and the biological processes governing resistance. A shortcoming is a lack of attention to human behaviour. Of particular importance is farmer adoption and compliance behaviour. Another shortcoming is their focus on the risk of resistance, without explicit consideration of the effect of IRM on conventional pesticide use, and the production value of *Bt* crops to farmers and industry, which are factors that relate more directly to the EPA's stated IRM objectives.

The purpose of this paper was to demonstrate how ignoring adoption and compliance behaviour can result in an inaccurate assessment of the efficacy of IRM policy. An example shows that the traditional assumptions of full adoption and compliance may over-estimate the risk of resistance. However, since adoption and compliance behaviour have countervailing effects, the result is sensitive to underlying assumptions of human behaviour that are not precisely specified at this time. The example also shows the risk of resistance can be a poor gauge of the success of alternative IRM policies.

Farmer adoption and compliance behaviour will play a key role in determining the successes and failures of IRM policy. To assure more successes than failures, farm-level data on adoption and compliance would be useful, so farmer behaviour can be more accurately specified. Another important issue not addressed by any IRM model is industry behaviour. Industry behaviour is important because it sets the price farmers pay for *Bt* crops and is required by the EPA to enforce IRM guidelines. In the present example, the price of the technology is held constant over time and industry enforcement of IRM guidelines is ignored. Since the introduction of ECB *Bt* corn, the price has fallen from around \$10/acre to about \$5/acre. In 2003, the industry instituted a new IRM-compliance assurance programme that includes on-farm monitoring and sanctions for non-compliant farmers. New data and models exploring industry pricing and enforcement behaviour under alternative IRM policies would provide additional insights into how to regulate *Bt* crops more effectively. Adaptive models of IRM are also starting to emerge (Andow and Ives 2002). With better surveillance techniques for monitoring the evolution of resistance, adaptive IRM policies become attractive because much of the information necessary to develop an effective policy can only be learned from commercial release and close scrutiny of *Bt* crops in the field.

References

Agricultural Biotechnology Stewardship Technical Committee, 2002. *Bt corn insect resistance management grower survey: 2001 growing season*. Pioneer Hi-Bred International, Johnston.

Alstad, D.N. and Andow, D.A., 1995. Managing the evolution of insect resistance to transgenic plants. *Science* (268), 1894-1896.

Andow, D.A. and Ives, A.R., 2002. Monitoring and adaptive resistance management. *Ecological Applications,* 12 (5), 1378-1390.

Bauer, L.S., 1995. Resistance: a threat to insecticidal crystal proteins of *Bacillus thuringiensis*. *Florida Entomologist,* 78 (3), 414-443.

Becker, G., 1968. Crime and punishment: an economic approach. *Quarterly Journal of Economics,* 76, 169-217.

Brammer, M., Dixon, F. and Ambrose, B., 2003. *Monsanto and genetic engineering: risks for investors*. Innovest Strategic Value Advisors, New York.

Caprio, M.A., 1998. Evaluating resistance management strategies for multiple toxins in the presence of external refuges. *Journal of Economic Entomology,* 91 (5), 1021-1031.

Caprio, M.A., 2001. Source-sink dynamics between transgenic and non-transgenic habitats and their role in the evolution of resistance. *Journal of Economic Entomology,* 94 (3), 698-705.

Carrière, Y., Ellers-Kirk, C., Sisterson, M., et al., 2003. Long-term regional suppression of pink bollworm by *Bacillus thuringiensis* cotton. *Proceedings of the National Academy of Sciences of the United States of America,* 100 (4), 1519-1523.

Clark, J.S. and Carlson, G.A., 1990. Testing for common versus private property: the case of pesticide resistance. *Journal of Environmental Economics and Management,* 19 (1), 45-60.

EPA, 1998. *The Environmental Protection Agency's white paper on Bt plant-pesticide resistance management*. US Environmental Protection Agency, Washington. [http://www.epa.gov/fedrgstr/EPA-PEST/1998/January/Day-14/paper.pdf]

Gahan, L.J., Gould, F. and Heckel, D.G., 2001. Identification of a gene associated with Bt resistance in *Heliothis virescens*. *Science,* 293 (5531), 857-860.

Gorddard, R.J., Pannell, D.J. and Hertzler, G., 1995. An optimal control model for integrated weed management under herbicide resistance. *Australian Journal of Agricultural Economics,* 39 (1), 71-87.

Gould, F., 1998. Sustainability of transgenic insecticidal cultivars: integrating pest genetics and ecology. *Annual Review of Entomology,* 43, 701-726.

Hueth, D. and Regev, U., 1974. Optimal agricultural pest management with increasing pest resistance. *American Journal of Agricultural Economics,* 56 (3), 543-553.

Hurley, T.M., Babcock, B.A. and Hellmich, R.L., 2001. Bt corn and insect resistance: an economic assessment of refuges. *Journal of Agricultural and Resource Economics,* 26 (1), 176-194.

Hurley, T.M., Secchi, S., Babcock, B.A., et al., 2002. Managing the risk of European corn borer resistance to Bt corn. *Environmental and Resource Economics,* 22 (4), 537-558.

Ives, A.R. and Andow, D.A., 2002. Evolution of resistance to Bt crops: directional selection in structured environments. *Ecology Letters,* 5 (6), 792-801.

Laxminarayan, R. and Simpson, R.D., 2002. Refuge strategies for managing pest resistance in transgenic agriculture. *Environmental and Resource Economics,* 22 (4), 521-536.

Liu, Y.B. and Tabashnik, B.E., 1997. Experimental evidence that refuges delay insect adaptation to *Bacillus thuringiensis. Proceedings of the Royal Society of London. Series B, Biological Sciences,* 264 (1381), 605-610.

Mason, C.E., Rice, M.E., Calvin, D.D., et al., 1996. *European corn borer ecology and management.* Iowa State University, Ames. North Central Regional Extension Publication no. 327.

McGaughey, W.H. and Beeman, R.W., 1988. Resistance to *Bacillus thuringiensis* in colonies of Indianmeal moth and almond moth (Lepidoptera: Pyralidae). *Journal of Economic Entomology,* 81 (1), 28-33.

Morin, S., Biggs, R.W., Sisterson, M.S., et al., 2003. Three cadherin alleles associated with resistance to *Bacillus thuringiensis* in pink bollworm. *Proceedings of the National Academy of Sciences of the United States of America,* 100 (9), 5004-5009.

NASS, USDA, 2002. *Crop production acreage supplement (PCP-BB).* National Agricultural Statistics Service, United States Department of Agriculture, Washington. [http://jan.mannlib.cornell.edu/reports/nassr/field/pcp-bba/acrg0602.pdf]

Onstad, D.W. and Gould, F., 1998a. Do dynamics of crop maturation and herbivorous insect life cycle influence the risk of adaptation to toxins in transgenic host plants? *Environmental Entomology,* 27 (3), 517-522.

Onstad, D.W. and Gould, F., 1998b. Modeling the dynamics of adaptation to transgenic maize by European corn borer (Lepidoptera: Pyralidae). *Journal of Economic Entomology,* 91 (3), 585-593.

Onstad, D.W. and Guse, C.A, 1999. Economic analysis of the use of transgenic crops and nontransgenic refuges for management of European corn borer (Lepidoptera:Pyralidae). *Journal of Economic Entomology,* 92 (6), 1256-1265.

Onstad, D.W., Guse, C.A., Spencer, J.L., et al., 2001. Modeling the dynamics of adaptation to transgenic corn by western corn rootworm (Coleoptera: Chrysomelidae). *Journal of Economic Entomology,* 94 (2), 529-540.

Peck, S.L., Gould, F. and Ellner, S.P., 1999. Spread of resistance in spatially extended regions of transgenic cotton: implications for management of *Heliothis virescens* (Lepidoptera: Noctuidae). *Journal of Economic Entomology,* 92 (1), 1-16.

Press, W.H., Teukolsky, S.A. and Vetterling, W.T., 1992. *Numerical recipes in C++: the art of scientific computing.* 2nd edn. Cambridge University Press, Cambridge. [http://lib-www.lanl.gov/numerical/bookcpdf.html]

Regev, U., Gutierrez, A.P. and Feder, G., 1976. Pest as a common property resource: a case study in alfalfa weevil control. *American Journal of Agricultural Economics,* 58 (2), 186-197.

Regev, U., Shalit, H. and Gutierrez, A.P., 1983. On the optimal allocation of pesticides with increasing resistance: the case of alfalfa weevil. *Journal of Environmental Economics and Management,* 10 (1), 86-100.

Roush, R. and Osmond, G., 1997. Managing resistance to transgenic crops. *In:* Carozzi, N. and Koziel, M. eds. *Advances in insect control: the role of transgenic plants*. Taylor and Francis, London, 271-294.

Tabashnik, B.E., 1994. Evolution of resistance to *Bacillus thuringiensis*. *Annual Review of Entomology,* 39, 47-79.

Tang, J.D., Collins, H.L., Metz, T.D., et al., 2001. Greenhouse tests on resistance management of Bt transgenic plants using refuge strategies. *Journal of Economic Entomology,* 94 (1), 240-247.

Taylor, C.R. and Headley, J.C., 1975. Insecticide resistance and the evaluation of control strategies for an insect population. *Canadian Entomologist,* 107 (3), 237-242.

Roush, R. and Campbell, G. 1997. Managing resistance to transgenic crops. In: Carozzi, N. and Koziel, M. (eds.) *Advances in insect control: the role of transgenic plants*. Taylor and Francis, London, 271-294.

Tabashnik, B.E. 1994. Evolution of resistance to *Bacillus thuringiensis*. *Annual Review of Entomology* 39: 47-79.

Tang, J.D., Collins, H.L., Metz, T.D., [illegible] 2001. Greenhouse tests on resistance management of Bt transgenic plants using refuge strategies. *Journal of Economic Entomology* 94: 240-247.

Taylor, C.E. [illegible] [illegible] *Journal of Economic Entomology* [illegible]

6b

Comment on Hurley: *Bacillus thuringiensis* resistance management: experiences from the USA

Rick Welsh[#]

In this comment I will deviate from the standard discussant format utilized thus far and not focus on the limitations of the paper or model. Terry does us the favour of doing a good job of laying these out in the paper and based on discussions with Terry and his presentation we know that he has undertaken the research required to begin to address the limitations. Also, there is an old adage of "don't bite my finger, look where I am pointing". So in that spirit I will focus on the issues the paper raises and hopefully this will stimulate a discussion of the implications of the paper. In general I found the paper helpful, primarily because of the introduction of producer behaviour and its measurement into the project of managing environmental risks from transgenic crops.

There are some significant implications of this approach. For instance, David Ervin and I argue in our paper (Environmental effects of genetically modified crops: Differentiated risk sssessment and management) that risk-assessment frameworks should be differentiated. We looked at type of genetic modification as the basis for the differentiation. Before reading Terry's paper, we were struggling with incorporating regional or geographic variation into our model. However, Terry has shown that there is another layer of complexity to incorporate potentially. This is also a point made earlier by our discussant (Willem Stiekema). At first, incorporating farmer behaviour into risk assessment seems problematic. However, current insect-resistance management (IRM) plans often rely on speculative assumptions about pest life cycles and new information is emerging continuously. Also risk assessment of the spread of disease among humans relies on research into expected human behaviour. So there are precedents.

Another issue raised is how the extensive literature on technology adoption in agriculture can inform Terry's work on IRM. There is a long history of measuring adoption of technology in economics, sociology and anthropology. A variety of variables have been found to be important in this regard, depending on the situation: farm structure, education, age and even membership in certain cultural groups. It would be interesting to discern if that literature informs this issue.

Finally, Terry's main argument is that regulatory policy is one-sidedly informed by entomology and especially the ecology of insects, and that an effective IRM policy has to be more balanced in favour of the type of information and theory social scientists can provide. He also suggests that ignoring this type of information requirement can lead to inaccurate assessment of IRM policy. I agree with Terry on this point but I also believe that this is a long-standing problem. Social scientists have been making this type of argument to the biological scientists for a long time and about a variety of environmental regulatory issues.

[#] Clarkson University, P.O. Box 5750, Potsdam, NY 13699-5750, USA. E-mail: welshjr@clarkson.edu

J. H. H. Wesseler (ed.), Environmental Costs and Benefits of Transgenic Crops, 95.

7a

Gene flow from crops to wild plants and its population-ecological consequences in the context of GM-crop biosafety, including some recent experiences from lettuce

Clemens van de Wiel[#], Mirella Groot[##] and Hans den Nijs[##]

Abstract

The public concern about the impact of genetically modified crops on the natural environment triggered a steady stream of research during the last decade. Among the possible impacts, the 'escape' of the transgene, either through dispersal of the crop plant outside the agricultural area or through hybridization with wild relatives, attracted a lot of attention, in particular in relation to the possibility of increasing 'weediness'. For gene flow through hybridization to occur, pollen grains must achieve fertilization and seeds must germinate and produce sexually mature plants. Subsequently, the first generation hybrids should be sufficiently fit to survive to sexual maturity and thus produce follow-up generations by which actual introgression into wild acceptor-species genomes could occur through repeated backcrossing. All these steps are reviewed in this paper. It will become evident that, in order to estimate a transgene's capacity to introgress and persist in wild relatives, all steps in the introgression process should be considered. Areas where still relatively little definite data has been published are i) assessing the extent to which genes, such as those conferring resistance to biotic as well as abiotic stresses, indeed enhance fitness in natural settings and the consequences of introgression of these for these environments; and ii) improving this assessment of fitness, e.g. by not only scoring relevant traits, such as those related to fecundity, but also monitoring them in realistic field situations. In this regard, more data on, for instance, the effects of the transgene insertion site on the introgression process and the importance of fitness of the intermediate stages (backcrosses) would be needed to reach a more general insight. In relation to co-existence of GMO and organic agriculture, crop-to-crop gene flow also needs to be controlled. Therefore, a wide variety of possible hybridization barriers, both physical and biological, are discussed. The technical limitations of assessing introgression from crop to wild avoiding the use of transgenic markers are discussed on the basis of work on lettuce.

Keywords: gene flow; hybrids; crops; wild relatives; fitness; GMO; biosafety; molecular markers; transgenes

---

[#] Plant Research International, PO Box 16, 6700 AA Wageningen, The Netherlands. E-mail: clemens.vandewiel@wur.nl

[##] University of Amsterdam, Institute for Biodiversity and Ecosystem Dynamics, PO Box 94062, 1090 GB Amsterdam, The Netherlands

J. H. H. Wesseler (ed.), Environmental Costs and Benefits of Transgenic Crops, 97–110.

Introduction

Gene flow, the movement of genes between populations of a species and between these populations and interfertile relatives, has recently received considerable attention in relation to the introduction of genetically modified (GM) crops. Gene flow from genetically modified (GM) crops to conventional crops and/or wild relatives could occur through dispersal of pollen or seeds, or vegetative parts capable of clonal propagation.

Even though genetically engineered crops could have a number of agronomical or environmental benefits, such as an increase in yield or a decrease of the use of pesticides or fertilizers, there are serious concerns about the possible consequences of the escape of transgenes into the environment. Examples of the risks mentioned in the context of gene flow from GM plants are: i) creation of new weeds resulting from an escape by the crop itself; ii) creation of superweeds by hybridization of a (wild/weedy) species with the transgenic crop; iii) genetic erosion (loss of original diversity of wild relatives).

A massive body of publications on environmental safety of GM crops has appeared (for recent general reviews see Nap et al. 2003; Conner, Glare and Nap 2003). In this paper, a short overview of the gene-flow topic illustrated with a few examples of crops from the literature will be given. These examples include an understudied crop, lettuce (*Lactuca sativa*), which is part of a recently started EU project, acronymed 'ANGEL" (http://www.plant.wageningen-ur.nl/projects/angel), in which the authors participate. First, a few words will be devoted to gene flow within the crop in relation to different styles of farming.

Gene flow from crop to crop

Gene flow from one crop to the other is mainly a concern of breeders, who wish to keep their varieties genetically pure, especially seed-production crops, in order to guarantee their quality, and of gene-bank curators, who want to safeguard the identity of their accessions. With regard to GMOs, there is particular concern with organic and other farmers who strive to produce harvests completely free from products of modern biotechnology. The EU regulation EEC 2092/91 explicitly excludes the use of GMOs in organic farming. Moreover, the supplementary regulation EC 1804/1999 prohibits the presence of GMO material in organic produce, for all practical purposes setting a threshold at the analytical limit of quantification of about 0.1%. Thus, unintended transfer of transgenes to an organic crop could lead to seriously diminishing the quality of seed or produce, leading to a serious setback in their marketability (Eastham and Sweet 2002).

Recently, IPTS (Sevilla) and the EC's Joint Research Centre produced a synthesis report on the co-existence of genetically modified, conventional and organic crops in European agriculture (Bock et al. 2002). They studied three examples, oilseed rape for seed production, maize for feed and potatoes for consumption. Even though there are already well established segregation schemes in place, e.g. for high-erucic-acid oilseed rape and starch potatoes, co-existence will be extremely difficult with a contamination limit of 0.1% as necessitated by the demands of organic farming. This even applies to the potato, where the only problems encountered are in the occurrence of groundkeepers and the possibility of post-harvest mixing. Suggested less stringent thresholds of 0.3% for allogamous seed-production crops like oilseed rape or 1% for feed use (maize) will lead to additional costs due to necessary changes in farming practice and the need for monitoring systems as well as additional insurance. These costs are in the order of 1-10% of the gross margin, but may go up to 41% in particularly difficult cases, such as occur in oilseed rape. Problems in oilseed rape not only relate to substantial insect-mediated pollen flow (20-30% outcrossing level on average), but also to the occurrence of volunteers in and outside of agricultural fields due to seed spillage. Volunteers are particularly problematic in the organic system with its less intensive weed control. In maize, problems are limited to wind-dispersed pollen and seed impurities. However, the difficulties in predicting long-distance pollen flow (see below) will at least necessitate co-operation between neighbouring farms. For a discussion of measures preventing gene flow, see below.

The problems with cross-pollination also apply to many other crops. Even in the as basically selfing regarded lettuce, outcrossing between varieties grown in each other's vicinity was shown up to a level of about 3% by Thompson et al. (1958).

In relation to contamination of genetic resources, a notorious case is the report of the occurrence of transgenic sequences in traditional land races of maize in the Oaxaca region of Mexico (Quist and Chapela 2001). The results were heavily criticized (Christou 2002; Metz and Fütterer 2002; Kaplinsky et al. 2002), and as a consequence, the Nature editor admitted in an editorial note that publication of the original paper was unwarranted on the basis of the evidence presented therein. In a reaction, CIMMYT (cf. http://www.cimmyt.org) stated that it had taken measures to keep its important maize gene-bank collection free of transgene contamination and that in line with this, no transgenes were found upon testing of accessions. As a wind-pollinated, outcrossing species, gene flow into land races of maize would come as no surprise. However, the whole affair highlighted the danger of premature publicizing in such a controversial area.

A consequence of crop-to-crop gene flow could be transgene stacking. For example, three-herbicide-resistant oilseed rape volunteers have been reported as a consequence of cross-pollination between single-resistant lines cultivated in the Alberta region of Canada (Hall et al. 2000). The appearance of such volunteers may lead to adaptation of herbicide treatments depending on the type of crop rotation used, e.g. whether there is a follow-up by other herbicide-tolerant crops, such as beet. On the other hand, weedy populations can also develop herbicide tolerance under selective pressure spontaneously, a phenomenon with which the agricultural industry is not unfamiliar (Conner, Glare and Nap 2003).

Gene flow from crop to wild

Gene flow basics

Ellstrand, Prentice and Hancock (1999) showed that 12 out of the 13 most important food crops of the world hybridize with wild relatives in some part of their agricultural distribution. Also other reviews suggest that gene flow from crop to wild relatives is a common phenomenon. In The Netherlands for instance, about one quarter of 42 reviewed species spontaneously hybridize with one or more species from the local flora (De Vries, Van der Meijden and Brandenburg 1992). De Vries, Van der Meijden and Brandenburg presented their results in the form of so-called botanical files, in which a D_{pdf} code (p for dispersal by pollen, d for dispersal by diaspores, f for frequency of dispersal) is given with each crop species for the benefit of regulatory authorities. According to Ellstrand, Prentice and Hancock (1999) the results on the universality of gene flow can be explained in an evolutionary context, because domesticated plants represent lineages that diverged from their progenitors only a few thousand years ago. Therefore, complete reproductive isolation would not be very likely to have occurred yet.

The key condition for gene flow through hybridization and consequent introgression is, of course, the presence of wild or weedy relatives in the vicinity of the crop. Generally, pollen flow from a specific source follows a leptokurtic distribution, implying that the great majority occurs over a short distance. Long-distance dispersal is rare, but also quite variable and thus it is, for instance, hard to predict what the maximum distance of occurrence is. For instance, in a re-analysis of previous data, Klinger (2002) showed that even for the obligatory outcrossing radish (*Raphanus sativus*), it is difficult to predict the actual rate of gene flow in the field; variability in the rate of pollination was high between experiments, particularly so at larger distances from the source. In wind-pollinated species, particularly trees (e.g. poplar, DiFazio et al. 1999), the picture might even be more complicated, depending on prevailing wind direction etc.

Furthermore, additional pre- and post-zygotic barriers can be found in a number of species and/or varieties. Lefol, Fleury and Darmency (1996) showed that, for example, between genetically modified *Brassica napus* and hoary mustard (*Hirschfeldia incana*) two types of pre-zygotic barriers occurred; the first barrier is pollen germination and growth on the papilla of the foreign species, and the second barrier is the weak attraction of foreign pollen tubes to the micropyles of ovules. With oilseed rape, an example of a post-zygotic barrier was apparent in a cross with wild radish (*Raphanus raphanistrum*): there was low seedling survival when the hybrid plants received their cytoplasm from oilseed rape as mother plant (Guéritaine et al. 2002).

Further introgression of crop genes (transgenes) will depend on the chances of survival of the F_1 hybrids and their fecundity. This will be discussed in the next section.

Fitness of first generation hybrids

Chances of introgression of crop traits into wild relatives will especially be high if fertile hybrids occur frequently (Ellstrand, Prentice and Hancock 1999). In this regard, it is sometimes assumed that crop genes, particularly the ones associated with domestication (and therefore also transgenes), will only be disadvantageous to wild populations or will have certain detrimental/fatal fitness effects on hybrid populations (Small 1984), and therefore they will not be able to spread. However, alleles encoding domestication traits often show a recessive character (cf. Doebley et al. 1990). As a consequence, first-generation hybrids will show more similarity to the wild parent, which will increase their chances of survival away from cultivation.

Nevertheless, examples of maladaptation occur, such as described by Hauser (2002) in hybrids between cultivated carrot (*Daucus carota* ssp. *sativus*) and wild *Daucus carota* ssp. *carota*: these hybrids inherited the sensitivity to frost from their cultivar parent and therefore had a survival rate which was significantly lower than that of the wild carrot and only slightly better than that of the cultivar.

However, in many taxa interspecific hybrids are found that are sometimes as fit as or even fitter than their parents. For example, Hauser, Shaw and Østergård (1998) found that hybrids between weedy *Brassica rapa* and cultivated *Brassica napus* were intermediate in fitness between their parents and significantly fitter than weedy *Brassica rapa* as determined by seed characteristics and survival in the field.

Generally, the transfer of (trans)genes from a crop to wild relatives is more likely to succeed if the trait itself is neutral or beneficial to the hybrid population. In

Cucurbita pepo, the F_1 generation produced from a doubly virus-resistant crop and wild *C. pepo* ssp. *ovifera* was vigorous enough to contribute to the gene pool of subsequent generations (Spencer and Snow 2001), but there was a lot of variation in fecundity between experiments. Whether the trait for virus resistance had any significant effect on hybrid fitness could not be established.

In some crop–wild combinations F_1 hybrids are assumed to be generally rare, and they can be sterile or have a very low fecundity. For example, fertilization of wild radish by oilseed rape was assessed using a transgenic herbicide-resistance marker under agronomical conditions by several groups and low, but widely varying success rates were reported: from none (Rieger et al. 2001), to 10^{-7} to 3.10^{-5} (Chèvre et al. 2000) or up to 6.10^{-5} to 2.10^{-3} (Darmency, Lefol and Fleury 1998). The varying results may be due to variation in experimental set-up, environmental conditions, or to genotypic differences; e.g., Guéritaine and Darmency (2001) showed wide individual variation in effectiveness of ovule fertilization in wild radish by oilseed-rape pollen under controlled conditions. In case of low hybrid fecundity, persistence of hybrid populations will depend on whether they are able to reproduce vegetatively or to recover fertility by repeated backcrossing to the wild progenitor. The possibilities of further introgression are the subject of the next section.

Introgression

In the case of poor compatibility between crop and wild species, fertility can be restored in hybrid offspring when the hybrids backcross to the wild relative. Thus, in hybrids between oilseed rape and wild radish, fertility could be restored by repeated backcrossing to the wild radish (Guéritaine et al. 2002). However, no stable integration of oilseed rape's transgenic herbicide resistance into the wild radish genome was found (Chèvre et al. 1997). It could not be excluded that this was a consequence of the position on the oilseed-rape genome of the herbicide-resistance transgene (Chèvre et al. 1998). Hauser, Jørgensen and Østergård (1998) found that backcrosses and F_2 hybrids between weedy *Brassica rapa* and oilseed rape (*Brassica napus*) had a reduced fitness relative to their parents for most of the fitness components measured, but there was a lot of variation related to genotypes used, and some of the hybrids were as fit as their parents. Therefore, in spite of the low fitness of second-generation hybrids, introgression of transgenes will not be completely hindered.

The other way around, backcrossing will also bring in fitness characters stemming from the wild parent, such as seed dormancy. Resulting progeny can adopt these traits from the wild relative, resulting in plants that are more adapted to the variable environmental conditions of wild populations. Landbo and Jørgensen (1997) found backcross seeds harvested from *Brassica napus* x *Brassica rapa* hybrids to show more weedy *B. rapa*-like behaviour, that is, more seeds were dormant. This result indicated that (trans)gene escape in time, through the persistence of dormant seeds in the soil, is likely to happen.

The chances of transfer of transgenes could be influenced by the integration site of the construct on the genome, as was implied by Metz et al. (1997). They found large differences in the transmission frequency of the transgene between the offspring of two transgenic lines of the amphidiploid *Brassica napus* (containing both A and C genomes) and non-transgenic *Brassica rapa* (containing only the A genome) during backcrossing. The authors suggested this to be due to the construct being situated on the C genome, which might limit the transfer to *Brassica rapa*. On the other hand, related species containing the C genome, e.g. *Brassica oleracea*, would then have an increased chance of introgression of the transgene. Tomiuk, Hauser and Bagger-Jorgensen (2000), however, using a simple population-genetic model, claimed that the results of Metz et al. (1997) could also be explained by selection against transgenic A chromosomes during backcrossing. In the end, the safety of using specific integration sites would depend on the amount of homologous and homeologous (between A and C genomes) recombination events.

In relation to the relevance of the position of the transgene on the parental genome, interesting observations came from studies of hybridization between wild species from the sunflower genus *Helianthus*. Rieseberg et al. (1996) experimentally resynthesized a known ancient hybrid between *H. annuus* and *H. petiolaris*, *H. anomalus*. The three different hybrid lineages created conformed well in their genomic composition to the hybrid species *H. anomalus*, as evidenced by a comparison of linkage maps made using molecular markers; so, only particular combinations of the parental genomes appeared to lead to viable hybrid offspring. This would mean that chances for a transgene to end up in introgressed wild populations could depend on the insertion site. On the other hand, it appears doubtful whether this would be a feasible approach to containment of transgenes in crops.

In conclusion, it is clear that, in order to estimate whether a transgene is capable of introgressing and persisting in wild relatives, all steps in the introgression process should be considered (Landbo and Jørgensen 1997). Furthermore, environmental consequences of introgression in the long run need to be taken into consideration, which is discussed in the next section.

Population-ecological effects

As Ellstrand, Prentice and Hancock (1999) already pointed out, gene flow between a crop and its wild relatives can have two potentially harmful ecological/evolutionary consequences: enhanced weediness in the wild/weedy relatives by introgression of certain crop traits and extinction of wild relatives through genetic swamping or outbreeding depression. For example, the evolution of enhanced weediness in one of the world's worst weeds, *Sorghum halepense*, is assumed to be a result of introgression from the crop *Sorghum bicolor*. Transgenes may enhance these chances, depending on the specific traits encoded.

With regard to the possibilty of 'swamping' of wild relatives, Rufener Al Mazyad and Ammann (1999) showed that the wild *Medicago falcata* is overtaken by the cultivated *Medicago sativa* and the hybrid between *M. sativa* and *M. falcata* (*Medicago x varia*) in Switzerland. This was in line with the observations that backcrosses of hybrids to cultivated *M. sativa* were considerably more vigorous than backcrosses to *M. falcata*. In some former populations of *Medicago falcata,* nowadays only introgressed forms with weakly represented traits of *M. falcata* are found. However, this phenomenon was limited to tetraploid forms of *M. falcata*, since these cross most readily to tetraploid *M. sativa*. Diploid *M. falcata* populations occurring in the eastern part of Switzerland were essentially free from introgression by *M. sativa*.

In contrast, the genetic variation of sea-beet populations (*Beta vulgaris* ssp. *maritima*) was found to increase slightly, for most parameters, by gene flow from sugar beet as well as red beet/Swiss chard (*Beta vulgaris* ssp. *vulgaris*) (Bartsch et al. 1999). In this regard, beet may be a special case, since beet cultivars show an equivalent level of genetic diversity as compared to their wild progenitors, which is highly unusual for a crop/wild combination (Bartsch et al. 1999). Beet also shows interesting complications as a consequence of the possibility of gene flow from wild to crop in the seed multiplication areas in Southern Europe. In Northern Europe, the biennial beet is usually harvested before the onset of flowering, except for the occasional occurrence of bolters. However, hybridization in the seed-production areas between sugar beet and annual/weedy forms of wild sea beet (*Beta vulgaris* ssp. *maritima*) can lead to contamination of seed lots. Through this route, annual weed beets have established in beet cultivation areas in Northern Europe and these weed beets may facilitate gene flow from the crop to wild sea beet where beet cultivation occurs close to the coastal habitat of sea beet in Northern Europe (Boudry et al. 1993). As far as has been ascertained up to now, exchange rates between weed beet and coastal sea beet were below detection level in French cultivation areas (Desplanque et al. 1999).

Recent preliminary analyses of ecological consequences of transgenes in wild populations of *Helianthus annuus* for the first time showed that backcrossed plants transgenic for the *Bacillus thuringiensis* (*Bt*) gene cry1Ac can have a considerably higher seed production (due to decreased herbivory) in a realistic field situation (Snow et al. 2003). This could lead to a high frequency of the *Bt* gene in wild populations and consequently influence the population structure of certain native herbivores.

Barriers to gene flow

Even though pollen flow quickly falls off with distance, distance alone will not completely prevent gene flow and hybridization between crop and wild relatives. In a specific set-up, Reboud (2003) showed that a gap between crops is an inefficient way of reducing pollination: an unplanted distance of 3-4 m had no more effect than one of 1m. Removing the field-border plants of the acceptor plot would be the most efficient way of removing undesired cross-pollinated seeds. Likewise, a non-transgenic barrier crop could act as an 'absorber' of GM pollen by planting it on the area immediately surrounding the crop and subsequently destroying it before seed set, as suggested by Morris, Kareiva and Raymer (1994). Yet, this would be a rather costly practice. Dense stands of shrubs or tree-sized vegetation will also act as traps for airborne particles, including pollen. Jones and Brooks (1952) showed for maize that a tree barrier was effective only for plants growing near to the barrier, but it was much less effective at greater distances.

In addition, engineering biological barriers to prevent gene flow has been widely suggested. Possibilities include genetic engineering the inhibition of flowering, male sterility or seed sterility, and inserting the transgene into the chloroplast genome. Since the chloroplasts are usually only maternally inherited (with notable exceptions, such as in pines), the latter would seriously impede gene flow through pollen dispersal. Seed sterility has not yet been adopted as a technique for genetic isolation because several aspects of these so-called GURTs (Genetic Use Restriction Technologies) or, popularly, 'terminator' technology are not yet totally reliable (Eastham and Sweet 2002). For this purpose, Kuvshinov et al. (2001) suggested a newly developed method called Recoverable Block of Function (RBF) to overcome this unreliability. Their construct will prevent sexual reproduction completely (thus pollen and seed flow is stopped) unless a trigger is applied, which then restores the ability to reproduce. Most recently, a repressible seed-lethal system was proposed, in which the repressor is inserted at the same locus as the seed-lethal system (to which the desired transgene can be linked) itself, but on a different chromosome. At outcrossing, the seed-lethal system (together with any other transgene linked to it) will become segregated from its repressor, thus leading to non-viable offspring

(Schernthaner et al. 2003). However, all these terminator technologies have also been challenged because of possible disadvantageous effects to biological diversity by keeping other breeders from further improvement of this germplasm and by harming subsistence farming by preventing farmers from multiplying the crop themselves for future croppings (Visser et al. 2001)

Assessing gene flow

In outcrossing crops with cross-compatible wild relatives, studying gene flow may be relatively straightforward using standard population-genetic methods of analysis (cf. Raybould and Clarke 1999). This will also be the case with well-circumscribed populations. Thus, Whitton et al. (1997) could assess persistence of crop alleles over five generations following a single hybridization event with the crop in a naturally occurring population of wild *Helianthus annuus* using cultivar-specific RAPD markers. However, ascertaining the occurrence of gene flow will be more difficult where outcrossing events might be expected to be relatively rare or in broader field surveys in which origins of crop alleles are not directly assignable. In an example of the latter on strawberry (*Fragaria* x *ananassa)*, Westman et al. (2001) showed the presence of crop-specific AFLP markers, both from cultivars used in the past and in the present, in wild populations of *F. virginiana* growing in the vicinity of areas of strawberry cultivation. However, the reliability of such observations depends strongly on the degree to which the crop-specificity of the AFLP markers can be ascertained.

In lettuce, Frietema-De Vries, Van der Meijden and Brandenburg (1994) performed an additional field study, because the botanical files study by De Vries, Van der Meijden and Brandenburg (1992) indicated uncertainty on the relationship to its closest wild relative, prickly lettuce (*Lactuca serriola*). They were shown to be essentially conspecific, based on morphology and crossability. Nevertheless, it remained unclear whether any gene flow occurred between crop and wild, for both are regarded as self-pollinating crops, with very limited knowledge available on any amount of cross-pollination. If there were any exchange, this would make it a potentially interesting case for studying the involvement of gene flow from the crop in increased invasiveness, since the wild *L. serriola* expanded its occurrence in Northwestern Europe enormously during the last decades.

One of the aims of the afore-mentioned EU project is filling in this knowledge gap by the use of molecular marker systems. When tracing evidence for past introgression from crop to wild by such methods, several problems arise. Even though markers can be selected that are specific for a panel of varieties of the cultivated form, rare occurrences of these seemingly crop-specific markers in the wild form may still be attributable to common ancestry at a considerably earlier stage of domestication, and not to recent introgression. On the other hand, wild relatives are widely used in breeding, e.g. for the introduction of disease-resistance genes, and thus, specific genomic segments from particular wild accessions have recently been introgressed into the cultivated form. In turn, these segments may also end up in other wild populations in the vicinity of lettuce cultivation.

To tackle these problems, three molecular methods are tested: in the first place, two 'neutral' multi-locus marker systems (AFLP and the retrotransposon-based SSAP). Markers will be selected by linkage mapping, to achieve a good coverage of the genome and thus to obtain maximum informativeness. In this way, it should, in principle, be possible to identify specific chromosomal segments shared by wild and cultivated populations, which could be taken as better evidence for recent introgression than random marker data. For the AFLP technique, this method was indicated to be workable in principle (Van de Wiel et al. 2003). The relatively novel system, SSAP, which is based on retrotransposons occurring ubiquitously in eukaryotic genomes, was also successfully developed for lettuce (Van de Wiel et al. 2003). It has the additional advantage of, in principle, providing markers of which the direction of change can be inferred from the way they move through the genome. Thus, shared unique insertions can be taken as strong evidence for a common origin of the chromosomal parts of the plants/populations in which they are found (for a review see Kumar and Hirochika 2001). In the second place, in order to be able to trace genomic regions most likely involved in breeding efforts, the third 'functional' multi-locus marker system, NBS-profiling, was implemented (Van de Wiel et al. 2003). NBS profiling is screening for variation in and around disease-resistance genes containing the conserved NBS (Nucleotide Binding Site) region, which make up the great majority of resistance genes known so far (Van Tienderen et al. 2002; Van der Linden et al. 2004). In lettuce, a considerable amount of breeding effort has been put into introducing resistance against pathotypes of the downy mildew, *Bremia lactucae*. At least one of these resistance factors, Dm3, has recently been shown to be encoded by a gene belonging to the NBS-LRR type of resistance genes (Meyers et al. 1998).

Conclusions

In recent years, many new data have been generated on the whole gene-flow process as relevant to the GMO debate. Nevertheless, knowledge gaps remain in the following areas: i) actual outcrossing and introgression rates in understudied crops, particularly the ones that have a self-pollination reputation, like lettuce; ii) persistence of crop (trans)genes in wild populations in relation to the extent to which they confer a fitness advantage; and iii) the population-ecological and evolutionary effects of the persistence of introgressed genes. Most important in this regard is the establishment of baseline data, that is, the impact that gene flow from crops already had on their wild congeners, to which the effects of GM crops could be weighed. The great majority of transgenes will confer characteristics not essentially different from their conventional counterparts. Because of the high variability in gene-flow results and the diversity of constructs used, it is often stressed that biosafety should be assessed using a case-by-case approach. This is exemplified by the observations of Linder and Schmitt (1995) that two oil-modification transgenes had a different effect on persistence of both feral *Brassica napus* canola and hybrids of *Brassica napus* and wild *Brassica rapa*. Thus, even transgenes with similar functions could lead to different environmental effects.

With regard to i) it is evident that gene-flow assessment by molecular marker technology can be technically demanding, wherever the use of transgene markers needs to be avoided, as is shown by the lettuce case. In relation to ii) it should be emphasized that still very little evidence was generated on the extent to which genes, such as those conferring resistance to biotic as well as abiotic stresses, enhance fitness in natural settings. Potential threats of the use of certain types of traits (e.g. apomixis genes or stress-tolerance genes) have hardly (or not at all) been studied yet. In addition, there is also a need for improving assessment of fitness, e.g. not only by scoring relevant traits, such as those related to fecundity, but also by demographic monitoring in realistic field situations. Demographic monitoring could establish the life-cycle stages most relevant to population survival and growth, and so put any of the fecundity traits measured in the proper perspective, that is, to what extent the pertaining traits are of critical importance in a taxon's life cycle and therefore, for its establishment, survival and spread in the field (Oostermeijer 2000; Luijten et al. 2002).

Furthermore, evolutionary effects of (trans)gene flow can not be evaluated for a particular region only, because the whole life history of crop and wild relatives has to be examined, in which not only regional variation may occur but which can also be spread out over different areas. This is exemplified by the occurrence of gene flow from wild to crop in seed-producing areas far removed from the ultimate cultivation areas, confounding any gene flow from crop to wild occurring in the latter, such as described for beet. Also, regional differences in the presence of crop relatives may cause a given crop introduction to have very different potential effects. In this respect, export of genetically modified (transgenic) crops to other countries could be a point. For instance, a genetically modified potato cultivar bred in Europe will not show any gene flow there, because of the lack of cross-compatible wild relatives. However, when exported to the potato's region of origin in South America hybridization partners will abound.

With the present uncertainties, the EU Directive 2001/18/EC on the deliberate release into the environment of GM organisms includes a set of outlines for monitoring after introduction in order to establish any unforeseen harmful effects. Since such effects, by definition, are difficult to predict, these claims might very well prove so costly that they seriously interfere with marketability of crop varieties, with a possible exception for the largest and most profitable crops, such as maize or soybean.

Acknowledgements

Part of the reviewing work presented in this article was performed in the framework of a report on gene flow assigned by COGEM (Committee on Genetic Modifcation), advisory body for VROM (Netherlands Ministry of Housing, Spatial Planning and the Environment). The lettuce work was performed in a project co-funded by the EU FP5 Quality of Life Programme, contract no. QLK3-2001-01657.

References

Bartsch, D., Lehnen, M., Clegg, J., et al., 1999. Impact of gene flow from cultivated beet on genetic diversity of wild sea beet populations. *Molecular Ecology,* 8 (10), 1733-1741.

Bock, A.K., Lheureux, K., Libeau-Dulos, M., et al., 2002. *Scenarios for co-existence of genetically modified, conventional and organic crops in European agriculture,* Institute for Prospective Technological Studies and Joint Research Centre of the European Commission IPTS-JRC. Sevilla. [ftp://ftp.jrc.es/pub/EURdoc/eur20394en.pdf]

Boudry, P., Morchen, M., Saumitou Laprade, P., et al., 1993. The origin and evolution of weed beets: consequences for the breeding and release of herbicide-resistant transgenic sugar beets. *Theoretical and Applied Genetics,* 87 (4), 471-478.

Chèvre, A.M., Eber, F., Baranger, A., et al., 1998. Characterization of backcross generations obtained under field conditions from oilseed rape-wild radish F1 interspecific hybrids: an assessment of transgene dispersal. *Theoretical and Applied Genetics,* 97 (1/2), 90-98.

Chèvre, A.M., Eber, F., Baranger, A., et al., 1997. Gene flow from transgenic crops. *Nature,* 389 (6654), 924.

Chèvre, A.M., Eber, F., Darmency, H., et al., 2000. Assessment of interspecific hybridization between transgenic oilseed rape and wild radish under normal agronomic conditions. *Theoretical and Applied Genetics,* 100 (8), 1233-1239.

Christou, P., 2002. No credible scientific evidence is presented to support claims that transgenic DNA was introgressed into traditional maize landraces in Oaxaca, Mexico. *Transgenic Research,* 11 (1), iii-v.

Conner, A.J., Glare, T.R. and Nap, J.P., 2003. The release of genetically modified crops into the environment. Part II. Overview of ecological risk assessment. *The Plant Journal,* 33 (1), 19-46.

Darmency, H., Lefol, E. and Fleury, A., 1998. Spontaneous hybridizations between oilseed rape and wild radish. *Molecular Ecology,* 7 (11), 1467-1473.

De Vries, F.T., Van der Meijden, R. and Brandenburg, W.A., 1992. *Botanical files: a study of the real chances for spontaneous gene flow from cultivated plants to the wild flora of the Netherlands.* Rijksherbarium RU Leiden, Leiden. Gorteria Supplement no. 1.

Desplanque, B., Boudry, P., Broomberg, K., et al., 1999. Genetic diversity and gene flow between wild, cultivated and weedy forms of *Beta vulgaris* L. (Chenopodiaceae), assessed by RFLP and microsatellite markers. *Theoretical and Applied Genetics,* 98 (8), 1194-1201.

DiFazio, S.P., Leonardi, S., Cheng, S., et al., 1999. Assessing potential risks of transgene escape from fiber plantations. *In:* Lutman, P.J.W. ed. *Gene flow and agriculture relevance for transgenic crops: proceedings of a symposium held at the University of Keele, Staffordshire, on 12-14 April 1999.* British Crop Protection Council, Farnham, 171-176. Symposium Proceedings British Crop Protection Council no.72.

Doebley, J., Stec, A., Wendel, J., et al., 1990. Genetic and morphological analysis of maize-teosinte F2 population: implications for the origin of maize. *Proceedings of the National Academy of Sciences of the United States of America,* 87 (24), 9888-9892.

Eastham, K. and Sweet, J., 2002. *Genetically modified organisms (GMOs): the significance of gene flow through pollen transfer.* European Environment Agency, Copenhagen. Environmental Issue Report no. 28. [http://reports.eea.eu.int/environmental_issue_report_2002_28/en/GMOs%20for%20www.pdf]

Ellstrand, N.C., Prentice, H.C. and Hancock, J.F., 1999. Gene flow and introgression from domesticated plants into their wild relatives. *Annual Review of Ecology and Systematics,* 30, 539-563.

Frietema-De Vries, F.T., Van der Meijden, R. and Brandenburg, W.A., 1994. *Botanical files on lettuce (Lactuca sativa): on the chance for gene flow between wild and cultivated lettuce (Lactuca sativa L. including L. serriola L., Compositae) and the generalized implications for risk-assessments on genetically modified plants.* Rijksherbarium RU Leiden, Leiden. Gorteria Supplement no. 2.

Guéritaine, G. and Darmency, H., 2001. Polymorphism for interspecific hybridisation within a population of wild radish (*Raphanus raphanistrum*) pollinated by oilseed rape (*Brassica napus*). *Sexual Plant Reproduction,* 14 (3), 169-172.

Guéritaine, G., Sester, M., Eber, F., et al., 2002. Fitness of backcross six of hybrids between transgenic oilseed rape (*Brassica napus var. oleifera*) and wild radish (*Raphanus raphanistrum*). *Molecular Ecology,* 11 (8), 1419-1426.

Hall, L., Topinka, K., Huffman, J., et al., 2000. Pollen flow between herbicide-resistant *Brassica napus* is the cause of multiple-resistant *B. napus* volunteers. *Weed Science,* 48 (6), 688-694.

Hauser, T.P., 2002. Frost sensitivity of hybrids between wild and cultivated carrots. *Conservation Genetics,* 3 (1), 73-76.

Hauser, T.P., Jørgensen, R.B. and Østergård, H., 1998. Fitness of backcross and F2 hybrids between weedy *Brassica rapa* and oilseed rape (*B. napus*). *Heredity,* 81 (4), 436-443.

Hauser, T.P., Shaw, R.G. and Østergård, H., 1998. Fitness of F-1 hybrids between weedy *Brassica rapa* and oilseed rape (*B. napus*). *Heredity,* 81 (4), 429-435.

Jones, M.D. and Brooks, J.S., 1952. *Effect of tree barriers on outcrossing in corn.* Oklahoma Agricultural Experimental Station, Stillwater. Oklahoma Agricultural Experimental Station Technical Bulletin no. T-45.

Kaplinsky, N., Braun, D., Lisch, D., et al., 2002. Maize transgene results in Mexico are artefacts. *Nature,* 416 (6881), 601.

Klinger, T., 2002. Variability and uncertainty in crop-to-wild hybridization. *In:* Letourneau, D.K. and Burrows, B.E. eds. *Genetically engineered organisms: assessing environmental and human health effects*. CRC Press, Boca Raton, 1-15.

Kumar, A. and Hirochika, H., 2001. Applications of retrotransposons as genetic tools in plant biology. *Trends in Plant Science,* 6 (3), 127-134.

Kuvshinov, V., Koivu, K., Kanerva, A., et al., 2001. Molecular control of transgene escape from genetically modified plants. *Plant Science,* 160 (3), 517-522.

Landbo, L. and Jørgensen, R.B., 1997. Seed germination in weedy *Brassica campestris* and its hybrids with *B. napus*: implications for risk assessment of transgenic oilseed rape. *Euphytica,* 97 (2), 209-216.

Lefol, E., Fleury, A. and Darmency, H., 1996. Gene dispersal from transgenic crops. II. Hybridization between oilseed rape and the wild hoary mustard. *Sexual Plant Reproduction,* 9 (4), 189-196.

Linder, C.R. and Schmitt, J., 1995. Potential persistence of escaped transgenes: performance of transgenic, oil-modified *Brassica* seeds and seedlings. *Ecological Applications,* 5 (4), 1056-1068.

Luijten, S.H., Kery, M., Oostermeijer, J.G.B., et al., 2002. Demographic consequences of inbreeding and outbreeding in *Arnica montana*: a field experiment. *Journal of Ecology,* 90 (4), 593-603.

Metz, M. and Fütterer, J., 2002. Suspect evidence of transgenic contamination. *Nature,* 416 (6881), 600-601.

Metz, P.L.J., Jacobsen, E., Nap, J.P., et al., 1997. The impact on biosafety of the phosphinothricin-tolerance transgene in inter-specific *B. rapa* × *B. napus* hybrids and their successive backcrosses. *Theoretical and Applied Genetics,* 95 (3), 442-450.

Meyers, B.C., Chin, D.B., Shen, K.A., et al., 1998. The major resistance gene cluster in lettuce is highly duplicated and spans several megabases. *Plant Cell,* 10 (11), 1817-1832.

Morris, W.F., Kareiva, P.M. and Raymer, P.L., 1994. Do barren zones and pollen traps reduce gene escape from transgenic crops? *Ecological Applications,* 4 (1), 157-165.

Nap, J.P., Metz, P.L.J., Escaler, M., et al., 2003. The release of genetically modified crops into the environment. Part I: Overview of current status and regulations. *The Plant Journal,* 33 (1), 1-18.

Oostermeijer, J.G.B., 2000. Population viability analysis of the rare *Gentiana pneumonanthe*: importance of genetics, demography, and reproductive biology. *In:* Young, A. and Clarke, G. eds. *Genetics, demography and viability of fragmented populations*. University Press, Cambridge, 313-334.

Quist, D. and Chapela, I.H., 2001. Transgenic DNA introgressed into traditional maize landraces in Oaxaca, Mexico. *Nature,* 414 (6863), 541-543.

Raybould, A.F. and Clarke, R.T., 1999. Defining and measuring gene flow. *In:* Lutman, P.J.W. ed. *Gene flow and agriculture relevance for transgenic crops: proceedings of a symposium held at the University of Keele, Staffordshire, 12-14 April 1999*. British Crop Protection Council, Farnham, 41-48. Symposium Proceedings British Crop Protection Council no. 72.

Reboud, X., 2003. Effect of a gap on gene flow between otherwise adjacent transgenic *Brassica napus* crops. *Theoretical and Applied Genetics,* 106 (6), 1048-1058.

Rieger, M.A., Potter, T.D., Preston, C., et al., 2001. Hybridisation between *Brassica napus* L. and *Raphanus raphanistrum* L. under agronomic field conditions. *Theoretical and Applied Genetics,* 103 (4), 555-560.

Rieseberg, L.H., Sinervo, B., Linder, C.R., et al., 1996. Role of gene interactions in hybrid speciation: evidence from ancient and experimental hybrids. *Science,* 272 (5262), 741-745.

Rufener Al Mazyad, P. and Ammann, K., 1999. The *Medicago falcata/sativa* complex, crop-wild relative introgression in Switzerland. *In:* Van Raamsdonk, L.W.D. and Den Nijs, J.C.M. eds. *Plant evolution in man-made habitats : proceedings of the VIIth international symposium of the International Organization of Plant Biosystematists, 10-15 August 1998, Amsterdam, The Netherlands*. Hugo de Vries Laboratory, Amsterdam, 271-286.

Schernthaner, J.P., Fabijanski, S.F., Arnison, P.G., et al., 2003. Control of seed germination in transgenic plants based on the segregation of a two-component genetic system. *Proceedings of the National Academy of Sciences of the United States of America,* 100 (11), 6855-6859.

Small, E., 1984. Hybridization in the domesticated-weed-wild complex. *In:* Grant, W.F. ed. *Plant biosystematics*. Academic Press, Toronto, 195-210.

Snow, A., Pilson, D., Rieseberg, L.H., et al., 2003. A Bt transgene reduces herbivory and enhances fecundity in wild sunflowers. *Ecological Applications,* 13 (2), 279-286.

Spencer, L.J. and Snow, A.A., 2001. Fecundity of transgenic wild-crop hybrids of *Cucurbita pepo* (Cucurbitaceae): implications for crop-to-wild gene flow. *Heredity,* 86 (6), 694-702.

Thompson, R.C. , Whitaker, T.W., Bohn, G.W., et al., 1958. Natural cross pollination in lettuce. *Proceedings of the American Society for Horticultural Science,* 72, 403-409.

Tomiuk, J., Hauser, T.P. and Bagger-Jorgensen, R., 2000. A- or C-chromosomes, does it matter for the transfer of transgenes from *Brassica napus*. *Theoretical and Applied Genetics,* 100 (5), 750-754.

Van de Wiel, C., Van der Linden, G., Den Nijs, H., et al., 2003. An EU project on gene flow analysis between crop and wild forms of lettuce and chicory in the context of GMO biosafety: first results in lettuce. *In:* Van Hintum, T., Lebeda, A., Pink, D., et al. eds. *Eucarpia leafy vegetables 2003: proceedings of the Eucarpia meeting on leafy vegetables genetics and breeding, Noordwijkerhout, The Netherlands, 19-21 March 2003*. Center for Genetic Resources, Wageningen, 111-116. [http://www.leafyvegetables.nl/download/19_111-116_Wiel.pdf]

Van der Linden, C.G., Wouters, T.C.A.E., Mihalka, V., et al., 2004. Efficient targeting of plant disease resistance loci using NBS profiling. *Theoretical and Applied Genetics,* 109 (2), 384-393.

Van Tienderen, P.H., De Haan, A.A., Van der Linden, C.G., et al., 2002. Biodiversity assessment using markers for ecologically important traits. *Trends in Ecology and Evolution,* 17 (12), 577-582.

Visser, B., Van der Meer, I., Louwaars, N., et al., 2001. The impact of 'terminator' technology. *Biotechnology and Development Monitor* (48), 9-12. [http://www.biotech-monitor.nl/4804.htm]

Westman, A.L., Levy, B.M., Miller, M.B., et al., 2001. The potential for gene flow from transgenic crops to related wild species: a case study in strawberry. *Acta Horticulturae,* 560, 527-530.

Whitton, J., Wolf, D.E., Arias, D.M., et al., 1997. The persistence of cultivar alleles in wild populations of sunflowers five generations after hybridization. *Theoretical and Applied Genetics,* 95 (1/2), 33-40.

7b

Comment on Van de Wiel, Groot and Den Nijs: Gene flow from crop to wild plants and its population-ecological consequences in the context of GM-crop biosafety, including some recent experiences from lettuce

Sara Scatasta[#]

Gene flow from crop to wild plants is one of the potential risks associated with the use of genetically modified organisms. The paper by van de Wiel et al. is an effort to understanding how far we are from answering the question:

"How likely is hybridization from transgenic crops to wild relatives to occur?"

To be able to answer this question is of fundamental importance for the future use of genetically modified organisms in agriculture. The potential evolutionary consequences of the use of transgenic crops have already raised a great deal of concern from consumers as well as producers and the government side of the market for agricultural products. From the consumer side the issues raised can be summarized in 3 broad categories:

1. Loss of biodiversity and environmental concerns;
2. Human health concerns;
3. Concerns related to the effect on prices of agricultural products of an imperfectly competitive biotech industry.

From the producer's point of view the concerns related to gene flow from crop to wild are of 3 main categories:

1. Liability for damages caused to other producers;
2. Ability to identify responsible parties for damages suffered;
3. Ability to calculate the risk of causing damages for which the producer is liable and the extent of these damages and costs of reducing this risk.

From the government point of view the concerns related to gene flow from crop to wild are also of 3 main categories:

1. Ability of setting standards to reduce the risk of damages caused by gene flow;
2. Choice of the optimal governance structure for these standards;
3. Ability to maintain friendly international relations with foreign countries that have adopted different standards.

The paper by van de Wiel et al. analyses the steps that are needed for gene flow from crop to wild to occur actually. In the paper these key elements may influence gene flow:

1. "The presence of wild or weedy relatives in the vicinity of the crop"
2. "The chance of survival of the F_1 hybrids and their fecundity"
3. "The integration site of the construct on the genome".

[#] Wageningen University, Social Sciences Group, Wageningen, The Netherlands. E-mail: sara.scatasta@wur.nl

J. H. H. Wesseler (ed.), Environmental Costs and Benefits of Transgenic Crops, 111–112.

The paper also describes potential barriers to gene flow:

1. Planting a non-transgenic barrier crop around the field planted with the transgenic crop could absorb GM pollen
2. Genetically engineered barriers such as "inhibition of flowering, male sterility, or seed sterility and inserting the transgene into the chloroplast genome".

With respect to lettuce, the paper by van de Wiel et al. describes some new methodologies that can be used to detect whether gene flow has indeed occurred and to what extent.

The paper concludes highlighting the following gaps in the actual knowledge on the issues:

1. The extent to which gene flows from crop to wild occurs in understudied crops such as lettuce;
2. The extent to which gene flows creates 'super weeds';
3. The ecological/evolutionary effects of gene flow.

The paper by van de Wiel et al. certainly confirms where we stand in answering the question asked at the beginning of this brief comment: "How likely is the occurrence of hybridization from transgenic crops to wild relatives?" The paper stresses that hybridization may occur but that further analysis is needed to understand how likely this is. It does, however, contribute greatly to the discussion by describing the steps we need to take now to be able to answer this question in the future.

In order to answer consumers', producers' and governments' concerns about the introduction of genetically modified organisms in European agriculture we need to have a clearer view on what the situation in Europe is with regard to those factors influencing gene flow from crop to wild highlighted in the paper. For example, are there genetically modified crops that are less likely to find wild or weedy relatives in the European Union? As stated by the American Biotechnology Industry Association, "In the United States, most crop species originated elsewhere, so there is very little environmental risk from gene flow for commercial crops such as corn, wheat, soybean, alfalfa, cotton, barley, dry bean and tobacco" (Biotechnology Industry Association 2003). Can we say the same thing for Europe?

Furthermore, is the characteristic mosaic structure of agricultural fields in Europe more likely to cause gene flows from crop to wild than an extensive agricultural-field structure such as that found in the US? When analysing the consequences of genetically modified organisms (GMOs) in European agriculture we need to keep in mind what the advantages of such cropping should be. If the farmer's point of view is taken into consideration, GMOs should increase the profitability of the field. If European farmers are to introduce GMO crops they will lose the premium paid by consumers to have GMO-free products. This means that farmers should be able to produce GMOs at a cost advantage with respect to their international competitors, and this cost advantage should be at least as great as the loss of the premium. The potential liability from damages caused by gene flow from crop to wild introduces a component of risk that should be quantified if the farmer is to base the choice of adoption of GMOs on the correct information.

References

Biotechnology Industry Association, 2003. *Weediness and gene flow*. Available: [http://www.bio.org/foodag/background/geneflow.asp] (June 25th, 2003).

8a

Irreversible costs and benefits of transgenic crops: what are they?

Matty Demont[#], *Justus Wesseler*[##] *and Eric Tollens*[#]

Abstract

The decision of whether to release transgenic crops in the EU is one subject to flexibility, uncertainty and irreversibility. We analyse the case of herbicide-tolerant sugar beet and estimate the maximum irreversible environmental cost that can be tolerated for this technology from a benefit–cost perspective. Among Member States, these costs range from an annual € 50 to € 212 per hectare planted to transgenic sugar beet, i.e. in the range of 27-80% of the reversible benefits. It is questionable whether the environmental cost of herbicide-tolerant sugar beet would exceed this threshold.
Keywords: irreversibility; uncertainty; biotechnology; externality; social benefits and costs; sugar beet

Introduction

The decision whether or not to release transgenic crops is one subject to uncertainty and irreversibility. This has been recognized by economists (Wesseler 2002; Demont, Wesseler and Tollens 2004; Laxminarayan 2003) as well as biologists (Gilligan 2003). Uncertainty related to the release of transgenic crops exists with regard to the future benefits of the technology as, in general, future output and input prices in agriculture are not known with certainty due to several factors including the microclimate, agriculture policies and technical change.

The irreversible effects of a release of transgenic crops include effects on: human health, due to changes in pesticide use; biodiversity, due to gene drift, impacts on unintended target organisms and on pest resistance; climate change, due to changes in greenhouse-gas emissions; investment in farm equipment due to changed seeding technology; and administrative costs due to new biosafety regulations.

Rejecting the existence of some or all of the effects or rejecting that they are irreversible has brought these irreversibility effects into question. We feel that clarification on the meaning of irreversibility from an economic point of view is important, so as to avoid misunderstanding among economists as well as between economists and biologists. A clarification on this point will not only improve communication, but will also be important for future assessments of the technology.

We proceed by providing a definition of irreversibility and discuss the implications for an economic assessment of biotechnology. We are able to show that the criticism

[#] Katholieke Universiteit Leuven, de Croylaan 42, B-3001 Leuven, Belgium. E-mail: matty.demont@agr.kuleuven.ac.be; eric.tollens@agr.kuleuven.ac.be

[##] Environmental Economics and Natural Resources Group, Wageningen University, The Netherlands. E-mail: justus.wesseler@wur.nl

J. H. H. Wesseler (ed.), Environmental Costs and Benefits of Transgenic Crops, 113–122.

against the irreversibility effect is due to a misunderstanding concerning the economic interpretation of the irreversibility effect.

What are irreversibilities?

The effects of irreversibilities on the value of a project, be it an investment by a single investor or a project financed by the government, were analysed in the seminal papers of Arrow and Fisher (1974) and Henry (1974). The basic message is that if one considers an investment with uncertain costs and benefits, irreversible costs and the possibility to postpone the investment (flexibility), then the investment should only be undertaken immediately if the benefits exceed the costs by a certain amount and not if they are equal to or greater than the costs as the standard net-present-value rule suggests. The amount by which the benefits have to exceed the costs under uncertainty, irreversibility and flexibility has been called the quasi-option value. The quasi-option value can be explained by the gains from waiting due to the arrival of new information over time. The concept of the quasi-option value is similar to the real-option value. The real-option value originated from financial economics. In the literature on real-option valuations, the opportunity to invest is valued in analogy to a call option in financial markets. Investors have the right but not the obligation to exercise their investments. This right, the option to invest (real option) has a value, which is a result of the option owner's flexibility. Chavas (1994) provided similar results in his application to investments in agriculture. Dixit and Pindyck (1994) suggest an application of the real-option approach not only to investment problems but to all kinds of decision-making under temporal uncertainty and irreversibility[1]. Recently, the approach has been applied in agriculture to, among others, the adoption of soil-conservation measures (Winter-Nelson and Amegbeto 1998; Shively 2000), marketing (Richards and Green 2000), wilderness preservation (Conrad 2000), agricultural labour migration (Richards and Patterson 1998) and investment in irrigation technology (Carey and Zilberman 2002). Applications related to agricultural biotechnology include studies by Demont, Wesseler and Tollens (2004), Knudsen and Scandizzo (2003), Morel et al. (2003) and Wesseler (2003). Leitzel and Weisman (1999) apply the real-option approach to the analysis of government reforms and argue that new government policies require investments in the form of training of government officials, hiring of additional workers and purchase of equipment. Part of these costs is irreversible and the success of the implemented policy is uncertain, which results under flexibility in a positive value of the option to delay the implementation of the policy.

Decision in the presence of irreversible costs

We consider the effects of irreversibility, uncertainty and flexibility in the context of releasing and adopting transgenic crops. Consider a sugar-beet farmer who wants to move from non-herbicide-tolerant sugar beets, n-htSB, to herbicide-tolerant sugar beets, htSB. To plant the htSB he needs a new planting machine, as he can increase the spacing of the beets[2]. The average gross margin of the n-htSB is about 1000 Euro per hectare. The gross margin of the htSB is expected to be about 1200 Euro per hectare due to higher yields and lower pesticide use. The expected incremental benefit is therefore 200 Euro per hectare and year received at the end of the year. The example will be kept simple by assuming the incremental benefits are certain and will remain constant forever. The discount rate is 10%. What is the value V of adopting htSB under these assumptions? This is simply the present value of the infinite incremental benefit stream, $V=\sum_{t=1}^{t=\infty}200\cdot(1.1)^{-t}=\frac{200}{0.1}=2000$. For the decision to invest in htSB the investment costs have to be deducted. Now, let us assume the farmer can sell his old sugar-beet planter for 500 Euro and buys a new one for 2100 Euro. The net investment costs I are 1600 Euro. The net present value, NPV, of an investment in htSB is $NPV = V - I = 2000 - 1600 = 400$. The NPV is positive and we can conclude that a profit-maximizing farmer adopts htSB. This example illustrates a decision under certainty.

Now, we introduce risk about the future incremental benefits. We will assume the incremental benefits can either be high at 300 Euro or low at 100 Euro depending on the price for sugar beets. The farmer will only know at the end of the year whether or not the price for sugar beets and, hence, the incremental benefits will be high or low. Both situations are equally likely and occur with a probability of $q = 1\text{-}q = 0.5$. As by assumption the farmer is risk-neutral, he would invest if the expected present value of the project is positive. The expect value, $E[V]$, of the project is the sum of the probability-weighted two states of nature:

(1) $$E[V]=0.5\cdot\sum_{t=1}^{t=\infty}300\cdot(1.1)^{-t}+0.5\cdot\sum_{t=1}^{t=\infty}100\cdot(1.1)^{-t}=2000.$$

The result is the same as before. Deducting the initial investment costs of 1600 Euro provides the same NPV of 400 as before.

What is the value, V, of adopting htSB, if the future incremental benefits are low? This is: $V_0=\sum_{t=1}^{t=\infty}100\cdot(1.1)^{-t}=1000$. In this case the value of the project does not cover the initial investment costs of 1600 Euro. This would not be a problem if the farmer could easily sell his planting machine after one year for 1600 Euro or a little less due to depreciation from using the machine. In this case the initial investment costs would be reversible.

In most cases it would be difficult to sell the planting machine. The low incremental benefits would not only effect one particular farmer but several and therefore, several farmers would want to sell their machines, whereas there would be almost no one interested in buying. Also, asymmetric information about the quality of the machine and the transaction costs of finding a buyer and negotiating the sale lower the net price of the machine.

In the case he is unable to sell the planting machine, the investment costs are totally irreversible. In the case he has to sell the machine at a price below 1600 Euro, the investment would be partially irreversible or sunk[3].

Now, we assume the farmer is flexible and can postpone his decision. Would this provide him with any additional gain? Yes, it would, if the investment costs are irreversible. Consider the following: the farmer postpones his investment by one period. In the case the gross margin increases, the NPV of the investment one year from now is: $NPV_1=-1600+\sum_{t=2}^{t=\infty}300\cdot(1.1)^{-t}=1400$ or in today's value $NPV_0=NPV_1/1.1=1273$. In case the gross margin decreases, the NPV of the investment one year from now is: $NPV_1=-1600+\sum_{t=2}^{t=\infty}100\cdot(1.1)^{-t}=-600$ or in today's value $NPV_0=NPV_1/1.1=-545$. In the latter case the farmer would not invest. The gain from waiting is the gain from avoiding losses of 545 Euro in present value. The economic gain from waiting can be calculated by comparing the expected

$E\left[NPV_0^I\right]$ of the immediate investment with the $E\left[NPV_0^P\right]$ from waiting one year. The $E\left[NPV_0^I\right]$ from immediate investment is 400 Euro. The $E\left[NPV_0^P\right]$ is:

$$(2)\qquad E\left[NPV_0^P\right]=\left[0.5\cdot\left(-1600+\sum_{t=2}^{t=\infty}300\cdot(1.1)^{-t}\right)+0.5\cdot(0)\right]\Big/1.1=636\,.$$

The $E\left[NPV_0^P\right]=636$ and is greater than the $E\left[NPV_0^I\right]$ of 400 Euro from immediate investment. In this case it would be worthwhile waiting. The economic gain from waiting is the difference between the two, i.e. 236 Euro.

At this point it is worthwhile noting the importance of the irreversibility effect. It only pays to wait when the investment costs are irreversible. This observation will be even more obvious if the incremental net benefit would be negative in the bad case. Then the farmer would immediately stop producing htSB and move back to planting n-htSB[4].

If the initial investment costs were not irreversible, immediate investment would be optimal[5]. Also, it would be optimal to invest immediately if the investment could not be postponed due to other circumstances, such as a contract for planting htSB only offered once.

A third important observation is the opportunity cost of waiting. Waiting pays as the veil of uncertainty will be removed after one year, but at the same time the benefits at the end of year one are foregone. These foregone benefits of expected 200 Euro are the opportunity costs of waiting.

Decision in the presence of irreversible costs and irreversible benefits

The benefits that have been discussed, the incremental benefits, are reversible. By stopping planting htSB, incremental benefits are also foregone. As there are irreversible costs there are also irreversible benefits. These are benefits that will continue to be present even if the action that has produced them stops. Consider, for example, a one-time subsidy of 500 Euro for planting htSB. There are other examples that will be discussed in more detail later. The $E\left[NPV_0^I\right]$ increases in this case by exactly 500 Euro and the $E\left[NPV_0^I\right]=900$. The $E\left[NPV_0^P\right]$ from waiting in this case is:

$$(3)\qquad E\left[NPV_0^P\right]=\left[0.5\cdot\left(-1600+500+\sum_{t=2}^{t=\infty}300\cdot(1.1)^{-t}\right)+0.5\cdot(0)\right]\Big/1.1=864\,.$$

The $E\left[NPV_0^I\right]>E\left[NPV_0^P\right]$ and there are no gains from waiting. The irreversible benefits reduce the irreversible cost, which leads in this case to an immediate investment.

In the last case, irreversible benefits were considered in the form of a grant. Now, consider the case where the subsidy is in the form of a loan and has to be paid back after ten years. In this case do irreversible benefits matter? Comparing the results of the subsidy as a grant with those as a loan will provide the information. The $E\left[NPV_0^I\right]$ of an immediate investment in htSB, where the subsidy has to be paid back after ten years, provides the following result:

$$(4)\qquad E\left[NPV_0^I\right]=-1600+500-\frac{500}{1.1^{10}}+0.5\cdot\left[\sum_{t=1}^{t=\infty}300\cdot(1.1)^{-t}+\sum_{t=1}^{t=\infty}100\cdot(1.1)^{-t}\right]=707\,.$$

The $E\left[NPV_0^I\right]$ of an immediate investment in this case is 707 Euro, which is more than in the case without the subsidy (400 Euro) and less than in the case with the subsidy as a grant (900 Euro).

The result for a postponed investment is the following:

$$(5)\qquad E\left[NPV_0^P\right] = \left[0.5\cdot\left(-1600+500-\frac{500}{1.1^{10}}+\sum_{t=2}^{t=\infty}300\cdot(1.1)^{-t}\right)+0.5\cdot(0)\right]\Big/1.1=776.$$

The $E\left[NPV_0^P\right]$ of an immediate investment in this case is 775 Euro, which is also in this case higher than in the case without the subsidy (636 Euro) and lower than in the case with the subsidy as a grant (864 Euro). We further observe that the optimal decision will be to postpone the investment, wait for one year and to invest if the incremental benefits increase and not to invest if they decrease. Again, we observe positive gains from waiting. The first case, of irreversible benefits, only is similar to the case where the adoption of transgenic crops reduces the use of pesticides that are harmful to human health.

Decision in the presence of irreversible benefits

Another interesting question related to the irreversible benefits is whether there are gains from waiting if only irreversible benefits and no irreversible costs are present or if the net irreversibility effect is positive. Under a positive net irreversibility effect there will be no gains from waiting, as there are no losses that can be avoided. The $E\left[NPV_0^I\right]$ in the case of irreversible benefits only is

$$(6)\qquad E\left[NPV_0^I\right]=500+0.5\cdot\sum_{t=1}^{t=\infty}300\cdot(1.1)^{-t}+0.5\cdot\sum_{t=1}^{t=\infty}100\cdot(1.1)^{-t}=2500$$

and in the case of the postponed investment:

$$(7)\qquad E\left[NPV_0^P\right]=0.5\left[500+\sum_{t=2}^{t=\infty}300\cdot(1.1)^{-t}+500+\sum_{t=2}^{t=\infty}100\cdot(1.1)^{-t}\right]\Big/1.1=2273.$$

The $E\left[NPV_0^I\right]$ under this scenario will always be greater than the $E\left[NPV_0^P\right]$ due to the discounting effect and therefore waiting does not provide an economic gain.

The important observations about the irreversible benefits are threefold. First, irreversible benefits reduce irreversible costs and this by the order of one. One unit of irreversible benefits compensates for one unit of irreversible costs. Second, a decrease in irreversible benefits over time, even up to a hundred percent, still has a positive impact on the value of the project. Third, a positive irreversibility effect does not provide economic gains from waiting.

The special case of pest resistance

An interesting effect to analyse in more detail is the possibility of pest resistance. The susceptibility of pests to control agents has been viewed by economists as a non-renewable resource, and hence the appearance of pest resistance as an irreversibility. Biologists and entomologists in particular argue that susceptibility to control agents, pesticides in particular, should be viewed as a renewable resource. That is, if pests become resistant to a control agent and consequently the use of the control agent stops, pest resistance breaks down after a while and pests become susceptible again. The important question within the context of this paper is whether or not an irreversibility effect exists. To show that an irreversibility effect indeed exists consider the following hypothetical example for *Bt* corn used against damages from the European Corn Borer (ECB). The incremental benefits from adopting *Bt* corn are assumed to be 200 at the beginning, period one, and due to price uncertainty increase to either 300 or 100 after one time period and remain at the level until the end of the fourth period. At the end of the fourth period the ECB becomes resistant to *Bt* corn and the incremental benefits decrease to zero from period five till the end of period seven. At the end of period seven, the ECB becomes susceptible again to *Bt* corn. To keep the example simple, we assume that the incremental benefits increase to 200 Euro until infinity as the ECB will also be susceptible till infinity. The example is illustrated in Figure 1. The costs of pest resistance in present value terms are 1600 Euro. These are extra costs beyond the lost incremental benefits of periods five, six and seven.

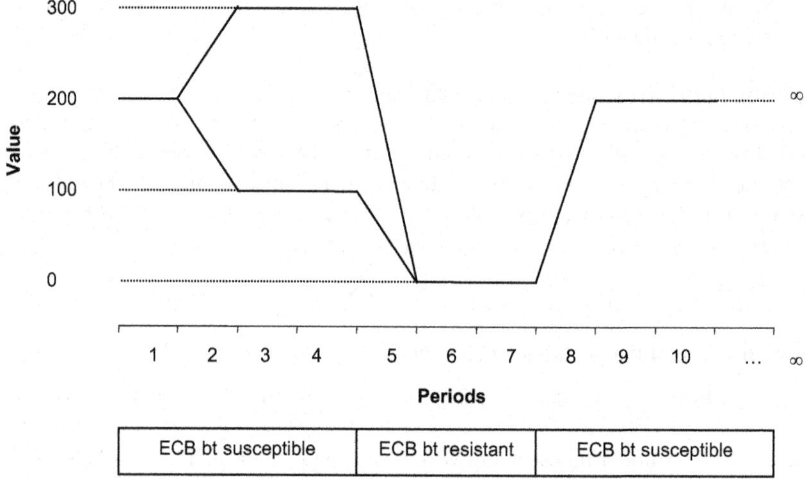

Figure 1. Example for appearance and breakdown of ECB resistance to *Bt* toxin

The value of *Bt* corn from immediate adoption is:

$$(8)\; E\left[NPV_0^I\right]=-1600+200\cdot 1.1^{-1}+0.5\cdot\sum_{t=1}^{t=4}300\cdot(1.1)^{-t}+0.5\cdot\sum_{t=1}^{t=4}100\cdot(1.1)^{-t}+\sum_{t=8}^{t=\infty}200\cdot(1.10)^{-t}=60\cdot$$

The result for a postponed adoption is:

$$(9)\qquad E\left[NPV_0^P\right]=\left[0.5\cdot\left(-1600+\sum_{t=2}^{t=5}300\cdot(1.1)^{-t}+\sum_{t=9}^{t=\infty}200\cdot(1.1)^{-t}\right)\right]=171\cdot$$

The above example illustrates that even though pest resistance can be reversible from a biological point of view, from an economic point of view an irreversibility effect may exist.

All the examples that have been discussed were constructed in a way that it was always optimal from an economic point of view to delay the adoption of transgenic crops. What is important to note is that while an irreversibility effect exists, it will not always be optimal to postpone the adoption. In cases where the irreversible costs are small or the incremental benefits are high, immediate adoption can be optimal.

Private and public irreversibilities

In the example we did not differentiate between irreversible benefits and costs. For the assessment of benefits and costs of transgenic crops and for the decision whether or not to release them, a distinction between private and social benefits and costs of transgenic crops has to be made. Private costs and benefits are important for the analysis of the adoption potential among farmers. This will provide information about the expected aggregated private net benefits from introduction. In addition, external benefits and costs have to be considered. These include, among others, climate-change effects, impacts on biodiversity and impacts on farmers' health. Further, the examples of the previous chapter illustrate the necessity of a differentiation between reversible and irreversible costs and benefits. A two-dimensional matrix (or three-dimensional one, if benefits and costs are added as an additional dimension) can be designed considering these differentiations for an *ex ante* analysis of social costs and benefits of transgenic crops as depicted in Figure 2. A complete *ex ante* analysis of economic benefits and costs of transgenic crops should consider all four quadrants of Figure 2.

Scope Reversibility	Private	External
Reversible	**Quadrant 1** Private Reversible Benefits (*PRB*) Private Reversible Costs (*PRC*)	**Quadrant 2** External Reversible Benefits (*ERB*) External Reversible Costs (*ERC*)
Irreversible	**Quadrant 3** Private Irreversible Benefits (*PIB*) Private Irreversible Costs (*PIC*)	**Quadrant 4** External Irreversible Benefits (*EIB*) External Irreversible Costs (*EIC*)

Figure 2. The two dimensions of an *ex ante* analysis of social benefits and costs of transgenic crops

As an example we use an *ex ante* assessment of herbicide-tolerant sugar beets (htSB) in Europe as explained in detail in Demont, Wesseler and Tollens (2004). The decision rule to release htSB is formulated as, to release htSB if the net reversible social benefits W, the sum of quadrant 1 and quadrant 2 in Figure 2, are greater than the net irreversible costs, the sum of quadrant 3 and quadrant 4, multiplied by a factor greater than one, the so-called hurdle rate η:

(10) $W \geq (I - R) \cdot \eta$.

As the social irreversible costs, $I=PIC + EIC$, and benefits, $R=PIB+EIB$, of transgenic crops are highly uncertain, instead of identifying the net reversible social benefits W required to release transgenic crops in the environment, the maximum tolerable social irreversible costs I^* under given net social reversible benefits W and social irreversible benefits R are identified:

(11) $I^* = R + W/\eta$.

The results are presented in Table 1. The estimated hurdle rates are entirely coherent with the expectations. We observe a bimodal distribution. Low-cost sugar-beet producers such as France, Belgium, the Netherlands, Germany, Denmark, the UK and Italy have low hurdle rates (1.25-1.82), while high-cost areas like Spain, Ireland, Austria, Sweden, Greece and Finland have higher ones (2.10-3.69), requiring higher values of *W* to justify a release of HT sugar beet.

The values of *W*, *R*, and *I** are presented as annuities of an infinite and continuous stream of benefits and costs, respectively, per hectare planted to transgenic sugar beet. *W* ranges from 121 Euro to 354 Euro with an average of 199 Euro per hectare. High-cost areas generally have high values for *W*, which can be explained by the EU sugar policy. Except a few outliers, estimates for *R* are low and range from 0.18 Euro to 3.36 Euro with an average of 1.59 Euro per hectare. This is due to the fact that we use conservative estimates from literature for the average external social cost of pesticide application. The maximum tolerable social irreversible costs range from 50 Euro to 212 Euro per hectare, i.e. in the range of 27-80% of the annual net private reversible benefits. For the EU as a whole this means that it should accept transgenic sugar beets as long as social irreversible costs do not exceed 121 Euro per hectare, totalling 103 million Euro per year. There is a large divergence between estimates for *R* and *I**. For the EU, e.g., *I** is 76 times larger than *R*. The social irreversible benefits *R* include impact of pesticide use on the environment, biodiversity and climate. As the social irreversible costs *I** include the same environmental effects, it is hard to believe that they are higher by a factor of 76. The total net private reversible benefits forgone, *W*, if the *de facto* moratorium is not lifted are in the order of 169 million Euro per year. On the other hand, the social reversible net benefits plus the social irreversible benefits are only about one Euro per household in the EU. If households put a value on the potential irreversible costs of transgenic crops of one Euro or more, than the *ex ante* net social benefits of htSB are negative and htSB should not be released.

Table 1. Hurdle rates and annual net private reversible benefits (*W*), social irreversible benefits (*R*), and maximum tolerable social irreversible costs (*I**) per hectare transgenic sugar beet

Member State	*W* (€/ha)	*R* (€/ha)	Hurdle rate	*I** (€/ha)	Total *I** (€)
Austria	251	3.36	2.88	91	1,842,164
Belgium & Luxembourg	168	2.09	1.26	135	5,852,023
Denmark	178	2.06	1.73	105	2,864,870
Finland	251	0.74	3.69	69	976,108
France	179	1.05	1.25	145	24,964,742
Germany	179	1.57	1.36	134	27,846,376
Greece	264	7.97[b]	3.12	93	1,771,502
Ireland	116	-0.96[b]	2.29	50	691,951
Italy	330	2.32	1.82	183	22,682,730
The Netherlands	121	0.83	1.31	94	4,630,433
Portugal	354	-0.65[b]	1.67[c]	212	615,218
Spain	252	0.53	2.10	121	7,258,219
Sweden	150	0.18	3.01	50	1,226,127
UK	127	1.78	1.76	74	5,135,522
EU	**199**	**1.59**	**1.67[a]**	**121**	**102,628,681**

[a] Sugar beet area-weighted average of the individual Member States' hurdle rates.
[b] The extreme estimates for Greece, Ireland and Portugal are probably due to data inconsistencies. These countries only cover 4% of total EU sugar-beet area, almost not affecting the EU average.
[c] No data on margins have been found for Portugal. We use the EU area-weighted average.
Source: Demont, Wesseler and Tollens (2004).

Conclusion

In this paper we have shown the multi-dimensional features of the irreversibility effect for the *ex ante* assessment of social benefits and costs of transgenic crops. We have demonstrated the irreversibility effect by using very simple examples. They illustrate the differences between irreversible benefits and irreversible costs. In addition, the example of pest resistance shows the difference between irreversibility at the biological and economic level. While pest resistance can be considered reversible from a biological point of view, it may nevertheless result in irreversible costs. The different types of irreversibilities are summarized in a two-dimensional matrix that we propose as a guideline for a complete *ex ante* analysis of social benefits and costs of transgenic crops. An application for the decision to release herbicide-tolerant sugar beets in the EU illustrates the use of the matrix.

References

Amram, M. and Kulatilaka, N., 1999. *Real options: managing strategic investment in an uncertain world.* Harvard Business School Press, Boston.

Arrow, K.J. and Fisher, A.C., 1974. Environmental preservation, uncertainty, and irreversibility. *Quarterly Journal of Economics,* 88 (2), 312-319.

Carey, J.M. and Zilberman, D., 2002. A model of investment under uncertainty: modern irrigation technology and emerging markets in water. *American Journal of Agricultural Economics,* 84 (1), 171-183.

Chavas, J.P., 1994. Production and investment decisions under sunk cost and temporal uncertainty. *American Journal of Agricultural Economics,* 76 (1), 114-127.

Conrad, J.M., 2000. Wilderness: options to preserve, extract, or develop. *Resource and Energy Economics,* 22 (3), 205-219.

Demont, M., Wesseler, J. and Tollens, E., 2004. Biodiversity versus transgenic sugar beet: the one Euro question. *European Review Agricultural Economics,* 31 (1), 1-18.

Dixit, A., 1989. Entry and exit decisions under uncertainty. *The Journal of Political Economy,* 97 (3), 620-638.

Dixit, A.K. and Pindyck, R.S., 1994. *Investment under uncertainty.* Princeton University Press, Princeton.

Gilligan, C.A., 2003. Economics of transgenic crops and pest resistance: an epidemiological perspective. *In:* Laxminarayan, R. ed. *Battling resistance to antibiotics and pesticides: an economic approach.* Resources for the Future, Washington, 238-259.

Henry, C., 1974. Investment decisions under uncertainty: the "irreversibility effect". *American Economic Review,* 64 (6), 1006-1012.

Knudsen, O. and Scandizzo, P.L., 2003. Environmental liability and research and development in biotechnology: a real options approach. *In:* Evenson, R.E. and Santaniello, V. eds. *The regulation of agricultural biotechnology.* CABI, Wallingford.

Laxminarayan, R., 2003. *Battling resistance to antibiotics and pesticides: an economic approach.* Resources for the Future, Washington.

Leitzel, J. and Weisman, E., 1999. Investing in policy reform. *Journal of Institutional and Theoretical Economics,* 155 (4), 696-709.

Merton, R.C., 1998. Applications of option-pricing theory: twenty-five years later. *American Economic Review,* 88 (3), 323-349.

Morel, B., Farrow, S., Wu, F., et al., 2003. Pesticide resistance, the Precautionary Principle and the regulation of Bt corn: real and rational option approaches to decision making. *In:* Laxminarayan, R. ed. *Battling resistance to antibiotics and pesticides: an economic approach.* Resources for the Future, Washington, 184-213.

Pindyck, R.S. and Rubinfeld, D.L., 1995. *Microeconomics.* 3rd edn. Prentic-Hall International, Englewood Cliffs.

Richards, T.J. and Green, G., 2000. Economic hysteresis in variety selection: why grow no wine before its time? *In:* Peters, G.H. and Pingali, P. eds. *Tomorrow's agriculture: incentives, institutions, infrastructure and innovations: proceedings of the twenty-fourth International Conference of Agricultural Economists, Berlin, 13-18 August 2000.* Ashgate, Aldershot.

Richards, T.J. and Patterson, P.M., 1998. Hysteresis and the shortage of agricultural labor. *American Journal of Agricultural Economics,* 80 (4), 683-695.

Shively, G., 2000. Investing in soil conservation when returns are uncertain: a real options approach. *In:* Peters, G.H. and Pingali, P. eds. *Tomorrow's agriculture: incentives, institutions, infrastructure and innovations: proceedings of the twenty-fourth International Conference of Agricultural Economists, Berlin, 13-18 August 2000.* Ashgate, Aldershot.

Special issue on irreversibility, 2000. *Resource and Energy Economics,* 22 (3).

Wesseler, J., 2002. The economics of agrobiotechnology. *In: Knowledge support for sustainable development.* EOLSS Publishers, Oxford. Encyclopedia of life support systems vol. 1, chapter 3.4.6.58.4.11.

Wesseler, J., 2003. Resistance economics of transgenic crops under uncertainty. *In:* Laxminarayan, R. ed. *Battling resistance to antibiotics and pesticides: an economic approach.* Resources for the Future, Washington, 214-237.

Winter-Nelson, A. and Amegbeto, K., 1998. Option values to conservation and agricultural price policy: application to terrace construction in Kenya. *American Journal of Agricultural Economics,* 80 (2), 409-418.

[1] Nobel laureate Robert C. Merton (1998) provides an overview of the application of the option-pricing theory outside financial economics. The book by Amram and Kulatilaka (1999) includes several case studies of real-option pricing. The special issue on irreversibilities of the journal Resource and Energy Economics, volume 22 (2000) includes application in the field of environmental and natural-resource economics.

[2] Note, this is a simplifying assumption and not necessarily correct for the case of sugar beets. The assumption has been made for convenience, as the empirical application in section 4 will be on htSB.

[3] The investment costs are sunk costs, as they are costs that cannot be recovered and do not affect future economic decisions ignoring the irreversibility effect (Pindyck and Rubinfeld 1995, p. 197). This changes if irreversibilities are considered (Dixit 1989).

[4] Assuming he does not need to buy a new planter and can still use the old one he used before switching to htSB.

[5] This would be the case if the planter could be used for n-htSB and htSB as well.

8b

Comment on Demont, Wesseler and Tollens: Irreversible costs and benefits of transgenic crops: what are they?

Meira Hanson[#]

This paper is an elucidating and useful contribution to a subject that is of interest not only to economists and biologists, but to anyone concerned with the meaning accorded to 'irreversibility' and the way the concept is employed in the debate on transgenic crops. That said, and reading the paper as an 'outsider' to this specific academic debate, there are several points that I believe ought to be clarified.

First, it is not clear whether the authors are discussing the concept 'irreversibility' or the 'irreversibility effect'. The former relates to those effects on human health, biodiversity, climate change, etc., which are (arguably) not reversible. The latter is the term used for the effect that the likelihood of more information about the nature (and/or possibility) of irreversible costs has on the valuation of a project. In other words, it relates to the value of information about irreversible costs under uncertainty rather than to some property of the widespread introduction of transgenic crops. It may be that this is the way economists relate to 'irreversibility', i.e. only to the degree that it effects the valuation of a project. However, if this is the case, it should be made clearer to the reader who is not an economist.

Second, considering the centrality of information to the 'irreversibility effect', some clarification is needed about the nature of the information expected. The illustrative examples provided by the authors deal mainly with uncertainty about the future benefits of a technology, i.e. those regarding output and input prices in agriculture "not known with certainty due to several factors including the microclimate, agriculture policies and technical change". While the authors mention that there is uncertainty about the irreversible costs and benefits of transgenic crops, it is less clear to me how this dimension of uncertainty figures in the 'irreversibility effect', according to the authors.

Furthermore, the value of further information has underlying it the assumption that such information is forthcoming and that uncertainty is to an extent reducible. However, is it not the case that much of the debate about the irreversible costs of transgenic crops deals with a condition more adequately characterized as 'ignorance' (i.e. not knowing what it is that we do not know)? And if this is indeed the case, does not the necessary information about the effects of transgenic crops ultimately depend on their widespread cultivation? Of course it may be that the way 'irreversibility' is discussed by economists precludes such questions, in which case we return to the necessary clarification of what irreversibility means to economists (vis-à-vis the 'irreversibility effect').

Finally, a central contribution of the paper is introducing and operationalizing the concept of irreversible *benefits*, which play an important role in a cost–benefit assessment of transgenic crops that recognizes the 'irreversibility effect'. The authors

[#] Department of Political Science, The Hebrew University of Jerusalem, Jerusalem, Israel

J. H. H. Wesseler (ed.), Environmental Costs and Benefits of Transgenic Crops, 123–124.

argue that when considering the 'irreversibility effect', irreversible benefits reduce irreversible costs by an order of one. However, while this may be the case where the benefits and costs are monetary, does it remain the case where the benefits and costs pertain to effects on health, biodiversity, climate change etc.? Does the inclusion of irreversible benefits not inevitably introduce a cost–benefit analysis into the 'irreversible dimension' of the equation and, as a consequence, the question of whether these irreversible benefits and costs are commensurable?

To conclude, while the economic understanding of 'irreversibility' is an important element in the debate on transgenic crops, at the end of the day, any decision – to introduce transgenic crops or not to introduce them – is irreversible (on various meanings of irreversibility, cf. Humphrey 2001). Either decision generates irreversible costs and benefits, and it is the likely distribution of these which remains the central political question.

References

Humphrey, M., 2001. Three conceptions of irreversibility and environmental ethics: some problems. *Environmental Politics,* 10 (1), 138-154.

9a

Ex post evidence on adoption of transgenic crops: US soybeans

Robert D. Weaver[#]

Abstract

Transgenic crops offer a complex new technology that is not universally dominant over alternatives. Instead, adoption decisions are conditional on incentives associated with alternative technologies and local conditions. These characteristics imply that transgenic crops can be expected to be adopted on a wide scale in existing cultural areas, if incentives are appropriate. Further, considerable potential exists for transgenic crops to be adopted in new areas where they offer advantages over alternative crops such as weed control and other management practices.
Keywords: transgenic crop; innovation adoption; diffusion; soybeans

Introduction

The potential of transgenic crops involves consideration of a technology that has several important features. First, as with many technologies that are involved with agricultural production, a shift to transgenic-crop production involves both private and public effects. Second, in each case, these effects involve both uncertain and, in some cases, irreversible costs and benefits. Third, while many technologies offer net benefits that render the innovation universally attractive to potential users, though actual diffusion is inhibited by imperfections in markets, transgenic-crop production involves a package of changes in practices, input mix, and basic opportunities for management of the crop. Because of the associated complex of private- and public-good changes, the attractiveness of these innovations may not be universal. The willingness-to-pay for transgenic crops is often conditioned by local, farm-specific conditions that result in what Weaver and Kim (2002) defined as local dominance of the technology. That is, rather than being universally adopted, an equilibrium is implied in which use of the technology is both incomplete and intertemporally unstable. Fourth, transgenic innovation has involved a change in perceived underlying attributes of food products derived from the crop. This change in attributes may or may not involve changes in the real functional value of the food product in consumption. In any case, this characteristic has proven to imply increased uncertainty with respect to the market value of the crop. Finally, the potential for public effects or changes in food attributes associated with transgenic crops has motivated national and regional regulatory responses ranging from prohibition of use to conditions for use creating a geographic fabric of varied experience.

[#] Dept. of Agricultural Economics, Pennsylvania State University, 207D Armsby Building, University Park, PA 16802, USA. E-mail: r2w@psu.edu

J. H. H. Wesseler (ed.), Environmental Costs and Benefits of Transgenic Crops, 125–140.

Within this context it is of interest to consider *ex post* evidence concerning the nature of adoption behaviour for these crops and, in particular, to provide a basis for drawing from that evidence any lessons that might be apparent concerning US experience and relevant to Europe. To proceed, the focus of this paper will be on herbicide-tolerant (ht) soybeans, leaving insecticidal-trait transgenic crops for another paper. Of particular interest in this paper is to consider United States (US) experience with adoption and to assess its implications for adoption in Europe, should current regulations be changed to allow growing of ht soybeans. Of subsidiary interest is whether adoption experience in the US appears to be consistent with well-established economic literature on the adoption of new technologies, or whether an alternative theoretical framework is needed to understand and predict adoption behaviour with respect to transgenic crops. Given that the assessment of environmental effects is considered elsewhere in the workshop, this paper will focus more sharply on adoption-behavioural implications of known or possible environmental effects.

To proceed, the outline of the paper will be to consider briefly the salient features of transgenic crops, though to defer consideration of regulation to another context. Next, implications of these features for adoption decisions will be considered. Third, evidence across US experience will be considered for the case of ht soybeans. In closing, conclusions will be drawn with a brief consideration of the relevance of US experience to Europe.

Salient features

Current status

Before proceeding, a brief consideration of the current status of genetically modified (GM) crops is in order. James (2002) estimates that globally 58.7 million hectares were planted to transgenic or GM crops in 2002, across 16 countries. He estimated that 27% of this area was planted in developing countries with Argentina and China accounting for the majority of this area. However, India, Columbia, and Honduras will rapidly claim a place on these charts as they shift to *Bt* cotton. Herbicide tolerance has dominated as a transgenic trait available in soybeans, corn and cotton, accounting for 75% of global transgenic area planted. Insecticidal traits of *Bt* claimed 17% of area with the remainder claimed by stacked genes delivering herbicide tolerance and insecticidal traits. Transgenic soybeans claimed 62% of global area accounting for 51% of soybean area while *Bt* corn claimed 13% of global transgenic area. These trends suggest that as a technology, transgenic crops might be distinguished as a universally dominant technology. That is, one that generates a benefit–cost stream that makes the technology dominant across a wide spectrum of heterogeneous producers.

The United States produces more soybeans than any other country involving approximately 380,000 farms in 29 states. A decade ago, in 1994, the US crop yielded 2.558 billion bushels of soybeans with an estimated farm-gate value of $13.813 billion and an average price of $5.40 per bushel; soybean production has expanded to cover 74.1 million acres (30.0 million hectares) in 2001, producing a record 2.891 billion bushels (78.68 million metric tons) of soybeans at an average farm price paid of $4.25 per bushel ($156 per metric ton) generating a crop value of $12.28 billion. Interestingly, soybean production in the US does not only take place in the wide, flat fields of the Midwest. Increasingly, soybeans are grown further north and in states such as Pennsylvania, where traditional crop rotations range across small grains, hay and forage, and corn. In these states, transgenic soybeans are being adopted by producers with no past experience with soybeans.

In fact, US experience indicates that two cases can be defined with respect to adoption experience. First, a substantial proportion of the US growing region can be viewed as having homogeneous growing conditions, farm scale and farm diversification. This would include what many would view as the traditional or primary areas where soybeans have been grown and accounted for a large percentage of principal crop area, e.g. Illinois, Iowa, Indiana, Missouri and Ohio. For these states, soybean area accounted for over 20% of principal-crop area before transgenic soybeans were available and adoption has continued to increase, though at a reduced rate (see Table 1). In addition to these areas, a secondary area is apparent. In the US, this area comprises states where soybeans accounted for less than 20% of area, however, more recently soybeans have expanded dramatically with the introduction of transgenic varieties. Nebraska, North Dakota, South Dakota, Wisconsin and Pennsylvania would be included in this category (see Table 1). In these secondary states, it is clear that soybean production is expanding through entry by farmers who previously did not produce soybeans, though with the availability of transgenic soybeans find the technology and crop attractive.

Private and public effects

The nature of innovation offered by transgenic crops is inherently complex. While most would agree that the innovation they offer is fundamentally different from a new variety or hybrid, the exact nature of the facets of the innovation are subject to substantial debate, uncertainty and variation across growing conditions and crops. One aspect of particular importance for consideration is the private and public effects of the innovation. With respect to private aspects, the yield effect is of particular interest. Yield impact of transgenic soy depends on weed control as well as the extent of adaptation of conventional varieties. Past studies have not found a striking difference in yields that cannot be unequivocally assigned to weed-control differences, or plant-growth or damage effects from application methods. Going a step further, transgenics provide opportunity for substantial changes in management practices, timing, flexibility and intensity of input use. Ervin et al. (2000) provide a thorough review of both the private and public effects.

Fernandez-Cornejo and McBride (2002) studied the private farm-level impacts of transgenic crops for the period 1996-98 focusing on yields, pest management and net returns. They found that use of ht cotton led to significant yield and net return increases, though no significant herbicide-use changes. Alternatively, for ht soy they found small increases in yield, no change in net returns, and significant decreases in herbicide use. However, their results varied substantially across farms and regions. In addition to changes in input mix or management practice, the incomplete knowledge of private and public effects of transgenics results in uncertainty concerning efficient input combinations, performance of the crop, costs and revenues. Although evidence from analysis of farm-level experience is not available, the flexibility in timing of use of herbicides and of field practices for ht soy would suggest that efficiency gains could be realized compared to conventional practices for soy.

Public benefits in the form of both short- and long-term environmental benefits are a potentially important dimension of the adoption of transgenics. Ervin et al. (2000) note that substantial environmental benefits may be associated with transgenic crops,

Table 1. Evolution of soybean area: US percentage of total crop acres planted to soybeans

Region	State	1994	1995	1996	1997	1998	1999	2000	2001	2002	2003
South central	AR	41.27	40.90	40.90	43.74	41.72	40.20	39.46	34.54	35.67	36.27
	KS	9.52	9.36	8.48	10.21	11.06	12.44	12.88	11.89	11.86	11.62
	LA	29.53	27.74	27.26	34.65	29.59	26.91	24.64	17.19	21.08	26.05
	MS	39.67	38.14	36.89	44.03	42.62	39.76	35.64	25.47	32.04	31.55
	OK	2.79	2.73	2.65	3.11	4.43	4.36	4.20	4.16	2.59	1.76
	TX	1.01	1.11	1.19	1.80	1.85	1.60	1.24	1.08	0.93	0.95
North central	IL	40.09	41.99	41.38	42.12	44.82	45.07	44.36	45.67	45.15	45.41
	IN	37.90	41.87	42.69	41.27	43.31	44.02	43.32	45.01	47.63	44.29
	IA	36.35	39.57	39.26	42.49	41.95	43.39	42.82	44.69	42.38	41.87
	MI	22.12	22.09	23.49	26.59	28.04	28.34	30.29	32.18	30.94	31.77
	MN	28.43	30.81	30.00	32.18	33.63	34.70	35.97	37.67	35.49	37.94
	MO	36.17	38.16	30.89	36.50	37.42	39.67	37.64	36.68	36.48	35.51
	NE	15.18	16.96	16.21	18.83	20.05	22.25	24.22	25.62	24.55	24.54
	ND	2.95	3.19	3.75	5.16	6.96	6.73	8.75	10.51	11.92	14.11
	OH	38.44	40.40	44.23	40.56	41.31	43.52	41.76	43.45	45.73	43.53
	SD	14.84	17.79	15.97	18.82	20.92	24.81	25.45	25.47	24.70	23.45
	WI	10.20	10.13	11.27	12.98	14.23	16.13	19.85	20.84	19.03	19.09
Northeast	DE	44.12	46.35	44.00	45.19	42.39	41.16	43.00	42.09	39.92	42.79
	MD	35.69	35.53	31.11	34.06	31.97	32.91	33.96	34.76	33.29	36.04
	NJ	32.75	30.97	28.10	30.57	25.56	25.24	27.17	30.12	28.57	30.49
	PA	7.71	7.72	7.00	8.69	9.20	8.61	9.20	9.91	9.74	9.30
	VA	18.58	16.84	17.03	17.43	17.06	16.14	17.24	18.03	16.80	19.63
Southeast	AL	13.73	10.89	14.19	14.75	15.09	10.77	9.11	6.26	8.05	9.27
	FL	4.13	2.80	3.16	4.32	3.11	1.82	1.81	0.93	0.91	1.03
	GA	12.18	7.55	9.23	9.05	7.42	5.70	4.35	4.27	4.11	4.73
	KY	20.69	20.49	20.53	21.19	20.80	20.65	20.32	22.64	23.44	20.35
	SC	29.44	27.83	28.41	28.57	28.39	26.86	26.87	26.33	25.33	30.85
	TN	22.56	21.46	23.00	25.18	25.86	25.44	23.31	21.08	23.27	23.80
	NC	29.60	24.79	26.28	28.48	29.41	28.31	28.52	27.90	27.41	30.10
US		19.02	19.64	19.20	20.95	21.83	22.37	22.61	22.80	22.53	22.64

Data source: USDA-National Agricultural Statistics Service: Crop-Production-Acreage Supplement; Crop Production–Annual Summary though they also note that ecological negative effects have been suggested by some research. Other chapters in this volume address this issue in detail, though for this paper it is sufficient to note that a review of US experience would reveal, for most readers, an absence of accepted scientific conclusions concerning the private and public effects of transgenics. The knowledge base is small and does not support consensus interpretation. As might be expected for any innovation, some uncertainty remains concerning the extent and nature of these effects. Recent USDA data shows that herbicide-tolerant seed slightly reduced the average number of active ingredients applied per acre, while slightly increasing the average amount applied per acre (Benbrook 2001).

In the long term, it was expected that transgenic crops would facilitate the introduction of pesticides that imply reduced environmental risk. While rapid adoption of such crops suggests that strong incentives may be in place to motivate these decisions, the interplay of private vs. public effects in these decisions is not clear. Based on ERS/USDA estimates, the expansion of ht soy in the US followed rapidly after 1997 and was accompanied by increased use of glyphosate (Economic Research Service USDA 1999b) and decreased use of other herbicides leading to a net reduction in total weight applied. ERS (1999a) indicated that ht soy allowed reduced active-ingredient application. Numerous other studies find that change in herbicide use overall has not been significant, e.g. Benbrook (2001), although the extent of impact varies by crop. Most recently, Conner, Glare and Nap (2003) noted that the variety of concerns raised relative to the impacts of GM crops on the environment remain open to debate. These include putative invasiveness, gene flows and ecological effects, and drift of GM material into other products (e.g. feeds to animals).

A more specific look at farm-level survey data suggests that changes in production practices for ht soy include a shift toward conservation tillage, a notable reduction in the number of active-ingredient herbicides used with a sharp focusing on glyphosate; and finally, that the planting window has become much wider offering substantial flexibility for timing of planting, weed control and movement toward no-till planting that eliminates tillage and other field preparation activities. From 1989 to 1998 the acreage of soybeans planted with conservation-tillage methods increased from 30% to 54%; see Carpenter and Gianessi (1999). These changes appear to offer reduction in fuel use.

Nonetheless, some evidence exists that supports the claim that changes in practice include shifts to no-till planting, pre-emergence herbicide use, change in the type of active ingredients used, and perhaps substantial changes in environmental impacts of crop practices. Benbrook (2001) provides evidence that herbicide-tolerant varieties have slightly reduced the average number of active ingredients used per acre while increasing the average pounds applied per acre. Carpenter and Gianessi reviewed shifts in practices citing in particular the role of transgenic soybeans as a natural extension of an evolution toward increased use of post-emergence herbicides, simplification of weed-control programmes, and improved effectiveness of active-ingredient applications; see e.g. Pike, McGlamery and Knake (1991). Importantly, this shift in practice had substantial implications for tillage practices that had focused on field preparation and post-emergence tillage. Given post-emergence herbicides, adoption of conservation tillage was facilitated, leading to over 50% adoption by 1998; see Kapusta and Krausz (1993) and Conservation Tillage Information Center (1999). This shift was further extended by introduction of herbicide-tolerant soybeans that allow post-emergence, broad-spectrum herbicide application at nearly any stage of plant growth. Second, improved post-emergence herbicides have allowed for a reduction in row spacing, significantly reducing cultivation, improved weed control due to canopy closure and increased land-area yield. The key innovation offered by transgenic soybeans is the reduction of crop damage (e.g. stunting, delayed canopy closure) from herbicide application (see Padgette et al. (1996)) and increased effectiveness of weed kill (see Rawlinson and Martin (1998)). This latter effect follows directly from tolerance that allows effective dosage to be determined with consideration of crop-damage relaxing constraints in conventional systems with respect to timing (early in weed emergence).

Some negative environmental effects that have been considered include impacts on soil structure (no till results in macropore development and exposure of groundwater to surface effluents) and reduced incorporation of plant residue or animal waste to amend the soil. In addition to these, concern regarding cross-pollination between GM (genetically modified) and non-GM soybean strains as well as the potential for evolution of herbicide-resistant weed varieties has been noted. That there are possible (unverified) negative effects of biotechnology products has been a social concern. In the United States, substantial concern with respect to BST use was initially raised among dairy producers, though after a period during which scientific evidence was interpreted and debated, these concerns subsided. A similar pattern of learning has been associated with transgenic soybeans. Importantly, the shift to reduced tillage has been credited with increased crop residue, reduced fuel, labour and machine time, reduced wind and water erosion; see American Soybean Association (2001).

Uncertainty and irreversibility

The productivity and market value of new products and production practices or technologies that result from innovation are inherently uncertain. In the case of transgenic crops, the extent of this uncertainty is extensive. This scope of uncertainty follows from uncertainty in applied science, the complexity of the production-process changes involved with transgenic crops, and the scope of private and public effects that may exist or are perceived along the supply chains through which value is created from these crops. In addition to this uncertainty, both private and public costs and benefits associated with transgenic-crop adoption may be or are perceived to be irreversible. On the uncertainty front, adoption of transgenics allows for or requires substantial change in production practices as reviewed above. This by definition introduces uncertainty that characterizes the adoption of most innovations. However, to the extent that the performance of transgenic crops is conditioned by local climate, soils and pest exposure, the extent and speed of resolution of this uncertainty through learning will be reduced. The suggestion of physical science is that site- and environmental-condition-specific characteristics can significantly affect the performance of ht soy. Nonetheless, the rapid and widespread adoption across the primary states in the US is consistent with experience with other technological change in agriculture and suggests that much of the technological uncertainty has been quickly resolved in these states.

Irreversibility is a key feature of many innovations. Typically, investment costs must be incurred for learning, change in management practices or acquisition of machinery or other services for assets. Where these are irreversible, adoption of the innovation is affected. In the case of transgenic crops, both irreversible costs and benefits have been cited including both private and public effects.

Transgenics: universal or locally dominant?

The characteristics of the technology and its impacts on production practices and input mixes suggest that transgenic seed constitutes a complex set of changes in the overall production technology, rather than a single augmentation of a particular input. Within this context, the role of heterogeneity across agents would be expected to be accentuated. Bullock and Nitsi (2001) found that the potential of transgenics varied with extent of pest exposure and the type of pest-control practices used. This confirms the physical science evidence concerning the complexity of the changes induced by the transgenic technology. In the presence of a substantial potential role of heterogeneity that might result in heterogeneity in adoption, adoption of transgenic soybeans has been rapid within what has been labelled the primary-states region of the US. However, a different story is apparent in the secondary-states region. In this section, we briefly note the distinction between these two cases.

To begin, past literature has considered technology adoption both at an individual-agent level and at the level of various aggregations. While the individual agent's decision to adopt is most often a binary one, its timing is conditioned by a variety of individual determinants that imply that adoption will not be instantaneous by all producers. Instead, over time the proportion of adoption in an aggregation of agents has been described as 'diffusion' process (see e.g. Karshenas and Stoneman 1995). The role of agent-specific factors implies that as these conditions evolve, the technology choice of particular agents will change. Recognition of the role of heterogeneity across agents in adoption decisions and timing has been expanded to include agent managerial characteristics, risk preferences, labour-market participation, exposure to uncertainty with respect to technological performance, and information access. Fernandez-Cornejo and McBride (2002) presented estimates of adoption of GM crops in the US that indicated that scale of operation (farm size) was not a significant determinant for ht soybeans, while indicators of operator characteristics (experience, operator risk aversion, use of marketing contracts) and general indicators of farm characteristics (limited resource, location in marginal crop region) were found statistically significant.

Feder and Umali (1993) note that factors affecting adoption may vary over the life span of the innovation (early vs. late phase).

Heterogeneity is also recognized in theargument that rationalizes why a technology does not immediately 'diffuse' as would be expected within competitive market settings with instantaneous information and costless adjustment; see most recently Fernandez-Cornejo and McBride's (2002) diffusion model for GM crops in the US. Importantly, this theory of adoption suggests implicitly that after some finite time period, for a particular population of producers, adoption will be complete or, in other words, the market for the new technology will be saturated. This interpretation translates into the postulation of ceiling or upper limit for adoption that may be conditioned by factors that characterize the population of producers. Fernandez-Cornejo, Klotz-Ingram and Jans (1999) and Fernandez-Cornejo and McBride (2002) use estimates of pest pressure. Below, we continue in this tradition and consider the case where producers are not homogeneous, implying that for particular incentive vectors and quasi-fixed factor positions, adoption may not be chosen by some producers. Note that this perspective differs from that of Moschini and Lapan (1997), who consider input-price adjustment as a reason for incomplete adoption.

Marketing and consumer preferences

The feasibility of marketing transgenic crops poses an important basis for distinguishing them from other crop innovations. Two issues deserve note. First, the geographic scope of transgenic crops appears to be changing. In this regard, the availability of marketing channels and opportunities are essential determinants of the feasibility of expansion of the crop into new locations. A second concern is consumer reaction to transgenic crops that might be used for or affect foods used for human consumption or for animal feed. In the soybean complex, three forms of products must be considered: beans, oil and meal. Traditionally, soybeans have been marketed either through local grain elevators or feed mills for animal feed, or to crushing plants that produce soybean meal and oil. While direct, local feed use of soybeans has evolved in secondary states such as Pennsylvania, sale to crushing plants remains constrained in these locations due to absence of nearby plants. The processing market as well as the large-scale feed-user procurement in the US operates through use of forward contracting and brokerage. This approach minimizes search costs and stabilizes procurement price risk. The question of consumer response to transgenic crops has been considered in depth elsewhere, though it is important to note that transgenic crops constitute an innovation for which the level and uncertainty of private and public effects results in consumer response. To the extent that consumers adjust preferences based on news or scientific announcements, consumer behaviour may be unstable until uncertainty is resolved and a consensus is formed concerning the nature of private and public effects.

Implications for adoption decisions

Static perspective

Based on the salient features of the transgenic crop innovation, the economics of adoption of the innovation deserve attention. To summarize, these salient features include private and public effects, uncertainty, irreversibility and regulation.

Clark (1999), Fernandez-Cornejo, Klotz-Ingram and Jans (1999) and Weaver and Kim (2002) note that transgenic-adoption decisions are complicated by factors that go beyond the private economics. Weaver and Kim (2002) note the complexity of effects on production practices that suggest the innovation goes beyond a single input augmentation. In contrast, transgenic technology has often been specified as a single factor augmenting technological change, see e.g. Moschini and Lapan (1997). Fernandez-Cornejo, Klotz-Ingram and Jans (1999), Nadolnyak and Sheldon (2001) and Neill and Lee (1999) discuss adoption of transgenic soybeans though they do not pursue actual empirical modelling. Fernandez-Cornejo and McBride (2002) present empirical results for models of adoption of GM crops as noted above. Further, they consider the possibility that adoption of GM technology and no-till technology is simultaneous. Based on estimates of structural equations for such simultaneous choice they find evidence that is consistent with the conclusion that simultaneity does not exist, however, adoption of GM seed is conditioned by use of no-till practices. Weaver and Kim (2002) present a table summarizing available literature. To consider the adoption decision as well as the nature of dominance of transgenic-crop technologies, the notation of Weaver and Kim (2002) is useful.

Define the production function for *c*th crop output y^{i}_{jc} (quantity per land area, e.g. hectare) from the *j*th farm operating the *i*th technology. Suppose that while a common technology is available across farms, the technology is conditioned by farm-specific

quasi-fixed and fixed input flows represented by a vector, θ_j. Suppose crop output is also conditioned by a stochastic shock, ε_j, generated by a density function $g(\varepsilon_j|0,1)$, and a vector of inputs, x^i_c. Note, this input vector includes inputs relevant for the *i*th technology, though some elements may also be relevant for other technologies. The technology-specific production function reflects unique technological attributes such as planting flexibility, management intensity, etc. Define the crop output per land-area production function as: $y^i_{jc} = y^i_{cj}(x^i_{cj}, \theta_j, \varepsilon_j)$ and producer profit per land area for crop *c* produced with technology *i* as:
$\pi^i_{jc} \equiv p_{jc}{}^i y^i_{jc} - r^i_{jc} x^i_{jc} - w^i_c \delta^i_{jc}$ where p^i_{jc} is the output price that is allowed to be technology- and farm-differentiated, r^i_{jc} is the input-price vector, δ^i_{jc} is the seeding rate per land area, and w^i_c is the price paid for *i*th seed type for crop *c*. The seed price w^i_c is uniform. The differences in seed prices at the farm gate are captured by the input price and input vector.

Based on this notation, sequential planting decisions will first select the optimal technology to operate for each crop that might be grown, and second select the area allocation and production plan across crops conditional on the optimal technology selected for each crop. The choice of optimal technologies follows from a consideration of optimal-value functions indicating the value per land-area unit for each crop *c* and technology *i*:

$$(1)\, V^i_{jc}(\rho^i_{jc}) = V^i_{jc}(p_{jc}{}^i, r^i_{jc}, w^i_c, \delta^i_{jc}, \theta_j) \equiv \max EU(\pi^i_{jc}) \quad where\, \pi^i_{jc} \equiv p^i{}_{jc} y^i_{jc} - r^i_{jc} x^i_{jc} - w^i_c \delta^i_{jc}$$

$$\text{s.t. } y^i_{jc} = y^i_{jc}(x^i_{jc}, \theta_j, \varepsilon_j).$$

Define the farm-specific vector of determinants of value as an 'incentives' vector $\rho^i_{jc} \equiv [p_{jc}{}^i, r^i_{jc}, w^i_c, \delta^i_{jc}, \theta_j]$ and the set *I* as the set of all economically feasible alternative technologies for crop *c* defined as those technologies *i'* for which $V^{i'}_{jc} > 0$ at the prevailing $\rho^{i'}_{jc}$. Based on this notation, the producer's relative net benefit for technology *i* versus technology *i'* for the same crop is: $\omega^{i}{}_{jc} = V^i{}_{jc} - V^{i'}{}_{jc}$.

The implications of heterogeneity across farms is clear by defining a technology *i* for crop *c* as *locally dominant* on farm *j* relative to other technologies $i' \in I$ if $\omega^{ii'}{}_{jc} > 0 \;\; \forall \;\; i' \neq i, i' \in I$, and is *universally dominant* relative to other technologies *i'* and for a set of *J* farms if $\omega^{ii'}{}_{jc} > 0 \;\; \forall \;\; i' \neq i, i' \in I, j \in J$. Thus, the dominance of a technology involves comparative evaluation of value across alternatives at the farm level. This implies that a wider scope of incentives become involved in the adoption decision than simply those that determine the value of the dominant technology. That is, it is clear that the choice of technology involves a comparative, though discontinuous, role for the incentive vectors $\rho^i_{jc}, \rho^{i'}_{jc}$ across the set of possible technologies, *I*. Further, the composition of the set *J* of farms for which technology is universally dominant is similarly conditional on and discontinuously related to the vectors $\rho^i_{jc}, \rho^{i'}_{jc} \;\; \forall \;\; j \in J$; especially noteworthy are the farm-specific characteristics embedded in these vectors. To continue the story, given the sequential nature of cropping decisions, dominance of technology *i* for crop *c* does not imply that the associated crop will be dominant. Thus, adoption or use of the dominant technology follows from dominance of the crop conditional on the dominance of the technology. In each case, the dominance condition involves discontinuous roles for incentives.

To consider choice of crops conditional on the set of dominant technologies, define local crop dominance of crop c for farm j as occurring for the locally dominant technology i $if\ \omega^{i}_{jcc'} \equiv V^{i}_{jc} - V^{i}_{jc'} > 0 \forall c' \neq c \in C$ where C is the set of all economically feasible alternative crops defined as those crops c' for which $V^{i}_{jc'} \geq 0$ at the prevailing $\rho^{i'}_{jc}$. It follows for area a and input vector x demands can be immediately derived and used to analyse or model these choices; see Weaver and Kim (2002). However, more relevant to this paper is that *relative willingness-to-pay* for technology i for crop c versus the second best alternative can now be defined as $\omega^{ii'}_{jcc'} \equiv V^{i}_{jc} - V^{i'}_{jc'} > 0 \quad \forall \quad [i,c] \neq [i',c'], c \in C, i \in I$.

Several propositions follow. First, the choice of the locally dominant technology is conditional on, though not continuous in, the vectors of determinants of value across all technologies, i.e. $\rho^{i}_{jc} = (p^{i}{}_{jc}, r^{i}_{jc}, w^{i}_{c}, \delta^{i}_{jc}, \theta_{j})$ as well as $\rho^{i'}_{jc} = (p_{jc}{}^{i'}, r^{i'}_{jc}, w^{i'}_{c}, \delta^{i'}_{jc}, \theta_{j}) \forall i'$. Further, the choice of crop is conditional on the vectors of determinants of value for dominant technologies across all crops, i.e. ρ^{i}_{jc} and $\rho^{i}_{jc'}\ \forall\ c'$. Despite this conditionality, the final choice of area allocated to and, therefore, demand for seed for crop c produced by technology i, is not a continuous function of these determinants of value for alternative technologies and crops. Instead, demand for seed is functionally continuous only in the vector of determinants of value of the dominant crop using the dominant technology, i.e. ρ^{i}_{jc}. That is, we define the demand for seed: $s^{i}_{jc} = s^{i}_{jc}(\rho^{i}{}_{jc}) \equiv a^{i}_{jc}(p^{i}{}_{jc}, r^{i}_{jc}, w^{i}_{c}, \delta^{i}_{jc}, \theta_{j})\delta^{i}_{jc}$. This last result has often led to confusion concerning the implications of a patent grant as noted by Weaver and Kim (2002) and Weaver and Wesseler (2003). Viewed alone, the demand function for seed seems to imply that a patent grant transfers monopoly power to the innovator to price the seed for technology i for crop c. However, it is clear from the above notation that this pricing power would only exist in the case where no other technologies or crops were economically feasible alternatives for the farmer.

The implications of this theory for empirical study of adoption are clear. The relative willingness-to-pay rule provides the basis for definition of a binary indicator of local crop dominance conditional on a particular technology:

(2) $\lambda^{i}_{jcc'} = 1 \quad if\ V^{i}_{jc} - V^{i}_{jc'} \geq 0\ otherwise\ \lambda^{i}_{jc} = 0$.

Providing further definition to the underlying functions motivates an empirical approach to estimating the probability of particular types of dominance by a given technology or innovation. For example, adding a stochastic error to the value function, $V^{i}_{jc}(\rho^{i}_{jc}) = v^{i}_{jc}(\rho^{i}_{jc}) + \omega^{i}_{jc}\ define \quad \lambda^{ii'}_{jc} = 1\ if \quad \omega^{i}_{jc} - \omega^{i'}_{jc} \geq -v^{i}_{jc}(\rho^{i}_{jc}) + v^{i'}_{jc}(\rho^{i'}_{jc})$

This motivates the probability of *local dominance* of technology i over alternatives i' on farm j (by generalization, a group of farms)

$$pr(\lambda^{ii'}_{jc} = 1\ \forall i') = \prod_{i' \neq i} pr(\omega^{i}_{jc} - \omega^{i'}_{jc} \geq -v^{i}_{jc}(\rho^{i}_{jc}) + v^{i'}_{jc}(\rho^{i'}_{jc}))$$

and the probability of *global dominance* of technology i on a group of farms J:

(3) $$pr(\lambda^{i}_{c} = 1\ \forall i', j) = \prod_{j=1}^{J} \prod_{i' \neq i} pr(\omega^{i}_{jc} - \omega^{i'}_{jc} \geq -v^{i}_{jc}(\rho^{i}_{jc}) + v^{i'}_{jc}(\rho^{i'}_{jc})).$$

By extension, crop choice can be similarly motivated, and by addition of parameterization to the underlying functions, we have the basis for a parametric approach to estimation of the dominance probabilities, or equivalently, the adoption probabilities for a particular technology or crop for a particular farm type or group of farms.

Dynamics perspective

The presence of uncertainty and irreversibility is an important feature of the setting in which transgenics are considered. These characteristics suggest that the static nature of the above framework could be fruitfully extended to consider the timing of adoption, or equivalently, the dynamics of dominance of a technology or crop. The theory for this problem has been developed by Weaver and Wesseler (2003) and illustrated with simulation. In essence, the framework above defines a relative willingness-to-pay that can be viewed as a return in the current period if the innovation were adopted. Adoption involves investment, a cost that can be viewed as irreversible if rental and resale markets are incomplete or sticky. The nature of this investment will vary across transgenic crops, though as an example, adoption of transgenic soybeans may involve a change in equipment used for tillage, planting, herbicide treatment, and herbicide handling and storage. From the perspective of learning, investment will be necessary to facilitate adjustment of management practices. Further, as claimed by some, and based on a certain amount of evidence, it is possible that adoption of transgenics results in private or public irreversible benefits. These may involve benefits such as improved long-term weed control, reduced soil erosion, or reduced surface water pollution. In each case, these benefits may be private and local, or public and go beyond the boundaries of the farm.

Weaver and Wesseler (2003) consider the adoption decision within the context of uncertainty and irreversibility from a real options perspective. They find that under uncertainty, irreversibility, and flexibility the relative willingness-to-pay for the new technology will be smaller than in the deterministic setting considered above. Further, the decision to adopt is shown to be delayed as uncertainty increases, and as either irreversible costs increase, or irreversible benefits decrease. Importantly, they show that as irreversible benefits are emphasized relative to irreversible costs, adoption is accelerated. This clarifies the importance of balanced research that is unbiased in its consideration of irreversible benefits as well as costs.

Evidence from US Perspective

Past studies of adoption of GM crops have taken a static approach and have not considered the role of the incentive vector for competing technologies. More generally, these studies have been motivated most frequently by a theory of diffusion that presumes the innovation is universally dominant for a subset of producers. On this basis a ceiling or maximum adoption proportion is defined. As noted above, this is not likely to be the case for transgenic crops. Past work has identified six types of factors affecting adoption: farm characteristics (farm size, field characteristics), experience and knowledge of the technology, market conditions (price risk, profitability, cost and yield effects) and environmental implications (e.g. decreased use of pesticides). Farm characteristics such as farm size (measured by acres planted or total gross farm income) and farm-operator demographics (farmer's age, experience, education), and tenure (share land owned) have been found to be statistically significant determinants; see Alexander, Fernandez-Cornejo and Goodhue

(2002). Their survey results show that farms with high total gross farm income and high education are more likely to adopt GM crops. Cameron (1999) found evidence of a role of accumulated knowledge of the performance of the GM crop that is consistent with learning theory. Market conditions such as the perceived ability to market GM crops, existence of premiums for non-transgenics, and consumer acceptance have been considered in surveys though not related to adoption. Perceived profitability has been decomposed into existence of unrealized cost savings (due to unsatisfied expectations), realized cost savings, premiums received, reduced pesticide cost, improved pest control and increased yield. Environmental factors such as perceived benefits from decreased use of pesticides to environment have also been considered; see Darr and Chern (2002). Results from past studies are summarized in Table 2 of Weaver and Kim (2002).

To close, results are presented for adoption of transgenic soybeans based on a static adoption model for a secondary state, Pennsylvania. Equation 3) motivates a probit model for this decision and results for a model of the adoption of transgenic soybean seed are presented in Table 2. Data are from a 1999 sample of Pennsylvania producers of soybeans based on the Agricultural Resource Management Survey (ARMS) implemented by the Pennsylvania Agricultural Statistics Service with funding from the National Agricultural Statistics Service, USDA. The survey includes information on the adoption of herbicide-tolerant soybean seeds, as well as an extensive set of data recording possible exogenous determinants of the adoption decision. Previous related studies include the Alexander, Fernandez-Cornejo and Goodhue (2002) consideration of Iowa GM corn, Darr and Chern's (2002) consideration of Ohio's soybean and corn experience from 1996 to 1999, and Fernandez-Conejo and McBride (2002).

Results reported in Table 2 are based on a data set of 158 observations. Descriptive statistics are reported in the right-hand side of this table. Given the observations are drawn from a common geographic region, the market environment across farm respondents can be assumed homogeneous and is not empirically described. About 46.6% of respondents cited increased yield as a reason for adoption of GM soybeans, 12.5% cited reduction in pesticide-input cost, 8.1% cited increased planting flexibility, and less than 1% cited perceived improvement in environmental effects of field practices as a reason, respectively. Eighty-two percent of respondents indicated farming as their major occupation. About 50.3% of respondents indicated farm gross value of sales between $50,000 and $250,000. Eighty percent of respondents had high school or less education. Respondents averaged 17.3 years of experience operating the farm. Over 76.6% of the respondents used post-emergence herbicides only, while 28.6% used pre-emergence herbicides. With respect to past experience with ht soy, only 5.6% of respondents planted ht soy the previous year (1998). Eighty-five percent of respondents indicated growing corn as the preceding crop to soybeans. While it would be of interest to explore the conditionality of adoption on use of no-till technology, data for this characteristic are not available for the sample and based on Fernandez-Conejo and McBride's results, its exclusion will result in inefficiency, though not bias estimates. However, data were available indicating whether tillage or cultivation was conducted in the soybean field for weed control during the growing season. Five point six percent of respondents indicated such a practice.

Results in Table 2 indicate that based on summary statistics, the model's fit can be interpreted as acceptable. Education level of the operator was found to have a

Table 2. Probit results
Adoption of transgenic soybeans Pennsylvania 1999 (=1)

Variable	**Definition & Type**	**(Mean, SD) Freq. %**	**Coefficient**	**t-stat**
Intercept			-1.51896	-4.65389**
Education	polychotomous =1 if < high school =2 if high school grad =3 if some college =4 if college grad =5 if grad school	 31.1% 50.3 9.3 7.5 1.9	-.07037	-1.48634
Gross value of sales	Polychotomous 1 if $1k<x<$2.5k 2 if 2.5<x<4.999 3 4 by $5k intervals 5 6 7 if 25k<x<39.999 8 if 40k<x<49.999 9 if 50k<x<99.999 10 if100k<x<249.999 11 if 250k<x<499.999 12 if > $500.000	 .6% .6 2.5 2.5 5.6 1.9 3.7 5.6 20.5 29.8 18.6 6.9	-.07104	-.14812*
Experience	Years as operator	(17.32 13.45)	.00296	.96726
GM used 1998	=1 if yes, =0 if no	5.6%	.01294	.31286
To increase yield	=1 if yes, =0 if no	46.6%	.51979	3.63506**
To decrease pesticide cost	=1 if yes, =0 if no	12.5%	.73090	4.30910**
To increase planting flexibility	=1 if yes, =0 if no	8.1%	.62445	3.51113**
Previous crop corn	=1 if yes, =0 if no	85.1%	-.31661	-2.48402**
Used pre-emerg herbicide	=1 if yes, =0 if no	28.6%	-.21260	-1.98350**
Used post-emerg herbicide	=1 if yes, =0 if no	77.6%	-.06791	-.45372
Major occup. 1= farm	1= farm, 2=hired manager, 3=other, 4=retired	82.0% 0 17.4 .6	-.01887	-.28994
Till/Cultivate	=1 if yes, =0 if no	5.6%	-.20075	-.95449

* Significant at 5% level; ** Significant at 10%.
Pearson goodness-of-fit chi square = 224.012 DF = 145 P = .000

negative, though statistically insignificant effect on adoption. Using a polychotomous indicator of gross value of crop sales as a measure of scale, the results indicate a negative relationship that is significant. Experience and occupation were both found insignificant. Past use of GM herbicide-tolerant seeds was found to be a positive factor though it is statistically insignificant. Use of tillage or cultivation during the growing season for weed control was found statistically insignificant. Estimates corroborate the low frequency of respondents reporting concern for environmental impacts as a motivation for adoption. Results indicate that hedonic factors played a key role. Specifically, an expectation of increased yield, decreased pesticide costs and increased planting flexibility were found to be positive and highly significant factors in predicting adoption. Use of pre-emergent herbicides was found to be a negative factor, as was previous planting of corn.

Overall the results confirm the importance of private economic performance as a determinant of GM soybean adoption. This is reflected in the significant role found for hedonistic rationale for adoption. Evidence was found that the technology is not scale-neutral, and a bias against large scale was found. Adoption was found to be strongly conditional on and negatively related to the previous crop being corn, perhaps reflecting continuous cropping of corn. Results emphasized the importance of perceived increase in yield as a factor that is interpretable as a hedonistic rationale, however, the interpretation of this result cannot go beyond yield increase given that no further explanation concerning the origin of such expected increases in yield was available. In general, increased yield could result from varietal performance, improved weed control, reduced plant suppression or damage due to herbicide intolerance, or increased seed density. The finding that planting flexibility and reduced pesticide costs also have a positive relationship with adoption is consistent with this interpretation.

While past work has focused on wide geographic areas, the results reported here are for a more geographically restricted area. It is also important to note that these results characterize a region in which soybeans are grown within a context of an agricultural system that is dominated by dairy farming with some limited animal feeding (beef and pork). Given that the predictions of physical science suggest that the appeal of GM soybeans will be field- and site-dependent, our results suggest that the economic appeal of this technology is dominant over substantial heterogeneity in field and site characteristics

Conclusions

At first consideration, US experience with respect to transgenic-crop adoption appears to suggest that this innovation involves a universally dominant technology motivating rapid adoption by all producers. However, at a deeper level it is clear that this type of innovation involves a multiple faceted technology, a complex set of changes in input mix, production practices and outputs. Further, the performance of the innovation is, in many settings, conditioned by local characteristics. This implies that the innovation is not universally dominant across the landscape of the US. This result is likely to hold for the E.U. as well. While the technology may be associated with public effects, either in the short term or irreversibly, results in hand do not suggest that these effects have been a driving force behind producer decisions to adopt. In this brief paper, an overview of representative results was presented, along with more recent results from the state of Pennsylvania where soybean cultivation has expanded. In this type of setting, transgenic crops may offer important flexibility in crop management as well as in efficiency of weed management. Empirical results reported here suggest that in this type of setting, producers remain focused on private net benefits associated with the technology as a determinant of their adoption decision.

References

Alexander, C., Fernandez-Cornejo, J. and Goodhue, R.E., 2002. Determinants of GMO use: a survey of Iowa maize-soybean farmers' acreage allocation. *In:* Santaniello, V., Evenson, R.E. and Zilberman, D. eds. *Market development for genetically modified foods*. CABI publishing, Wallingford.

American Soybean Association, 2001. *ASA study confirms environmental benefits of biotech soybean.* American Soybean Association, Saint Louis. [http://www.soygrowers.com/newsroom/releases/2001%20releases/r111201.htm]

Benbrook, C., 2001. Do GM crops mean less pesticide use? *Pesticide Outlook,* 12 (5), 204-207.

Bullock, D.S. and Nitsi, E.I., 2001. Roundup Ready soybean technology and farm production costs: measuring the incentive to adopt. *American Behavioral Scientist,* 44 (8), 1283-1301.

Cameron, L.A., 1999. The importance of learning in the adoption of high-yielding variety seeds. *American Journal of Agricultural Economics,* 81 (1), 83-94.

Carpenter, J. and Gianessi, L., 1999. Herbicide tolerant soybeans: why growers are adopting roundup ready varieties. *AgBioForum,* 2 (2), 65-72. [http://www.agbioforum.org/v2n2/v2n2a02-carpenter.htm]

Clark, E.A., 1999. *Ten reasons why farmers should think twice before growing GE crops*. Plant Agriculture, University of Guelph, Guelph. [http://www.plant.uoguelph.ca/faculty/eclark/10reasons.html]

Conner, A.J., Glare, T.R. and Nap, J.P., 2003. The release of genetically modified crops into the environment. Part II. Overview of ecological risk assessment. *The Plant Journal,* 33 (1), 19-46.

Conservation Tillage Information Center, 1999. *Conservation tillage survey data.* Available: [http://www.ctic.purdue.edu/Core4/CT/CT.html] (June 4, 1999).

Darr, D.A. and Chern, W.S., 2002. Estimating adoption of GMO soybeans and maize: a case study of Ohio, USA. *In:* Santaniello, V., Evenson, R.E. and Zilberman, D. eds. *Market development for genetically modified foods*. CABI publishing, Wallingford.

Economic Research Service USDA, 1999a. *Genetically engineered crops for pest management*. Economic Research Service, US Department of Agriculture, Washington. [http://www.biotechknowledge.com/biotech/knowcenter.nsf/ID/C217513C7EDCB48886256AFB0067D273]

Economic Research Service USDA, 1999b. *Impacts of adopting genetically engineered crops in the United States*. Economic Research Service, US Department of Agriculture, Washington. [http://www.ers.usda.gov/emphases/harmony/issues/genengcrops/genengcrops.htm]

Ervin, D., Batie, S., Welsh, R., et al., 2000. *Transgenic crops: an environmental assessment*. Wallace Center for Agricultural and Environmental Policy. Policy Studies Report. Wallace Center for Agricultural and Environmental Policy. Winrock International.

Feder, G. and Umali, D.L., 1993. The adoption of agricultural innovations: a review. *Technological Forecasting and Social Change,* 43, 215-239.

Fernandez-Cornejo, J., Klotz-Ingram, C. and Jans, S., 1999. Farm-level effects of adopting genetically engineered crops in the USA. *In: Transitions in agbiotech: economics of strategy and policy: proceeding of NE-165 conference, June 24-25, 1999, Washington DC.* 57-74. [http://www.umass.edu/ne165/conferences99/ta_program.html]

Fernandez-Cornejo, J. and McBride, W.D., 2002. *Adoption of bioengineered crops.* Economic Research Service, US Department of Agriculture, Washington. Agricultural Economic Report no. 810. [http://www.ers.usda.gov/publications/aer810/aer810.pdf]

James, C., 2002. *Preview: global status of commercialized transgenic crops.* ISAAA, Ithaca. ISAAA Briefs no. 27. [http://www.isaaa.org/kc/Publications/pdfs/isaaabriefs/Briefs%2027.pdf]

Kapusta, G. and Krausz, R.F., 1993. Weed control and yields are equal in conventional, reduced-, and no-tillage soybean (Glycine max) after 11 years. *Weed Technology,* 7 (2), 443-451.

Karshenas, M. and Stoneman, P., 1995. Technological diffusion. *In:* Stoneman, P. ed. *Handbook of the economics of innovation and technical change.* Blackwell, Oxford.

Moschini, G. and Lapan, H., 1997. Intellectual property rights and the welfare effects of agricultural R&D. *American Journal of Agricultural Economics,* 79 (4), 1229-1242.

Nadolnyak, D.A. and Sheldon, I.M., 2001. Simulating the effects of adoption of genetically modified soybeans in the US. *In: AAEA annual meeting, August 6-8 2001, Chicago.* AAEA, Ames.

Neill, S.P. and Lee, D.R., 1999. *Explaining the adoption and disadoption of sustainable agriculture: the case of cover crops in northern Honduras.* Cornell University, Ithaca. Working Paper Department of Agricultural, Resource, and Managerial Economics, Cornell University no. 1999-31.

Padgette, S.R., Re, D.B., Barry, G.F., et al., 1996. New weed control opportunities: development of soybeans with a Roundup Ready gene. *In:* Duke, S.O. ed. *Herbicide-resistant crops: agricultural, environmental, economic, regulatory and technical aspects.* CRC Press, Boca Raton, 53-84.

Pike, D.R., McGlamery, M.D. and Knake, E.L., 1991. A case study of herbicide use. *Weed Technology,* 5 (3), 639-646.

Rawlinson, J. and Martin, A., 1998. *Weed management strategies in soybeans.* Unpublished Manuscript. University of Nebraska, Lincoln.

Weaver, R.D. and Kim, T., 2002. Incentives for R&D to develop GMO seeds: a reconsideration of alternative pricing mechanisms. *In: 6th International conference on agricultural biotechnologies: new avenues for production, consumption and technology transfer, Ravello (Italy), from July 11 to 14, 2002.* International Consortium on Agricultural Biotechnology Research, Rome.

Weaver, R.D. and Wesseler, J., 2003. Restricted monopoly R&D pricing: uncertainty, irreversibility, and non-market effects. *In: 7th International conference on public goods and public policy for agricultural biotechnology, Ravello (Italy), from June 29 to July 3, 2003.* International Consortium on Agricultural Biotechnology Research, Rome.

9b

Comment on Weaver: *Ex post* evidence on adoption of transgenic crops: US soybeans

W.J.M. Heijman[#]

In this interesting paper the author aims at analysing the adoption of transgenic soybeans in areas in the US. He concludes that transgenic crops can be expected to be adopted in existing cultural areas if incentives are appropriate and that potential exists for transgenic crops to be adopted in new areas where they offer advantages over alternative crops in aspects such as weed control and other management practices.

After an overview of the current status of GM crops, the author continues with describing the private and public effects of transgenic crops. The public effects include the potentially important short- and long-term environmental benefits, such as the introduction of pesticides that would reduce environmental risk. The private benefits refer to increased yields and reduction of costs. In addition to that the author states that the uncertainty of private costs and benefits of GM crops is extensive.

"It is especially important to note that transgenic crops constitute an innovation for which the level and uncertainty of private and public effects results in consumer response".

Further, the degree of irreversibility connected with the investments in this new technology is high. The high uncertainty connected with the irreversibility of the adoption decision leads to a high risk for producers.

An important conclusion of the paper is that in the US transgenics is not a *universally* dominant new technology. It is likely that this conclusion is also true for the EU. According to the author the adoption of transgenics means:

"...a complex set of changes in the overall production technology, rather than a single augmentation of a particular input".

This accentuates the importance of local circumstances and characteristics of the producers in the adoption and diffusion of the new technology.

In the quantitative part of the paper the author describes four types of factors affecting adoption: farm characteristics (farm size, field characteristics), experience and knowledge of the technology, market conditions (price risk, profitability, cost and yield effects) and environmental implications (e.g. decreased use of pesticides). In the regression analysis only two variables have a significant coefficient: increased yield and the past use of herbicide-tolerant seed. All other factors are more or less irrelevant.

As the author rightly concludes, this means that in their adoption decision of new technologies such as GM crops the producers remain focused on private net benefits.

[#] Wageningen University, Wageningen, The Netherlands. E-mail: wim.heijman@wur.nl

J. H. H. Wesseler (ed.), Environmental Costs and Benefits of Transgenic Crops, 141–142.

Public consequences (such as gene flow and other environmental consequences) are considered important only when they lead to lower costs (e.g. the decreased use of pesticides). This stresses the public responsibility for the public costs and benefits of transgenics. It seems that the public awareness of these risks is developed more extensively in the EU than in the US.

10a

Spatial and temporal dynamics of gene movements arising from deployment of transgenic crops

Christopher A. Gilligan[#], David Claessen[#, ##] and Frank van den Bosch[##]

Abstract

Many of the pressing questions about whether or not to release a genetically modified crop can be resolved into population-dynamics questions concerning invasion and persistence of the transgenic crop itself, or of hybrids with other crops and wild relatives. Progress in assessing risk demands a coherent theoretical framework within which questions such as the reversibility of breakdown in pest resistance or 'escape' of herbicide-tolerance genes can be phrased and tested. Here we begin to sketch out a framework that provides some answers while taking account of the inherent spatial and temporal variability of agricultural and semi-natural systems. After a brief summary of the risks associated with the deployment of novel crop varieties, we define what we mean by invasion and persistence in variable environments. This leads to a discussion of stochasticity and spatial scales in which we distinguish between different sources of variability that affect the probability of invasion and persistence times. We illustrate this with the distributions of global extinction times of crop plants or hybrids that persist as local patches in the landscape following the introduction of a new variety. Empirical studies of persistence of transgenes are next discussed. We begin with pollen dispersal, which leads to the dispersal profiles from point sources from which separation distances can be calculated to minimize contamination between transgenic and conventional crops. Three important aspects that emerge from some work on the dynamics of feral populations of transgenic crops are identified: feral patches of some transgenic crops are transient with short persistence times; there is large environmentally-driven variability in ecological performance; the seed bank is an important reservoir from which plants may emerge even after a patch appears to have gone extinct. Next we show how such variability can be incorporated into stochastic life history models, from which it is possible to identify intrinsic, genetically controlled properties of crop plants that favour persistence. These analyses suggest how to design new crop varieties with characters that minimize the risk of spread and persistence in the landscape. Finally, we introduce landscape-dynamic models in which we show, using

[#] Department of Plant Sciences, University of Cambridge, Downing Street, Cambridge, CB2 3EA, UK. E-mail: cag1@cam.ac.uk

[##] Biomathematics Unit, Rothamsted Research, Harpenden, Herts, Al5 2JQ, UK. E-mail: david.claessen@bbsrc.ac.uk; frank.vandenbosch@bbsrc.ac.uk

J. H. H. Wesseler (ed.), Environmental Costs and Benefits of Transgenic Crops, 143–161.

individual-based percolation and metapopulation models, how the spatial arrangement of crop plants and feral patches of 'escaped' plants in the landscape influence invasion and persistence.

Keywords: Invasion; persistence; stochastic variation; life-history parameters; metapopulation analysis

Introduction

The deployment of any novel crop variety is seldom risk-free. The objective of much recent economic and biological modelling is to assess these risks in order to decide whether or not it is profitable and environmentally safe to release a novel crop. Recent advances in molecular biology that allow greater genetic control in the design of modern crop varieties promoting enhanced yield, greater pest and disease resistance, tolerance to herbicides as well as novel crops for biofuel, biopolymers, nutrients and biopharmaceuticals all lead to increased urgency for a robust theoretical framework for decision-making. In this chapter we focus on the spread of genetically modified plants and their introduced foreign genes (transgenes) into the environment. We argue that stochasticity plays an important role in successful invasion and persistence of these transgenes. We also show how, using ecological analyses of the population dynamics under variable environments, it is possible to identify which life-history parameters favour invasion and persistence and, hence, which crop characteristics should be targeted in breeding programmes so as to minimize the risk of unwanted spread and persistence.

Our analyses hold for the deployment of any novel crop variety, whether conventionally bred or genetically manipulated by transformation. Conventional plant breeding has been remarkably successful in steadily increasing yields of agricultural crops (Evans 1998). The demand for yet more progress continues to rise, however, with global population growth, so that the global demand for cereals is expected to increase by 30% by 2020, with developing countries accounting for two thirds of this demand (Rosegrant et al. 2001). Recent developments in molecular biology are likely to have an important impact on breeding technology. Thus, instead of selective breeding and reassortment of 50 to 100 thousand genes with up to ten recurrent selective back-cross generations to recover an elite line, modern methods of marker-assisted breeding allow the identification of many fewer genes that must be changed to achieve a successful line. Although this may involve transfer of genes between species more often than not it will occur between cultivars within the same species. Still, there remains uncertainty about the economic and environmental benefits and costs of releasing these newly developed varieties into commercial agriculture. Will, for example, a new variety continue to out-yield the conventional varieties it replaces? How does this change from year to year? Will novel pest or disease resistance remain effective or will it be necessary to supplement weakened genetic control with pesticides?

Some consequences of the deployment of novel varieties are classified as 'irreversible' by economists but still subject to uncertainty from a biological – and economic – perspective. These include benefits such as reduced risk of resistance to pesticides or to accumulation of toxic pesticide residues in soil, water or crops, following the release of varieties that are genetically resistant to pests or pathogens. Still greater uncertainty surrounds the irreversible costs associated with the release of a new variety. The principal concerns here are focused on:

- unexpected human-health risks from genetically-modified crops in the food chain;
- loss of biodiversity or an unfavourable change in ecological balance because of enhanced persistence or invasiveness of a genetically-modified crop;
- escape of transgenes to other crops or wild relatives.

Irreversible costs may also include squandering of resistance or toxin genes by promoting premature build-up of counter measures in a pest population so preventing further use of the resistance or toxin gene.

Many of these irreversible costs can be resolved into population-dynamic questions concerning invasion and persistence.

- Will a gene for pest resistance or herbicide tolerance 'escape' from a crop plant into a wild relative and if it does will the resulting hybrid invade?
- Will it replace the wild type?
- Will both persist?
- Is the process irreversible whereby one replaces the other and what is the time to extinction of one or the other form and hence the time-frame for irreversibility?

Gilligan (2003) has recently discussed the epidemiological perspective of pest resistance for transgenic crops in the context of options analysis advocated by Morel et al. (2003) and Wesseler (2003). He showed that: (i) theoretical progress can be made in predicting the risk of invasion and persistence of resistant pests and diseases; (ii) deterministic models are useful in identifying crude criteria for invasion but stochastic population models are essential to understand the risk of invasion and to identify strategies to minimize risk. The focus there is on the effects of new varieties on the dynamics of microbial and pest populations. Here we consider the spread of transgenes from crop plants. The spread of genes from these crops can originate from hybridization with wild or cultivated relatives or from transgenic seeds that disperse and potentially establish a persisting feral population of transgenic plants. Essentially both depend on the fitness of the resulting population irrespective of whether or not it arose by pollen or seed flow. Accordingly, we present below some analyses that focus on the effects of fitness on invasion and persistence and how transgenic crops may be designed to minimize the unwanted spread. We do this without proposing recourse to 'terminator genes' which themselves pose environmental threats (Masood 1998).

We first explain what we mean by invasion and persistence in variable environments and introduce some concepts of demographic and environmental stochasticity. Next we summarize some important empirical studies of pollen dispersal, with particular emphasis on separation distances between crops to minimize the risk of contamination and introgression, followed by a brief discussion of empirical studies of seed persistence. The importance of chance and space is then described for life-history models which leads naturally to landscape-dynamic models in which we consider the percolation of transgenes through a dynamically changing landscape.

Invasion, persistence, variability and spatial scales

Invasion and persistence of transgenic plants occur at a range of scales from the field through the farm to the regional scales. This resolves into a hierarchy of patches for which we distinguish two types of patch. A feral patch occurs on sites that have not previously grown the transgenic crop. Feral patches are often located around field or road margins. They are typified by relatively small groups of plants following accidental deposition of grain as it is transported from fields to processing sites (Figure 1). They also occur at sites of cross-pollination with wild relatives. A volunteer patch arises, by contrast, from shed grain at sites within fields that have been sown to a transgenic crop. Volunteer patches often form large reservoirs of seed in the soil but they are regularly disturbed by agricultural cultivation, bringing buried seeds to the surface, where they may germinate or die. Invasion is determined by whether or not there is an increase in density of a transgenic plant or introgressed offspring in a region of interest following introduction. There are two processes, enhanced colonization of previously occupied feral and volunteer sites and colonization of new sites. Hence some analyses may treat the plant as an individual while others treat the patch as an individual. The identification of discrete patches leads naturally to the concept of a metapopulation comprising asymmetrically-sized volunteer and feral patches within a dynamic landscape of sites available for feral colonization (Gaston et al. 2000; Fahrig 1992) that may be supplied by seed and pollen from continued cultivation of a transgenic crop. Within sites, it is usual to define a 'pre-set threshold' density of plants above which a population is assumed to have become established. The identification of criteria for invasion in metapopulations is currently the subject of both empirical and theoretically-motivated research. Within metapopulations, criteria involve analogous threshold densities for occupied patches or for the balance between local extinctions and probabilities of transit between patches. Future work will examine the effects on invasion of geometrical balance and connectivity between field crops, volunteer and feral patches in the landscape. In this chapter, we focus on the stochastic description of patch size, distinguishing between volunteer and feral patches. This leads naturally to consideration of persistence and to the description and analysis of extinction times (Figure 2). Here by adopting a stochastic approach it is possible to derive a distribution for extinction times of a population of feral or volunteer patches (Figure 2). This yields substantially more management information than a simple mean, allowing us to estimate the proportion of patches that might persist for long periods of time.

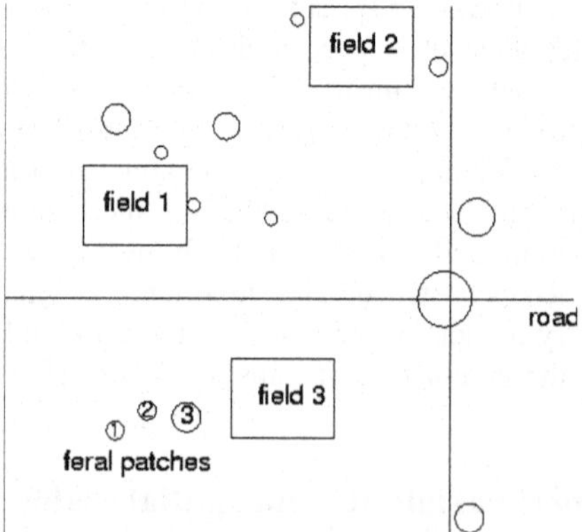

Figure 1. Schematic representation of the local mosaic for persistence of transgenes within fields as crops or as volunteer patches and as feral patches on field boundaries and along roads

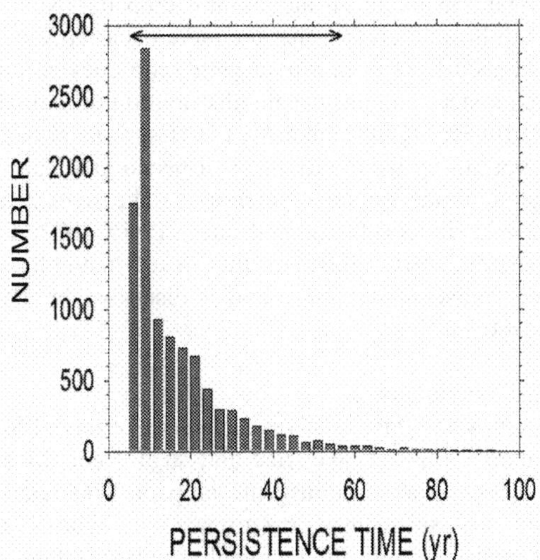

Figure 2. Distribution of persistence times in 10,000 model runs, measured as the time elapsed until the total population abundance falls below a pre-defined extinction threshold, given an initial condition of 100 seeds in a seed bank at t = 0. Extinction is defined as the total abundance < 1 individual (i.e., sum of above ground plants and seeds in seed bank). The arrow indicates the median 95% range of persistence times (6 yr – 57 yr), mean persistence time is 16 yr. The data were computed for a model similar to that in Box 1, with the probability of annual disturbance of p=0.15

Two broad types of variability affect the spatial and temporal dynamics of gene movement. One is demographic stochasticity, whereby the probabilities of individual seed germination, growth and subsequent seed production vary according to a set of probability-density functions that arise from the birth and death processes themselves. These have parameters that are fixed for a given set of environmental conditions. The second is environmental stochasticity, in which the underlying parameters for germination, growth and seeding fluctuate with environmental conditions. While this is largely driven by weather variables, it can also include other factors, notably disturbance of a volunteer site by cultivation or of a feral site by mowing, grazing or soil disturbance by rabbits or invertebrates.

Empirical studies of persistence

Empirical studies have been used to quantify the dispersal and persistence of transgenes following introduction in agricultural crops. They include quantification of dispersal gradients for the two important pathways of potential transgene spread via seed dispersal or via pollen dispersal (Wolfenbarger and Phifer 2000). They also involve measurements of seed survival as volunteer and feral patches within fields, along field boundaries or road margins as well as empirical tests for the feasibility of hybridization with other crop or weed species. The results of farm-scale field trials in the U.K. on the impacts of transgenic crops on biodiversity (Firbank et al. 2003; 1999) have recently been published (Squire et al. 2003). Here we focus on pollen and seed dispersal and the fate of feral populations, for which two important results emerge. The first is the importance of stochastic variation in determining whether or not a transgene persists after release in an agricultural crop (Crawley et al. 2001; 1993; Hails et al. 1997). It follows therefore that the mere occurrence of hybridization with another species or of seed shed from a transgenic crop does not guarantee long-term persistence. We need instead to consider the distribution of extinction times (and the corollary persistence times: Figure 2) in order to assess the *risk* of persistence. Second is the challenge of scaling-up from individual dispersal gradients around single sites (comprising transgenic fields or feral patches) to the dynamics of invasion and persistence in a mosaic of loosely-coupled sites (Perry 2002; Colbach, Clermont-Dauphin and Meynard 2001a; 2001b). This leads naturally to the theory of metapopulations and landscape dynamics and is discussed briefly in the section on Landscape-dynamic models.

Pollen dispersal and hybridization

Hybrids of GM crops and wild relatives have been observed for a number of crop species, including maize, oilseed rape, rice and sugar beet (Messeguer 2003). The transfer of a transgene into a non-transgenic crop or wild relative is mediated by pollen transfer and the subsequent occurrence of hybrids finally leading to introgression of the transgene into the recipient population. Four broad risks can be identified with differing economic consequences and public acceptability.

(i) Small-scale field testing of newly developed transgenic crops is usually done under semi-contained conditions to minimize dispersal of transgenic pollen, and the risk is considered to be small.

(ii) Consumer preference increasingly stipulates that products from non-transgenic crops should not be contaminated with transgenic material. The main source of such contamination is influx of transgenic pollen into non-transgenic crops. The Advanta Seeds contamination incident (Advanta Seeds UK 2000), whereby large amounts of seed from non-transgenic crops was rejected because of contamination by transgenic hybrids, showed that this risk is real and has large financial side effects.

(iii) Pollination of wild relatives with transgenic pollen can lead to hybridization. This has, for example, been observed in crosses between the crop species *Brassica napus* and the wild plant species *B. rapa* (Hauser, Shaw and Ostergard 1998). The hybrids have a low fitness but subsequent backcrosses with the wild plant species can lead to the development of a population where the transgene has introgressed into the wild species (Hauser, Shaw and Ostergard 1998). Introgression of transgenes into wild plant species is considered unwanted.

(iv) Wild plant species that have introgressed a transgene might be more fit than their relatives increasing the risk of weed invasion.

In order to quantify the risk of pollen transfer and the risk of invasion of the transgene through hybridization with non-transgenic crops or wild relatives we need to know the dispersal pattern of transgenic pollen as well as the probability of hybridization and fitness of the offspring. Stochasticity enters via the dispersal range, viability of pollen, viability and fitness of the progeny as well as the relative spatial distributions of donor and recipient sites in the landscape. In this section, we focus on dispersal from a single site and summarize some of the principal considerations to analyse persistence.

Most work to date has focused on the empirical description of pollen dispersal gradients, usually for non-transgenic crops, and, in particular, on the tail of the distribution to assess the furthest extent of transport. This leads naturally to the concept of a *separation distance* by which criteria are set for the separation of transgenic from other sensitive crops in order to reduce the probability of cross-pollination below a certain defined threshold (Perry 2002). There has therefore been considerable discussion in the general biological literature about the shape of dispersal kernels, particularly with respect to the thickness of the tail which can affect the probability of rare long-distance events and of the shapes of invasion wave-fronts (Ferrandino 1993; Shaw 1995; Kot, Lewis and Van den Driessche 1996; Shigesada and Kawasaki 1997). Two broad mathematical approaches have been used, one based on empirical description of dispersal gradients, the second on physical models for dispersal of particulates (Csanady 1973; Nieuwstadt and Van Dop 1982). Physical models include the effect of wind, turbulence, deposition and in some cases elements of landscape structure. These models have been applied to the dispersal of fungal spores, encysted bacteria and pollen (e.g. Aylor 1986) but so far relatively little exploitation has been made of the stochastic description of these models from which we may be able to ascertain risk of dispersal under a range of environmental conditions. Most attention has instead been directed towards the use of empirical models to describe dispersal profiles. Practical distinction of thick-tailed (typified by power-law, Pareto or Cauchy distributions) and thin-tailed (usually exponential) dispersal kernels, remains difficult, however, because of the problems of low detection rates at these long distances. These give few data-points in the critical range with which to distinguish models so that there may be considerable error in the estimation of rare distant events. Other factors can bias estimation. Recent reports of long-distance pollen transport leading to transgenic contamination of maize plants in Mexico (Quist and Chapela 2001; 2002) is now thought to have been due to illegal planting of GM seed by growers (Perry 2002). We argue below (Landscape-dynamic models section) that the tail of the distribution is rather less important than might be thought in predicting whether or not a transgene invades through a heterogeneous landscape, for which the magnitude of dispersal in the mid-ranges may be more important.

The nature of the gradient depends on the method of dispersal. Distances range from 100 to 200 m for short- to medium-distance dispersal to approximately 10 km for long-distance dispersal by insects. The corresponding ranges for wind dispersal are < 1 km for short to medium distances but the long distances extend to hundreds of kilometres. Some crops have more than one dispersal mechanism, leading to bimodality in dispersal ranges. Simple dispersal profiles for pollen transfer can be obtained by sampling the air. Recent developments in molecular diagnostics and biosensing promise improved specificity in detection and automatic quantification. Mere arrival of the pollen is not enough to guarantee hybridization. In practice, trap plants are usually used to sample pollen yielding dispersal profiles for successful hybridization that automatically take account of the uncertainty in viability of pollen. Yet more uncertainty arises after hybridization, since Hauser, Shaw and Ostergard (1998) have shown that there is considerable variability in the fitness of F1 hybrids and further backcrosses with the wild relative. We conclude that stochasticity is therefore of key importance in pollen dispersal and hybridization and must be given greater prominence in evaluating risks of pollen dispersal.

Separation distance

Consideration of the risks described above, in combination with the pollen-dispersal models, has led to the idea of a *separation distance,* the distance between the transgenic pollen source crop and the nearest non-transgenic crop or population of wild relatives. Methods are currently being developed to decrease the risks below a threshold that is perceived acceptable by regulations on minimum separation distance (Perry 2002; Firbank et al. 1999). For small-scale field testing of newly developed transgenic crops separation distances of 50 to 400 m and greater have been required in the past (Scheffler, Parkinson and Dale 1995). These separation distances were mainly based on those commonly used by plant breeders in the production of certified or basic seed. Experiments with *B. napus* showed that these distances of 400m result in hybridization ratios (fraction of hybrid seeds in the yield) of 0.3 - 4.0% (Scheffler, Parkinson and Dale 1995), which exceeds the threshold of 0.1% level that has been suggested to be publicly acceptable (Meacher 2001).

Perry (2002) discussed the relation between separation distance and the fraction of the agricultural land that can be used to grow a non-transgenic crop that has a hybridization ratio below the acceptable level, or conversely the fraction of agricultural land available to grow a transgenic crop given existing non-transgenic crops that must have this low hybridization ratio. He showed that the increase in separation distance currently under review by the UK government would have serious implications for future coexistence of non-contaminated non-transgenic crops and transgenic crops. A more involved treatment of the same problem including realistic pollen-dispersal models and realistic crop field spatial patterns can be found in Colbach, Clermont-Dauphin and Meynard (2001a; 2001b).

Hybridization rates between transgenic crops and wild relatives have been intensively studied for a variety of species when they are in close proximity (Wilkinson et al. 2000; 2003) Very little is however done on the relation between separation distance and the rate of hybridization. Moreover, the spatial spread of wild plant species with an introgressed transgene has not been studied so far.

Feral population dynamics

Many crops, such as oilseed rape and sunflower, are annuals that were originally selected from weedy ephemeral plants (Linder and Schmitt 1995). They both have a high seed yield but low competitive ability so that recruitment depends critically on disturbance of the vegetation cover to provide a site for invasion. The frequency of environmental disturbance therefore plays an important role in the invasion and persistence of these populations. Most European work has been focused on oil-seed rape and sugar beet. Crawley and Brown (1995) mapped the occurrence of feral populations of oilseed rape (i.e., patches of flowering plants) along the M25 motorway orbiting Greater London. Their results suggest that, in the absence of disturbance, the typical fate of a population of oilseed rape is local 'extinction' (i.e., absence of flowering plants) within 2 – 4 years, due to overgrowth by perennial grasses. However, absence of flowering plants does not imply true extinction of the population because viable seeds may still persist in the seed bank. In France, flowering plants from feral roadside populations were identified as an old cultivar which had not been cultivated for at least 8 – 9 years (Pessel et al. 2001). Unfortunately, it is unclear whether these populations had flowered every year or whether survival in the seed bank had allowed the populations to persist. Yet, seed-burial experiments show that oilseed-rape seeds, once dormant, have a high survival rate in the seed bank of >60% per year (P.J.W. Lutman, pers. comm.), suggesting that long-term persistence of the seed bank is possible. Hence disappearance of adult plants from a patch in one or more successive years does not necessarily mean that the seed bank is extinct.

Nevertheless, current evidence suggests that feral populations of oilseed rape are ephemeral. Further evidence for this and other crops comes from a large-scale field experiment in which Crawley et al. (2001; 1993) compared the performance of transgenic and conventional lines of oilseed rape, sugar beet, maize and potato in a range of 12 natural habitats in the UK over ten successive years. None of the transgenic traits was expected to increase fitness in the given environmental conditions. For oilseed rape, the field experiment was complemented with a seed-burial experiment in which the fate of transgenic seeds in the seed bank was compared with that of the conventional line (Hails et al. 1997). Overall, the transgenic lines did not out-perform the conventional line. Rather, there was a tendency for the transgenic lines that do perform less well than the conventional one, in particular in the seed-bank survival experiment (Hails et al. 1997).

Estimates of the population growth rate varied by more than two orders of magnitude between years and sites, but varied much less between conventional and transgenic lines in the same year and the same site (Crawley et al. 1993). This shows not only the importance of extrinsic factors such as disturbance for these crops, but also the importance of genotype x environment interactions. All seeded plots in the study by Crawley et al. (2001) went extinct within the 10 years of the monitoring period, most of them in two or three years. A smaller-scale field experiment with *Bt*-insecticidal oilseed rape shows that *Bt* transgenes can increase fitness when the plants are subjected to insect predation (Stewart et al. 1997). The differences between the results of Hails et al. (1997) for oilseed rape and Stewart et al. (1997) for maize show that the risk lies typically in those cases where the transgene *does* have an effect on life-history traits which increase fitness and favour persistence (see Life-history models section).

Three important aspects of the dynamics of feral populations of crops emerge from the work by Crawley:

(i) there is large environmentally-driven variability in ecological performance;

(ii) many feral patches of transgenic crops are transient with relatively short persistence times;

(iii) the seed bank is an important reservoir for long-term survival from which plants may emerge even after a patch has appeared to become extinct.

It follows from these empirical analyses that because there is rapid local extinction of feral patches, then global persistence of a transgenic crop can only occur if there is continual input to the system. Without this, feral patches, at least of oilseed rape, will die out. Reinvasion occurs through continual cultivation of the transgenic crop yielding a reservoir of volunteer patches from unharvested seed and a spatially dynamic reservoir of spillage from harvested seed to establish new feral patches.

Life-history models

Empirical studies show that variability is an important aspect of the spread and persistence of transgenes, both via seed dispersal and via pollen dispersal. Field-release experiments are the most reliable way to assess the fate of specific GM plants in cultivated or natural environments. Yet, such experiments do not get at mechanisms by which the risk can be controlled. One approach is to use life-history models to identify which features of transgenic plants ought to be selected to inhibit invasion and persistence. The formulation of a simple life-history model is summarized in Box 1.

Box 1. The matrix population model

Deterministic formulation

Here we formulate a simple matrix model that relates the state of the population in year t, denoted by $\mathbf{n}(t)$, to the population next year, $\mathbf{n}(t+1)$. The elements of the vector $\mathbf{n}(t)$ represent the number of adult plants, $n_1(t)$, and the number of seeds in the seed bank, $n_2(t)$, respectively. The projection from year to year can be represented in vector-matrix notation as

$$\mathbf{n}(t+1) = \mathbf{A}\mathbf{n}(t)$$

(Caswell 2001). The projection matrix **A** summarizes the contribution from each category to the population structure next year and is the mathematical representation of the life cycle of the studied population. In our model (Claessen et al. in press)

$$\mathbf{A} = \begin{pmatrix} F(1-\mu)[d\sigma_2 G + (1-d)\sigma_1]S & \sigma_2 GS \\ F(1-\mu)d\sigma_2(1-G)s_2 & \sigma_2(1-G)s_2 \end{pmatrix}$$

and the interpretation of the matrix **A** corresponds to a description of our assumptions on the life cycle: adult plants each produce F seeds, of which a fraction μ disperses. Of the remaining fraction $(1 - \mu)$, a fraction d is incorporated in the seed bank. In the seed bank, seeds have a probability of σ_2 to survive the winter while a fraction G germinates in spring. Of the seeds that did not enter the seed bank, i.e. the fraction $(1-d)$, a fraction σ_1 survives the winter as seedlings. Seedlings have a probability S to reach the flowering stage. Together, these assumptions are translated in the element a_{11} of **A**: the contribution of adult plants to new adult plants. In addition, new adult plants can emerge from seeds in the seed bank that survive the winter, germinate and reach the flowering stage, i.e. $\sigma_2 GS$, which equals element a_{12} of **A**. Element a_{21} represents the seeds produced by the current adults which enter the seed bank and do not germinate next spring, while element a_{22} represents current seeds in the seed bank that survive and remain in the seed bank.

Stochastic formulation

We incorporate chance by letting some life history processes depend on a stochastic environmental variable. First, fecundity F is assumed to be high in meteorologically favourable years ('good' years), and low in others. Good and bad years are assumed to occur with equal probability but this can easily be adjusted to reflect different environments. Second, germination G and seedling survival S are assumed to be high in years when the habitat is disturbed, with probability p, which reduces competing vegetation, and low in others. The combinations of good and bad years, and of disturbed and undisturbed habitat, produce four different environmental conditions, each occurring with a fixed probability. For each condition we can write down the corresponding projection matrix by substituting the values of F, G and S. The stochastic matrix model is hence equivalent to drawing randomly a matrix out of four possible matrices, at each time step, and can thus be written down as

$$\mathbf{n}(t+1) = \mathbf{A}_t\mathbf{n}(t),$$

in which environmental stochasticity enters via the transition matrix.

Deterministic formulation

Transgenes in plants affect life-history traits such as fecundity, plant survival and survival in the seed bank (Wolfenbarger and Phifer 2000). Some changes are intentional, such as increased plant survival and fecundity conferred by a *Bt* transgene (Stewart et al. 1997) while others are unintended 'side effects' of the transgene, such as increased seed-bank survival conferred by a transgene for oil modification (Linder and Schmitt 1995). In the face of genetic mechanisms such as recombination, however, it remains to be seen how stable such side effects are in successive generations.

The expected rate of population growth, log λ, can be estimated from knowledge of the life-history traits using matrix population models (Caswell 2001). Assuming a small population size, as is realistic for an invading species, log λ is equivalent to fitness (Metz, Nisbet and Geritz 1992). It indicates whether a population is growing (if log $\lambda > 0$) or declining (if log $\lambda < 0$). In addition, with elasticity analysis the sensitivity of population growth to model parameters can be computed. The latter analysis identifies life-history traits which, if modified by genetic engineering, have the largest impact on population growth rate (Claessen et al. in press; Bullock 1999). Modifying such traits can hence either produce an invasive crop, or a GM crop with a reduced capacity to invade and persist.

Several studies have used mathematical models to estimate λ of transgenic cultivars in a particular habitat (Crawley et al. 1993; Bullock 1999; Parker and Kareiva 1996). The fact that these investigations use deterministic models is a major shortcoming, however, because the dynamics of feral-crop populations are inherently influenced by unpredictable environmental variability.

Introducing chance

Here we focus on environmental stochasticity because it introduces and captures the natural variability in weather and disturbance from season to season. Apart from influencing the long-term population growth rate, environmental stochasticity also affects the persistence of a population. A run of bad years may drive a population with a positive expected population growth rate to extinction while, in contrast, a run of good years may result in long persistence of a population which is expected to go extinct.

We studied a stochastic matrix population model of feral oilseed rape, taking into account that in any given year the vegetation cover in the habitat may be disturbed or not, as well as that seed production is high in favourable years and low in others (Claessen et al. in press). Short-lived plant species like oilseed rape are typically structured into a seed bank and established plants above ground. The matrix model projects the number of individuals above ground and in the seed bank in one year, to the number of individuals above ground and in the seed bank in the next year. The contribution of each individual to the next year is determined by the life-history parameters given the current environmental conditions. For example, the fraction of seeds in the seed bank that germinates and emerges is larger if the habitat is disturbed. Seedling survival is also positively affected by disturbance, while survival in the seed bank is assumed to be unaffected by environmental conditions. The model is parameterized for oilseed rape in a feral habitat, but can be adapted to represent other crops or other habitats, by choosing appropriate values of the life history parameters.

In a stochastic matrix model the population growth rate is computed as the geometric mean of annual growth factors, and denoted log λ_s (Caswell 2001). Invasion and establishment happens with a non-zero probability only if log $\lambda_s > 0$ (Metz, Nisbet and Geritz 1992). If log $\lambda_s < 0$, extinction is certain, although the timing of extinction is uncertain. Persistence can be measured as the time it takes to go extinct given a certain initial condition. For feral oilseed-rape populations the predicted persistence times, depending on log λ_s, are shown in Figure 3a. The 95% confidence interval spans many decades for most values of log λ_s, indicating that persistence is highly variable.

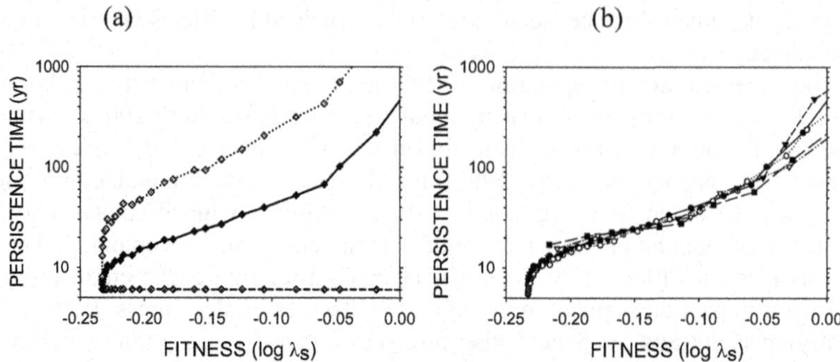

Figure 3. Persistence time computed as quasi-extinction time, versus fitness measured as log λ_s. (a) Mean persistence time and the 95% confidence interval for the default parameter set for a feral oilseed-rape population. (b) Mean persistence time for five different parameter settings

Importantly, persistence depends on overall fitness and not on the details of the life-history parameters (Figure 3b). For the issues of invasion and persistence the basic parameter that matters is therefore the population growth rate. Figure 4 summarizes the relation between invasion, persistence and fitness.

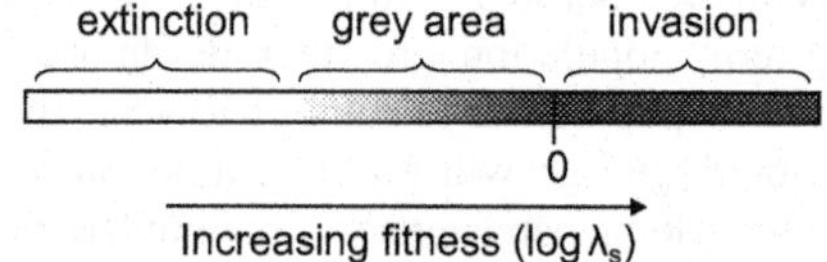

Figure 4. Schematic representation of the relation between fitness, extinction and invasion. If fitness is positive, invasion followed by establishment happens with a certain probability but environmental stochasticity may drive the population to extinction. If fitness is negative extinction is certain. However, if fitness is negative but close to 0 the time to extinction can be very large. In this 'grey area' transient persistence occurs

The determining role of log λ_s means that we can use elasticity analysis to refine the analysis, because it allows us to identify life-history parameters with large impact on fitness and hence on both invasion and persistence. From elasticity analysis we conclude that for oilseed-rape survival in the seed bank is of over-riding importance, while fecundity and plant survival appear less critical (Claessen et al. in press).

Landscape-dynamic models

Whether or not a transgene invades and persists in the landscape leads naturally to consideration of landscape-dynamic models in which spread occurs through a dynamic mosaic of habitable sites. Here we think of a dynamically changing mosaic of habitable sites and occupied sites. Some simple models that underlie the ideas are summarized in Box 2.

Persistence of a transgene in feral or volunteer patches within the landscape depends on the balance between the rate of occupation and extinction of these patches. These suggest simple criteria for invasion (Box 2) but they do not take explicit account of the spatial arrangement of patches nor of stochasticity in extinction and colonization.

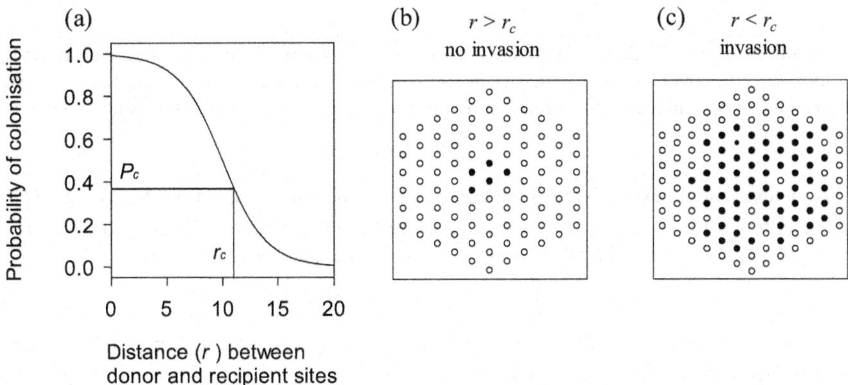

Figure 5. Use of percolation thresholds to predict whether or not a transgene invades across a lattice of contiguous sites. (a) Dispersal profile for transmission from donor to recipient sites. For a given lattice shape (here shown as triangular) there is a characteristic probability (P_c) for percolation that can be used to predict the critical distance (r_c) between sites on the lattice for percolation to occur in a population of sites. (b) Example of restricted spread for inter-site distances $r > r_c$. (c) Percolation through the landscape for $r < r_c$. The hollow sites are uncolonized and the dark sites colonized

The patch-occupancy rate depends upon the density of transgenic crops within a region, from which volunteer or feral patches arise. The extinction rate is influenced by the fitness of the plants as well as disturbance rate and weather. Although it is arguably a useful, if simplistic, starting point this model is naïve in that it assumes β_i (Box 2) to be one constant value for all patches. The model thus does not take into account that movement throughout the landscape is restricted. The probability of transmission of both seed and pollen decreases away from the source, so that dispersal is localized. This leads to a stepping-stone model for patch occupancy, whereby the transgene spreads through a spatially defined set of sites. One appealing simplification of this approach is to consider the system first as movement through a lattice, adapting models from statistical physics and epidemiology (Gilligan 2002; Grassberger 1983). Under static conditions (in which the locations of habitable sites do not change), percolation theory predicts that there is a critical probability (P_c) that corresponds with a threshold inter-patch distance above which the system is connected and invasion occurs (Figure 5). Below the threshold, contiguous patches may be colonized but the patches occur in isolated fragments and invasion is restricted. (At the notional threshold, theory predicts that the system will be self-similar, having clusters of all sizes from single sites right up to connect the entire lattice (Stauffer and Aharony 1991).) Hence the critical probability marks a phase transition between isolated sites with a very low probability of invasion below the threshold, while above the threshold the probability abruptly increases. Clearly habitable patches do not occur on a lattice but the concept can be extended to spread on a random graph. Moreover, the dependence of the phase transition on values for $p > 0.35$ (depending on the nature of the lattice) rather than on the tail of the distribution shows again the importance of considering the mass of the dispersal profile well above the tail.

Box 2. Patch-occupancy models

One of the simplest models for patch occupancy derives from the first metapopulation model proposed by Levins (1969) in which each site is either occupied (*I*) or unoccupied (1- *I*) and there is a colonization rate (β) and an extinction rate (δ), giving for continuous time,

$$\frac{dI}{dt} = \beta I(1-I) - \delta I \ .$$

This yields a simple expression for the equilibrium density of occupied patches, $\hat{I} = 1 - \delta / \beta$, with a criterion for invasion $\beta/\delta > 1$. Many systems have two sources of occupation, one from transgenic crops (β_1) and the other from feral or volunteer patches (β_2), giving

$$\frac{dI}{dt} = (\beta_1 I + \beta_2)(1-I) - \delta I \ ,$$

and an equilibrium density of occupied patches of $\hat{I} = m \pm (m^2 + n)^{1/2}$, where $m = (\beta_1 - \beta_2 - \delta)/2\beta_1$ and $n = -\beta_2 / \beta_1$.

Here the number of patches is treated as fixed. Keymer et al. (2000) circumvented this by adapting earlier models for landscape dynamics and patch occupancy in which patches appear and disappear. They have three classes of patch, 'habitable and empty' (*S*), 'habitable and occupied' (*I*) and 'non-habitable' (*R*). (The variables *S,I,R* are used because of analogy with epidemiological models, where *S,I,R* refer to susceptible, infected or removed states.) Keymer et al. (2000) introduced two parameters for patch dynamics, representing the rates of patch creation (λ) and patch extinction (*e*) as well as two parameters for biological dynamics of colonization (β) and local extinction (δ), analogous to those above. The mean-field form of the model is now given by:

Empty habitable patches $$\frac{dS}{dt} = \lambda R - \beta SI + \delta I - eS \ ,$$

Occupied habitable patches $$\frac{dI}{dt} = \beta SI - (\delta + e)I \ ,$$

Non-habitable patches $$\frac{dR}{dt} = e(S+I) - \lambda R \ .$$

Analysis of this simple mean-field model shows that, when allowance for patch dynamics is made, the criterion for invasion becomes $R'_0 = R_0 \bar{n} \gamma > 1$, where $R_0 = \beta / \delta > 1$ is the criterion derived above for fixed patches (and $\beta_2 = 0$) above. The new criterion depends on the inherent life history ($R_0 = \beta / \delta$, the balance of colonization to extinction), the long-term proportion of habitable sites ($\bar{n} = \bar{S} + \bar{I} = \lambda / (\lambda + e)$) and $\gamma = \delta / (\delta + e)$, which is the ratio of intrinsic extinction to effective extinction because of patch extinction in addition to biological extinction (Keymer et al. 2000). Moreover, it is possible to identify the minimum amount of suitable habitat (n_{min}) that a dynamic landscape needs in order to support the colonizing agent:

$$n_{min} = \frac{1}{\beta}(\delta + \frac{1}{\bar{\tau}}),$$

where $\bar{\tau} = 1/e$ is the average life span of a habitable patch. Stochastic analyses show that while the mean field model captures the qualitative behaviour of the spread it grossly underestimates the time taken to achieve dispersal (Keymer et al. 2000).

Patches, of course, are not fixed. They are ephemeral. Habitable patches appear and disappear, thereby making and breaking local connections across which a transgene can spread, and now the problem becomes one of directed percolation. The approach of Keymer et al. (2000) offers a promising way forward to examine extinction in this framework. Fahrig (1992) had earlier shown that when the rate of creation and extinction of patches is fast, the details of dispersal distance and inter-patch distance are relatively unimportant in influencing persistence. That work considered the density of plants within patches. Keymer et al. (2000) instead consider simple patch occupancy. By analysing the model first as a mean-field (Box 2) and then as a stochastic, patch-based model with dispersal amongst nearest neighbours Keymer et al. (2000) were able to show how the criterion for invasion was affected by the appearance of patches. This in turn allows prediction of the amount of suitable habitat that is required for persistence of a transgene in the landscape.

So far we have been considering presence or absence of the transgene at a habitable site. Clearly, the density of plants carrying the transgene will influence the chance of local extinction as well as the force of colonization for new sites. This leads naturally to consideration of structured metapopulations (Gyllenberg, Hanski and Hastings 1997) in which the stochastic dynamics within and between patches are explicitly modelled in order to identify criteria for invasion as well as persistence times. Recent work on the analogous problem of invasion of animal and plant disease in epidemiology, has shown how the number and size of patches affect the time to extinction of the entire population (Park, Gubbins and Gilligan 2001; 2002; Swinton et al. 1998). The theory shows that median extinction times pass through three regimes as patch population size increases. For small patch population sizes, few individuals make it to another patch and the extinction time scales with the log of population size. As population size increases, so spread occurs to some but not all available patches and median extinction times (across the entire population) tend to be longer but very variable. In the third regime, extinction time scales (T_E) with $T_E = a + bn^{1/2}$, where a is the local extinction time, b is the transit time from arrival in one patch till spread to the next and n is the patch population size (Park, Gubbins and Gilligan 2002).

The models introduced above all assume some overt over-simplification such as distributions of habitable sites on a lattice, or if in free-space, then that the patches are uniform in size. Colbach, Clermont-Dauphin and Meynard (2001a; 2001b) took a more challenging route to modelling reality in devising a detailed simulation model to map fields and feral patches. Our preference is to search for parsimonious metapopulation models in which we simplify the rules for spread in order to predict the risk of persistence. This may involve restriction of dispersal to local contiguous populations and habitable patches or perhaps dual sources of spread with a predominance of short-distance dispersal and some occasional long-distance dispersal analogous to small world mixing (Strogatz 2001). Some support for restricted dispersal in the regional spread of transgenic hybrids from oilseed rape comes from empirical work by Wilkinson et al. (2000). *Brassica napus* can hybridize with several species, notably *B. rapa*, which tends to be found in the U.K. near river courses, and *B. oleracea*, which is found close to coasts. Earlier work has shown that hybrids with *B. rapa* may exceed 90% when the recipient occurs as a weed in an oilseed-rape crop but this plummets to < 2% in natural populations where most pollen is from *B. rapa* (Scott and Wilkinson 1998). Wilkinson et al. (2000) concluded that hybridization would therefore be restricted to sympatric populations. An initial survey of a 100 x 120 km region in Southeast England revealed that only two *B. rapa* populations were found adjacent to *B. napus* fields, and out of the 305 plants in those populations, only one individual appeared to be a hybrid (Wilkinson et al. 2000). Subsequent work at the national scale has suggested that this inference may have been overly conservative and that widespread, relatively frequent hybrid formation is inevitable from male-fertile GM rapeseed in the UK (Wilkinson et al. 2003). Hybridization, however, does not necessarily lead to invasion, not least because of stochastic effects on spatially isolated events. We conclude that, because of this, the metapopulation framework, with allowance for stochastic extinctions, is well suited to the analysis of landscape-dynamic models of transgene invasion and persistence.

Conclusions

Our intention has been to outline a theoretical framework for the spatial and temporal dynamics of gene movements arising from the deployment of transgenic or other novel crops. In doing so we have sought to emphasize the importance of stochasticity in invasion and persistence of transgenic crops and wild hybrids. We have also tried to keep models sufficiently parsimonious so that on balance they shed more 'light' than 'darkness' on the dynamics of transgene spread. Consideration of empirical data for pollen dispersal and the persistence of feral patches of transgenic crops underlines the importance of chance variation in influencing invasion and persistence, although data remain scarce. We have distinguished environmental from demographic stochasticity. Environmental stochasticity is characterized by year-to-year and site-to-site variation as well as the frequency of disturbance of feral patches. Demographic stochasticity reflects chance variation in seed set, seed survival or in hybridization with another crop or wild relative, when relative population densities of donor and recipient become important in influencing successful reproduction and survival. One of the most important results of stochastic analysis is the derivation of probability distributions for persistence times for a population distributed amongst feral, volunteer or other patches (Figures 2 and 3). These allow assessment of the risk of long- or short-term persistence. Despite the complexities associated with plant growth in variable environments, we have shown that it is possible to model the population dynamics of transgenic plants. The use of life-history models, in which the growth, reproduction and survival of transgenic plants are characterized by relatively few parameters, shows how transgene-induced life-history changes affect persistence times. This suggests a challenging and attractive possibility of designing crop varieties that would have low persistence as feral colonizers. Such design of crop ideotypes is not new (Donald 1968). Finally, landscape models, including percolation and metapopulation models provide a framework together with stochastic life-history models not only to assess risk but ultimately to minimize risk by optimizing the spatial deployment of novel crop plants in the landscape.

Acknowledgements

The authors gratefully acknowledge the support of the Biotechnology and Biological Research Council for funding this work.

References

Advanta Seeds UK, 2000. *Minutes of evidence for the eighth report (Genetically modified organisms and seed segregation) from the House of Commons Agricultural Select Committee in the 1999-2000 session, 3 August 2000.* Her Majesty's Stationery Office, London. [http://www.parliament.the-stationery-office.co.uk/pa/cm199900/cmselect/cmagric/812/0071801.htm]

Aylor, D.E., 1986. A framework for examining inter-regional aerial transport of fungal spores. *Agricultural and Forest Meteorology,* 38 (4), 263-288.

Bullock, J.M., 1999. Using population matrix models to target GMO risk assessment. *Aspects of Applied Biology,* 53, 205-212.

Caswell, H., 2001. *Matrix population models: construction, analysis and interpretation.* 2nd edn. Sinauer Associates, Sunderland.

Claessen, D., Gilligan, C.A., Lutman, P., et al., in press. Which traits promote persistence of feral GM crops in a stochastic environment? *Oikosi.*

Colbach, N., Clermont-Dauphin, C. and Meynard, J.M., 2001a. GENESYS: a model of the influence of cropping system on gene escape from herbicide tolerant rapeseed crops to rape volunteers. I. Temporal evolution of a population of rapeseed volunteers in a field. *Agriculture Ecosystems and Environment,* 83 (3), 235-253.

Colbach, N., Clermont-Dauphin, C. and Meynard, J.M., 2001b. GENESYS: a model of the influence of cropping system on gene escape from herbicide tolerant rapeseed crops to rape volunteers. II. Genetic exchanges among volunteer and cropped populations in a small region. *Agriculture Ecosystems and Environment,* 83 (3), 255-270.

Crawley, M.J. and Brown, S.L., 1995. Seed limitation and the dynamics of feral oilseed rape on the M25 motorway. *Proceedings of the Royal Society of London. Series B. Biological Sciences,* 259, 49-54.

Crawley, M.J., Brown, S.L., Hails, R.S., et al., 2001. Transgenic crops in natural habitats. *Nature,* 409 (6821), 682-683.

Crawley, M.J., Hails, R.S., Rees, M., et al., 1993. Ecology of transgenic oilseed rape in natural habitats. *Nature,* 363 (6430), 620-623.

Csanady, G.T., 1973. *Turbulent diffusion in the environment.* Reidel, Dordrecht.

Donald, C.M., 1968. The breeding of crop ideotypes. *Euphytica,* 17, 385-403.

Evans, L.T., 1998. *Feeding the ten billion: plants and population growth.* Cambridge University Press, Cambridge.

Fahrig, L., 1992. Relative importance of spatial and temporal scales in a patchy environment. *Theoretical Population Biology,* 41 (3), 300-314.

Ferrandino, F.J., 1993. Dispersive epidemic waves. I. Focus expansion within a linear planting. *Phytopathology,* 83 (8), 795-802.

Firbank, L.G., Dewar, A.M., Hill, M.O., et al., 1999. Farm-scale evaluation of GM crops explained. *Nature,* 399 (6738), 727-728.

Firbank, L.G., Heard, M.S., Woiwod, I.P., et al., 2003. An introduction to the Farm-Scale Evaluations of genetically modified herbicide-tolerant crops. *Journal of Applied Ecology,* 40 (1), 2-16.

Gaston, K.J., Blackburn, T.M., Greenwood, J.J.D., et al., 2000. Abundance-occupancy relationships. *Journal of Applied Ecology,* 37 (suppl. 1), 39-59.

Gilligan, C.A., 2002. An epidemiological framework for disease management. *Advances in Botanical Research,* 38, 1-64.

Gilligan, C.A., 2003. Economics of transgenic crops and pest resistance: an epidemiological perspective. *In:* Laxminarayan, R. ed. *Battling resistance to antibiotics and pesticides: an economic approach.* Resources for the Future, Washington, 238-259.

Grassberger, P., 1983. Asymmetric directed percolation on the square lattice. *Journal of Physics A. Mathematical and General,* 16 (3), 591-598.

Gyllenberg, M., Hanski, I. and Hastings, A., 1997. Structured metapopulation models. *In:* Hanski, I. and Gilpin, M.E. eds. *Metapopulation biology: ecology, genetics and evolution.* Academic Press, San Diego, 93-122.

Hails, R.S., Rees, M., Kohn, D.D., et al., 1997. Burial and seed survival in *Brassica napus* subsp *oleifera* and *Sinapis arvensis* including a comparison of transgenic and non- transgenic lines of the crop. *Proceedings of the Royal Society of London. Series B. Biological Sciences,* 264 (1378), 1-7.

Hauser, T.P., Shaw, R.G. and Ostergard, H., 1998. Fitness of F-1 hybrids between weedy *Brassica rapa* and oilseed rape (*B. napus*). *Heredity,* 81 (4), 429-435.

Keymer, J.E., Marquet, P.A., Velasco-Hernandez, J.X., et al., 2000. Extinction thresholds and metapopulation persistence in dynamic landscapes. *American Naturalist,* 156 (5), 478-494.

Kot, M., Lewis, M.A. and Van den Driessche, P., 1996. Dispersal data and the spread of invading organisms. *Ecology,* 777 (7), 2027-2042.

Levins, R., 1969. Some demographic and genetic consequences of environmental heterogeneity for biological control. *Bulletin of the Entomological Society of America,* 15, 237-240.

Linder, C.R. and Schmitt, J., 1995. Potential persistence of escaped transgenes: performance of transgenic, oil-modified *Brassica* seeds and seedlings. *Ecological Applications,* 5 (4), 1056-1068.

Masood, E., 1998. Monsanto set to back down over 'terminator' gene? *Nature,* 396 (6711), 503.

Meacher, M., 2001. *Letter to Chair of Agricultural and Environment Biotechnology Commission.* May 21, 2001.

Messeguer, J., 2003. Gene flow assessment in transgenic plants. *Plant Cell Tissue and Organ Culture,* 73 (3), 201-212.

Metz, J.A.J., Nisbet, R.M. and Geritz, S.A.H., 1992. How should we define fitness for general ecological scenarios? *Trends in Ecology and Evolution,* 7 (6), 198-202.

Morel, B., Farrow, S., Wu, F., et al., 2003. Pesticide resistance, the precautionary principle and the regulation of Bt corn: real and rational option approaches to decision making. *In:* Laxminarayan, R. ed. *Battling resistance to antibiotics and pesticides: an economic approach.* Resources for the Future, Washington, 184-213.

Nieuwstadt, F.T.M. and Van Dop, H., 1982. *Atmospheric turbulence and air pollution modelling: a course held in The Hague, 21 - 25 September, 1981.* Reidel, Dordrecht.

Park, A.W., Gubbins, S. and Gilligan, C.A., 2001. Invasion and persistence of plant parasites in a spatially structured host population. *Oikos,* 94 (1), 162-174.

Park, A.W., Gubbins, S. and Gilligan, C.A., 2002. Extinction times for closed epidemics: the effects of host spatial structure. *Ecology Letters,* 5 (6), 747-755.

Parker, I.M. and Kareiva, P., 1996. Assessing the risks of invasion for genetically engineered plants: acceptable evidence and reasonable doubt. *Biological Conservation,* 78 (1/2), 193-203.

Perry, J.N., 2002. Sensitive dependencies and separation distances for genetically modified herbicide-tolerant crops. *Proceedings of the Royal Society of London. Series B. Biological Sciences,* 269 (1496), 1173-1176.

Pessel, F.D., Lecomte, J., Emeriau, V., et al., 2001. Persistence of oilseed rape (*Brassica napus* L.) outside of cultivated fields. *Theoretical and Applied Genetics,* 102 (6/7), 841-846.

Quist, D. and Chapela, I.H., 2001. Transgenic DNA introgressed into traditional maize landraces in Oaxaca, Mexico. *Nature,* 414 (6863), 541-543.

Quist, D. and Chapela, I.H., 2002. Maize transgene results in Mexico are artefact: reply. *Nature,* 416 (6881), 602.

Rosegrant, M.W., Paisner, M.S., Meijer, S., et al., 2001. *Global food projections to 2020: emerging trends and alternative futures.* International Food Policy Research Institute IFPRI, Washington. [http://www.ifpri.org/pubs/books/globalfoodprojections2020.htm]

Scheffler, J.A., Parkinson, R. and Dale, P.J., 1995. Evaluating the effectiveness of isolation distances for field plots of oilseed rape (*Brassica napus*) using a herbicide-resistance transgene as a selectable marker. *Plant Breeding,* 114 (4), 317-321.

Scott, S.E. and Wilkinson, M.J., 1998. Transgenic risk is low. *Nature,* 393 (6683), 320.

Shaw, M.W., 1995. Simulation of population expansion and spatial pattern when individual dispersal distributions do not decline exponentially with distance. *Proceedings of the Royal Society of London. Series B. Biological Sciences,* 259, 243-248.

Shigesada, N. and Kawasaki, K., 1997. *Biological invasions: theory and practice.* Oxford University Press, Oxford.

Squire, G.R., Brooks, D.R., Bohan, D.A., et al., 2003. On the rationale and interpretation of the Farm Scale Evaluations of genetically modified herbicide-tolerant crops. *Philosophical Transactions of the Royal Society of London. Series B. Biological Sciences,* 358 (1439), 1779-1799.

Stauffer, D. and Aharony, A., 1991. *Introduction to percolation theory.* Taylor and Francis, London.

Stewart, C.N., All, J.N., Raymer, P.L., et al., 1997. Increased fitness of transgenic insecticidal rapeseed under insect selection pressure. *Molecular Ecology,* 6 (8), 773-779.

Strogatz, S.H., 2001. Exploring complex networks. *Nature,* 410 (6825), 268-276.

Swinton, J., Harwood, J., Grenfell, B.T., et al., 1998. Persistence thresholds for phocine distemper virus infection in harbour seal *Phoca vitulina* metapopulations. *Journal of Animal Ecology,* 67 (1), 54-68.

Wesseler, J., 2003. Resistance economics of transgenic crops under uncertainty. *In:* Laxminarayan, R. ed. *Battling resistance to antibiotics and pesticides: an economic approach.* Resources for the Future, Washington, 214-237.

Wilkinson, M.J., Davenport, I.J., Charters, Y.M., et al., 2000. A direct regional scale estimate of transgene movement from genetically modified oilseed rape to its wild progenitors. *Molecular Ecology,* 9 (7), 983-991.

Wilkinson, M.J., Elliott, L.J., Allainguillaume, J., et al., 2003. Hybridization between *Brassica napus* and *B. rapa* on a national scale in the United Kingdom. *Science,* 302 (5644), 457-459.

Wolfenbarger, L.L. and Phifer, P.R., 2000. Biotechnology and ecology: the ecological risks and benefits of genetically engineered plants. *Science,* 290 (5499), 2088-2093.

10b

Comment on Gilligan, Claessen and Van den Bosch: Spatial and temporal dynamics of gene movements arising from deployment of transgenic crops

Ekko van Ierland[#]

Introduction

The paper by Gilligan, Claessen and Van den Bosch offers a very interesting and transparent analysis of the modelling of gene movements based on stochastic life-history models of plants, on metapopulation models (for assessing whether a plot will be occupied by a species or not) and on landscape models that may show, for instance, the impact of distances between plots on the dispersion of transgenic crops. The models presented in the paper consider many interactions and complexities like the stochastic behaviour of weather or the disturbance of a natural environment, which is relevant for the germination of seeds. It is interesting to see how concepts of stochastic metapopulation models, after appropriate modification, can be applied to the new research questions on the potential spatial distribution of transgenic crops.

Issues of concern

a. The paper discusses the application of metapopulation models in the analysis of the spatial distribution of transgenic crops. This raises the question whether sufficient information is available to identify the relevant parameters for the metapopulation models. It is well known that the extinction rates or the recolonization rates for various plots are difficult to estimate for traditional species. It will be even more complicated to assess these parameter values for transgenic crops, particularly if little experience is available on how these species will compete with other species under potentially very different circumstances. It would be worthwhile to analyse these complications in more detail, because they will be extremely relevant for the *ex ante* assessment of the spatial distribution of transgenic crops.

b. The paper raises the question of how in the process of modification the characteristics of the plants can be chosen in such a way that the risk of undesirable spatial distribution of genetically modified crops can be reduced or minimized. This raises the question what lessons can be learned from the analysis for the desirable characteristics of transgenic plants, for instance with regard to germination or cross-pollination.

c. The paper analyses the distances that should be considered in order to separate GMO crops sufficiently from non-GMO crops. Although the analysis seems

[#] Environmental Economics and Natural Resources Group, Wageningen University, P.O. Box 8130, 6700 EW Wageningen, The Netherlands. E-mail: ekko.vanierland@wur.nl

J. H. H. Wesseler (ed.), Environmental Costs and Benefits of Transgenic Crops, 163–164.

scientifically correct, the question may be raised whether in practice farmers will always comply with the regulations, and whether undesired dispersion of the GMO crops may occur, despite the precautionary measures. If only one or a few farmers – for whatever reason – plant GMO crops in the neighbourhood of traditional crops the dispersion of the GMO crop may already occur and the traditional crop will be affected with GMOs.

d. Finally, I would like to indicate that the paper focuses on the *ex ante* analysis of the risk of dispersion of transgenic crops under rather 'normal' circumstances. For a proper risk assessment it will be the combination of some very unlikely events that may have tremendous negative impacts on the natural environment as a result of the introduction of transgenic crops. This very small probability of a highly undesirable or catastrophic event is one of the most complicated factors to assess in making decisions on the introduction of GMOs. Despite the fact that the paper contributes a lot to resolving some of the urgent questions in this domain, this fundamental issue seems – at least to me – to remain unsolved.

11a

Minimum distance requirements and liability: implications for co-existence

Claudio Soregaroli[#] *and Justus Wesseler*[##]

Abstract

The co-existence of conventional and transgenic products in the food chain introduces new elements in the evaluation of the profitability of transgenic crops and, consequently, on the farmer's adoption decision. In particular, one emerging problem farmers are facing in Europe is related to the legal liability of transgenic-crop cultivation. In Europe, a mixture of *ex-ante* regulations and *ex-post* liability rules governing transgenic crops emerges.

One of the predominant *ex-ante* regulations discussed at the EU level is a minimum-distance requirement to neighbouring fields in order to avoid cross-pollination. The *ex-post* liability rules differ. They depend on the legal frameworks of individual members of the EU. The current interpretation of, for example, Italian and German law does not exclude *ex-post* liability for farmers planting transgenic crops in the case of cross-pollination.

In this paper, we analyse the value of planting transgenic crops when farmers face *ex-ante* regulatory and *ex-post* liability costs under irreversibility and uncertainty. The regulatory instrument analysed is the minimum distance to neighbouring fields. First results indicate that under irreversibility and uncertainty the value of cultivating transgenic crops presents a trade-off between *ex-ante* regulatory and *ex-post* liability costs with respect to farm size. From this, it is not possible to conclude *a priori* the net effect on the size of the adopting farms, if, *ceteris paribus*, a minimum distance regulation is adopted within the EU and farmers can be held liable *ex-post*.

Keywords: co-existence; *ex-ante* regulation; *ex-post* liability; irreversibility; uncertainty

Introduction

The cultivation of biotech crops is continuously expanding worldwide. According to the International Service for the Acquisition of Agri-biotech Applications (James 2002), 58.7 million hectares were planted with genetically modified (GM) organisms in 2002, an increase of 12% over the previous year. This involved nearly 6 million farmers. The United States (66.4%), Argentina (23.0%), Canada (6.0%) and China (2.1%) have the largest world share of transgenic crops. However, new countries are

---

[#] Istituto di Economia Agro-alimentare, Università Cattolica del Sacro Cuore, Italy
[##] Environmental Economics and Natural Resources Group, Wageningen University, The Netherlands. E-mail: justus.wesseler@wur.nl

J. H. H. Wesseler (ed.), Environmental Costs and Benefits of Transgenic Crops, 165–181.

emerging: for example in 2002 India, Colombia and Honduras introduced biotech crops in their fields.

Considering the different commodities, soybean production has the largest world share with 36.5 million hectares in 2002. Cultivation of this crop has expanded so much in the last years that today transgenic soybean represents more than half of the total world production of these seeds. This crop is followed in importance by maize with 12.4 million hectares, cotton with 6.8 million hectares and canola with 3 million hectares.

Considering that in 1996 the total area of GM crops was less than 3 million hectares, the adoption has undoubtedly been rapid and massive. The economic reasons are mainly considered in farmers' expectations on the profitability of transgenic crops, in particular as regards yield and/or cost savings. However, as reviewed by Demont and Tollens (2001) studies often do not show a significant difference in profitability between conventional and GM crops when yields and costs are considered. One important factor in determining the choice of transgenic crops is the convenience given to the farmer (Marra 2001). These crops allow for a greater flexibility in growing practices, which reduces the time specificity of labour and capital. This can translate into increased labour productivity and impact on farm restructuring. Moreover, as the report of the Directorate-General for Agriculture of the European Union underlines, "…the effective profitability of a GM crop can only be properly assessed on the basis of several years of cultivation and commercialization. Several years have to be considered for two main reasons. First, many other factors have an impact on profitability. In particular, there are important yearly fluctuations in yields and prices. Second, effective profitability depends on developments on the supply and on the demand side" (European Commision 2000, chapter 3).

Initial development and adoption of transgenic crops were supply-driven. The target of biotech firms was the farmer, and the crops produced had agronomic traits that favoured farm practices and output volumes. However, upstream effects from the demand side started to be important as the consumers' awareness and knowledge of GM organisms increased. Consumer concern about the possible negative effects of biotech on health and the environment induced part of the population, even in countries where GM cultivation is largely allowed, to demand GM-free products (Winston 2002). As a consequence, a double channel may develop, one including GM products and the other GM-free. This implies the segregation of agricultural products along the vertical food chain and the eventual development of an Identity Preservation (IP) and a traceability system together with opportune labelling requirements (Gaisford et al. 2001). The co-existence of conventional and transgenic products in the food chain introduces new elements in the evaluation of the profitability of GM crops and, consequently, in the farmer's adoption decision. In particular, one emerging problem farmers are facing is related to the legal liability of GM cultivation. As will be discussed in the next section, farmers face the risk of litigation with neighbouring farmers, biotech companies and public institutions. This introduces a new element in the farm adoption process. The value of the new technology will not only depend on the incremental changes in revenues and variable production costs but also on costs from *ex-ante* regulations and *ex-post* liability.

However, while it is possible to enumerate the factors influencing the farmer's actions when considering the adoption of biotech crops, the direction and interrelation of their effects is not clearly defined. For example, what is the influence of an *ex-ante* regulation imposing standards for GM cultivation? It probably increases field costs while reducing legal liability.

Given that, at least at the EU level, there is much debate on the type and level of regulation that should be adopted in order to govern co-existence, the purpose of the study is to highlight the implication on farm's GM adoption of an *ex-ante* regulation setting standards for GM farm cultivation practices in combination with *ex-post* liability rules. Awareness of these effects can help to evaluate the consequences of policy actions and, eventually, to identify the impact of those policy instruments on adoption of the new technology.

The paper is structured as follows. First, legal issues related to co-existence are discussed, followed by a model that values transgenic crops including *ex-ante* regulations and *ex-post* liability rules at farm level. Third, the effects of changes in regulations and liability rules on adoption are discussed before we conclude.

Legal issues of co-existence

There are different sources of litigation that can hold a GM farmer liable for his cultivation practices. Actions are likely to be taken by non-GM farmers who consider their business damaged when GM contamination occurs. This is summarized in the following paragraphs:

Product depreciation

A farmer growing GM crops can be sued by neighbouring non-GM farmers who find their crops contaminated by GM material. The mixing of GM and non-GM material can result from cross-pollination or volunteers (self-sown plants) carried by different agents (wind, animals…) from the GM field to the neighbouring soils (Kershen 2002; Schmidt 2002). Contamination causes harm to non-GM farmers since they risk not being able to sell their products as GM-free, with negative consequences for the value of the product[1]. This is especially true in the case of organic production, where the utilization of GM products is excluded. Financial losses to organic farmers can also be higher if the adventitious GM contamination implies the loss of their organic status: in this case the access to important markets could be precluded for several years.

Litigation is not limited to neighbouring farmers but can also occur between landlords and tenants. A landlord can claim that the loss of organic status due to GM cultivation has a negative impact on the land value and that the tenant did not comply with the rules of 'good husbandry' included in many tenancy agreements (Network of Concerned Farmers NCF 2003). As the Network of Concerned Farmers (NCF) underlines, depreciation of land can also cause concerns to banks, when secured loans are linked to the land value. Hence, if GM cultivation is proved to have effects on land values, landlords and banks could play a role in engaging legal disputes with the responsible GM farmer.

Legal actions by biotech companies

Companies producing GM seed invest a great deal of effort in protecting their property rights on the use of transgenic crops. Especially for those crops, such as oil-seed rape, where it is possible for the farmer to use seeds kept from the harvest of the previous year, contracts between farmers and agro-biotech companies explicitly state that the farmers cannot use as seed their own harvested GM crops (and, of course, they cannot give or sell seeds to other farmers). Moreover, companies reserve the right to control and take samples from harvested crops of the farms in the following years.

The behaviour of agro-biotech companies has consequences both on GM and non-GM farms as recent court cases demonstrate. Well-known is the case of a Canadian farmer (Schmeiser) sued by Monsanto and held liable by the Federal Court of Canada (Monsanto Canada Inc. vs. Schmeiser 2001). The farmer was found guilty of knowingly growing canola containing a gene patented by Monsanto (gene Roundup®-tolerant). One of the motivations of the judge that held the farmer liable was that "*...a farmer whose field contains seed or plants originating from seed spilled into them, or blown as seed, in swaths from a neighbour's land or even growing from germination by pollen carried into his field from elsewhere by insects, birds, or by the wind, may own the seed or plants on his land even if he did not set about to plant them. He does not, however, own the right to the use of the patented gene, or of the seed or plant containing the patented gene or cell*" (Monsanto Canada Inc. vs. Schmeiser 2001, p. 92). The farmer was considered guilty because he used seeds knowing they included the Monsanto gene: as the judge writes "*his infringement arises not simply from occasional or limited contamination of his Roundup-susceptible canola by plants that are Roundup-resistant. He planted his crop for 1998 with seed that he knew or ought to have known was Roundup-tolerant*" (Monsanto Canada Inc. vs. Schmeiser 2001, p. 125).

The above case highlights the importance of the effects of contamination. Legal actions by biotech companies put pressure on farmers to identify GM contamination and to take action to eliminate such plants from the field. However, this influences the farming practices and there is no unanimous opinion on the effect on costs and the effectiveness of the control[2].

Damages to the downstream vertical chain

Labelling of food products is an important issue related to biotechnology. Bodies such as the European Union (EU) require labelling of food products as GM if they contain more than 1% of transgenic material. Also, firms can voluntarily certify their products as GM-free or organic with a guarantee of a certifying institution. One key element in this system is the maintenance of a separate channel of GM-free products and the possibility to trace back the food components up to the producing farmer. This implies that, if GM contamination is found in the food product, liability can be transferred to the responsible non-GM farmer (if any). Hence, the cost of contamination would not only be the depreciation of the product, but also the payment for damages caused to the downstream food chain. According to NCF (Network of Concerned Farmers NCF 2003), this would expose farmers to high liability levels with difficulties also in obtaining an insurance coverage. These facts would be a further incentive for the non-GM farmer to sue neighbouring GM producers, with an overall increase of legal disputes.

In conclusion, farmers adopting transgenic crops face the risk of being held liable if they plant transgenic crops. Also, the introduction of transgenic crops often includes regulations, such as the refuge areas for *Bt*-corn in the United States. In the EU, minimum distance requirements to avoid cross-pollination are discussed. The distance requirements discussed range from a few meters up to several thousand meters (Agnet 2002; Bock et al. 2002). However, the risk of being held liable depends on the specific liability systems of the different countries. This will be discussed in the next section.

Legislation and specificities of different countries

GM farmers' liability is likely to be included in the category of 'damage to property' (so-called traditional damage). More specifically, this can be distinguished in two main categories: negligence and strict liability. Negligence, in the biotech case, occurs when the farmer fails to take adequate action in order to avoid GM contamination. If it can be proved that the GM farmer did not provide sufficient care (did not meet the standards established by law) to avoid contamination and that this caused prejudice to the non-GM farmer, the former could be held liable by the court. On the other hand, strict liability does not require fault or negligence by the person who caused harm. Hence, a farmer can be held liable simply because his activity is causing damages.

In order to evaluate the relationship between *ex-ante* regulation and *ex-post* liability on the farmer's GM adoption decision, the relevant question is whether strict liability is indeed applicable to the GM farmer[3]. The answer is different depending on the countries considered: as will be discussed below, in the US strict liability is not likely to occur, whereas there are more possibilities in the European Union.

US legislation

In the US regime of legal liability, biotechnology is regulated by laws that are generally applicable to agricultural products. Biotechnology in agriculture is considered by US legislation to be equivalent to other agricultural breeding practices (Office of Science and Technology Policy OSTP 1986, General Recommendation 2).

In order to claim, strict liability damages have to be demonstrated. For example, in the case of organic production the organic standards are set by the United States Department of Agriculture (USDA) under the federal law given by the National Organic Program (NOP). This programme explicitly states that the use of GM organisms is excluded for organic production. However, it is a process-based standard. USDA does not set specific tolerance levels for the presence of GM material. It is sufficient to respect the production standards to obtain the certification. Therefore, "*...organic producers may face significant difficulties in proving that the farmer growing transgenic crops caused damage*" (Kershen 2002, p.7). Even if the organic farmer complies with stricter private standards (from a non-governmental institution) the court under the US legislation could consider it an "*abnormally sensitive character*" of the plaintiff's activity (Kershen 2002).

Moreover, as underlined by (Kershen 2002, p.12-13), in the US "*...courts are unlikely to endorse* [claims] *that insist on zero tolerance of pollen flow or volunteer plants. Courts expect neighbors to have reasonable tolerances toward one another as the court engages in balancing of gravity of the harm against the social utility of each neighbor's use and enjoyment of their own land*". The basis of the court's view is the substantial equivalence, in the US case, of the biotech cultivation to the traditional farming practices.

From the above discussion it appears that the possibility of legal actions against a GM farmer is limited according to the US legislation over biotech practices. Excluding negligence, the non-GM farmer will have a hard task demonstrating that transgenic cultivation caused significant damage.

EU legislation

In the EU, damage to property is not covered by EU legislation (or proposals) and is left to the civil liability systems of the Member States (European Commision 2002). Moreover, from the policy debate on co-existence, it seems that much will be left to the specific legislations of Member States. In this regard, the following paragraphs will discuss the examples of Italian and German legislation.

Italian and German legislation

In this context, the starting point of the Italian legislation is article 844 of the Civil Code stating: "a farmer cannot impede the emission of smoke, heat, odours, noise, vibrations and similar propagations originating from a neighbouring field, if they do not exceed the normal tolerability while taking into account the conditions of the area". The emission can be forbidden if it is 'intolerable', that is it has to be over any reasonable tolerance of its external effects (Germanò 2002). A similar position can be found in Germany. §903 of the German Civil Code is similar to the article 844 of the Italian Civil Code. In combination with §906 of the German Civil Code organic farmers have to tolerate cross-pollination as long as this does not impose important constraints on their freedom to farm and if cross-pollination cannot be avoided by methods that are tolerable from an economic point of view. However, there is no precise definition of reasonable cost and considerable economic losses. Hence, the point is to define the nature and the level at which an emission can be considered 'intolerable'.

Different EU legislations pose constraints to non-GM farmers. For example, under regulation 2092/91 as amended by regulation 1804/99, organic farmers can receive certification for their products only if they avoid GM products in their farming practices. In Italy, farmers receiving subsidies within the framework of the regional Rural Development Plan (RDP) are often required to produce GM-free products. Also, the Protected Designation of Origin (PDO) certification requires practices that do not allow the use of GM products. However, although PDO and organic certification are process-based, the eventual contamination of GM material would preclude their labelling status if the GM content were over a threshold level. This has been defined at 1% for food (EC 49/2000) and 0.1% for organic products[4]. In the first case food would lose its organic status, while in the second case the product has to be labelled as GM. Hence, the key difference with respect to the US legislation is the mandatory labelling system: the definition of intolerable emission is likely to depend on its size and could be determined by the court on a case-by-case basis.

The European case illustrates that farmers planting transgenic crops risk *ex-post* liability costs, even if *ex-ante* regulations are implemented. The legal framework in the United States reduces the risk of *ex-post* liability costs. While *ex-post* liability will be less relevant for adoption of transgenic crops in the US, it may pose important additional adoption costs to farmers in the EU.

In the next step we model the value of transgenic crops at farm level including *ex-ante* regulatory and *ex-post* liability costs. The regulation we use is the minimum distance a farmer planting transgenic crops has to keep between his field and neighbouring fields to reduce cross-pollination. We show that indeed, *ex-post* liability adds additional costs that reduce the likelihood of adoption and, what is also important, that they will not be scale-neutral.

Theoretical framework for the GM farmer

As underlined in the previous section, the farmer's decision on whether to adopt a GM crop is not a simple one. The value of adopting a transgenic crop depends not only on the incremental profit from growing the transgenic crops, but also on *ex-ante* regulatory and *ex-post* liability costs.

Ex-ante regulation and *ex-post* liability

The starting point of the conceptual framework is the definition of the GM farmer's value function. The value of the option to adopt the GM crop can be defined as the expected value of the difference between the extra profit obtainable from the GM cultivation (as compared to the conventional one) considering the sole cultivation practices (Π) and the costs related to liability and its control (L):

$$V = E(\Pi - L)\,. \tag{1}$$

If, for the moment, we assume that the farmer does not face any reduction in costs related to the adoption of a transgenic crop, he/she is assumed to adopt the transgenic crop when V is equal to or grater than zero. The expected costs related to liability are the sum of the costs of respecting *ex-ante* regulations (C) and the value of tort liability (TL):

$$L = E(C + TL)\,. \tag{2}$$

Following Kolstad, Ulen and Johnson (1990) the above relation can be reformulated as

$$L = C + \mu * D * R \tag{3}$$

where μ is the probability of causing an accident (for example, contamination of the neighbouring non-GM fields), D is the monetary value of the accident, and R is the probability that the injurer will pay the damages. In our case, R can be interpreted as a function of the court view and the probability of being sued by the neighbour who has suffered damage.

From the previous equations the value function for the GM farmer can be formulated as follows:

$$V = E[\Pi(p,y,c,s) - C(s,reg) - \mu(s,reg)D(s,reg)R(law)] \tag{4}$$

where p is a vector of output prices, y is the vector of the per-hectare yields, c is the vector of the cost of inputs, s is the size of the field, *reg* is the enforced GM legal standard for the country, *law* is the tort-liability system of the country and E the expectation operator.

The above framework can be used to assess the impact of regulation standards on a farm's adoption of GM crops. One possibility is to evaluate the effect of the variable *reg* on the 'relevant' farm size. The relevant farm size for the given problem is the dimension at which the cultivation of the GM crop starts to be convenient, that is the value function is greater than or equal to zero.

Assuming the farm is a single field and interpreting the variable *reg* as the minimum distance (*d*) between the GM crop and the farm's external limits, it is possible to evaluate the relationship between the minimum adoption size ($\underline{s}$) and the severity of regulation. Assuming all of the other variables are constant, $\underline{s}$ can be solved from the following equation:

$$V = \Pi(s) - C(s,d) - \mu(s,d)D(s,d)R = 0 \tag{5}$$

where

$\partial\Pi(s)/\partial s > 0$ and $\partial^2\Pi(s)/\partial s^2 \leq or \geq 0$;
$\partial C(s,d)/\partial s > 0$ and $\partial^2 C(s,d)/\partial s^2 \leq 0$;
$\partial C(s,d)/\partial d > 0$ and $\partial^2 C(s,d)/\partial d^2 \leq 0$;
$\partial\mu(s,d)/\partial s > 0$ and $\partial^2\mu(s,d)/\partial s^2 \leq 0$;
$\partial\mu(s,d)/\partial d < 0$ and $\partial^2\mu(s,d)/\partial d^2 \geq 0$;
$\partial D(s,d)/\partial s > 0$ and $\partial^2 D(s,d)/\partial s^2 \leq 0$;
$\partial D(s,d)/\partial d < 0$ and $\partial^2 D(s,d)/\partial d^2 \geq 0$.

From the implicit function theorem it is possible to write

$$\partial\underline{s}/\partial d = -\frac{\partial V/\partial d}{\partial V/\partial s},$$

hence applying the above relation to equation (5) the resulting expression is

$$\partial\underline{s}/\partial d = -\frac{-\partial C/\partial d - R(\mu(s,d)\partial D(s,d)/\partial d + D(s,d)\,\partial\mu(s,d)/\partial d)}{\partial\Pi(s)/\partial s - \partial C/\partial s - R(\mu(s,d)\partial D(s,d)/\partial s + D(s,d)\,\partial\mu(s,d)/\partial s)}. \tag{6}$$

Given that at the break-even point $\underline{s}$, an increase in size determines a higher increase in the extra profit than in the implementation and liability costs, the denominator of the above equation can be assumed to be positive around $\underline{s}$. Hence, the discussion can be focused on the numerator of the equation. For simplicity, ignore the denominator and rewrite equation (6) as

$$\text{sign}\left[\partial\underline{s}/\partial d\right] = \text{sign}\left[\partial C/\partial d + R\{\mu(s,d)\partial D(s,d)/\partial d + D(s,d)\,\partial\mu(s,d)/\partial d\}\right]. \tag{7}$$

The first term on the right-hand side of the above equation is positive, while the second term is negative. With an increase in distance there is a trade-off between an increase in implementation costs and a decrease in the expected value of liability. This implies that the effect on the minimum adoption size of a policy that poses higher standards on distances between GM and non-GM fields is uncertain. Hence, it is not possible to conclude that an increase in distance will exclude smaller farms. Indeed, if the decrease in expected liability is superior to the increase in implementation costs it is possible to have smaller farms adopting the technology. Or rather, the more severe legislation would not have any impact, given that those smaller farms were already adopting the technology imposing higher crop distances on themselves.

The case of irreversibility and uncertainty

In the previous discussion it was assumed that incremental profits Π are certain and the farmer did not face reduced costs while deciding to adopt the GM technology. However, some of the costs could be irreversible: for example, the transgenic crop may require specific machinery, or as discussed in the introduction, the GM cultivation could make it difficult for the farmer to switch back to the non-GM status. These difficulties could include additional practices for the control of volunteers or a required minimum number of years of non-GM cultivation for a field to be considered for producing non-GM products. The multi-period time frame also adds uncertainty to the farmer's adoption decision as future yields, prices and costs are not known with certainty.

In the presence of net-irreversible costs, uncertainty and flexibility, the value of a GM crop is not simply the difference between the present value of future benefits and costs, as from equation (1), but the sum of this difference plus the value of the option to plant transgenic crops (Wesseler 2003). More formally, when some costs are irreversible, costs and benefits are uncertain and the decision to invest can be postponed, the farmer maximizes the option value of the investment. Hence, equation (1) can be reformulated as follows

$$F(V) = max\, E\left[\left(V(\Pi, C, TL) - I\right)e^{-\rho T}\right] \tag{8}$$

where $F(V)$ is the value of the investment opportunity, $V(\Pi, C, TL)$ is the value of the reversible net-benefits, and I are the net-irreversible costs of the investment.

As the time frame gets longer than a sowing season, the benefit of using a GM crop becomes uncertain. Profit from farm practices can change over time and there is always the risk of liability. It is possible to represent this uncertainty by the following stochastic process

$$d(\Pi - C) = \alpha(\Pi - C)dt + \sigma(\Pi - C)dz + (\Pi - C)dq \tag{9}$$

where (Π-C) evolves under a combined geometric Brownian motion and Poisson process. The first two terms are common for modelling incremental benefits of transgenic crops (e.g. Demont, Wesseler and Tollens 2002; Morel et al. 2003; Wesseler 2003). α is the drift of the Brownian motion, dz is the increment of a Wiener process, dt is the marginal increment in time and dq is the increment of a Poisson process. The third term represents tort liability modelled as the risk of a jump in the profit when the farmer is held liable. More precisely,

$$dz = \varepsilon_t \sqrt{dt}\text{, and}$$

$$dq = \begin{cases} 0 & \text{with probability } 1 - \lambda dt \\ -\phi & \text{with probability } \lambda dt \end{cases}$$

where ε_t is normally distributed with zero mean and unit standard deviation, λ is the mean arrival rate of a Poisson process, and ϕ the percentage of the *ex-post* liability costs of (Π-C).

From the above equation and the opportune boundary conditions, as shown in Appendix 1, it is possible to obtain the following relation defining the rule for the investment decision, assuming ϕ = 1:

$$(\Pi - C)^{*} = \left(\frac{\beta_1}{\beta_1 - 1}\right)(\rho - \alpha + \lambda)I \tag{10}$$

where

$$\beta_1 = \frac{1}{2} - \frac{\alpha}{\sigma^2} + \sqrt{\left(\frac{\alpha}{\sigma^2} - \frac{1}{2}\right)^2 + \frac{2(\rho + \lambda)}{\sigma^2}} > 1. \tag{11}$$

From the last two equations it is possible to evaluate the effect of a change in the regulation regarding co-existence. Taking again as an example the case of the distance between GM and non-GM fields it is possible to see the effect of an increase in distance on the hurdle rate, assuming $\partial I/\partial d = 0$. The same approach used in the case without irreversibility can be used to compare the effects of a change in the regulation on the minimum adoption size of the farm (s). This can be solved rearranging equation (10) and applying the implicit function theorem leading to the following derivative:

$$\partial \underline{s}/\partial d = -\frac{\dfrac{\partial(\Pi - C)}{\partial d} - \left[\dfrac{\partial\left(\dfrac{\beta_1}{\beta_1 - 1}\right)}{\partial d}(\rho - \alpha + \lambda) + \left(\dfrac{\beta_1}{\beta_1 - 1}\right)\dfrac{\partial(\rho - \alpha + \lambda)}{\partial d}\right]I}{\dfrac{\partial(\Pi - C)}{\partial s} - \left(\dfrac{\beta_1}{\beta_1 - 1}\right)(\rho - \alpha + \lambda)\dfrac{\partial I}{\partial s}} \tag{12}$$

Given that we are observing a break-even point, as in the case under certainty it is possible to assume that the denominator or the above equation is positive. Hence, equation (12) can be rewritten as

$$\text{sign}\,\partial \underline{s}/\partial d = \text{sign} - \underbrace{\frac{\partial(\Pi - C)}{\partial d}}_{(+)} + \left[\underbrace{\frac{\partial\left(\frac{\beta_1}{\beta_1 - 1}\right)}{\partial d}(\rho - \alpha + \lambda)}_{(+)} + \underbrace{\left(\frac{\beta_1}{\beta_1 - 1}\right)\frac{\partial(\rho - \alpha + \lambda)}{\partial d}}_{(-)}\right]I \tag{13}$$

Results

The result in equation 13 is similar to the case without irreversibility and uncertainty. The first term in the square brackets indicates the effect on the hurdle rate, which is positive. This is the positive effect of an increase in the future value of transgenic crops due to an increase in distance requirements, which can be explained by the decrease in *ex-post* liability costs (the option to wait is worth more). The second term in the square brackets is the effect on the annualized hurdle rate. This captures the effect on the reversible value of the transgenic crop. This effect is negative as an increase in the distance reduces directly the probability of *ex-post* liability, which increases the actual value of adopting transgenic crops. The overall sign of the terms in the square brackets cannot be determined and will depend on the specific parameter values (see the solution in Appendix 2). However, numerical examples show a very robust negative sign of the square bracket. This means that with an increase in distance requirement the greater value of the investment opportunity is outweighed by the increase in the actual value of the project. This is shown in Figure 1.

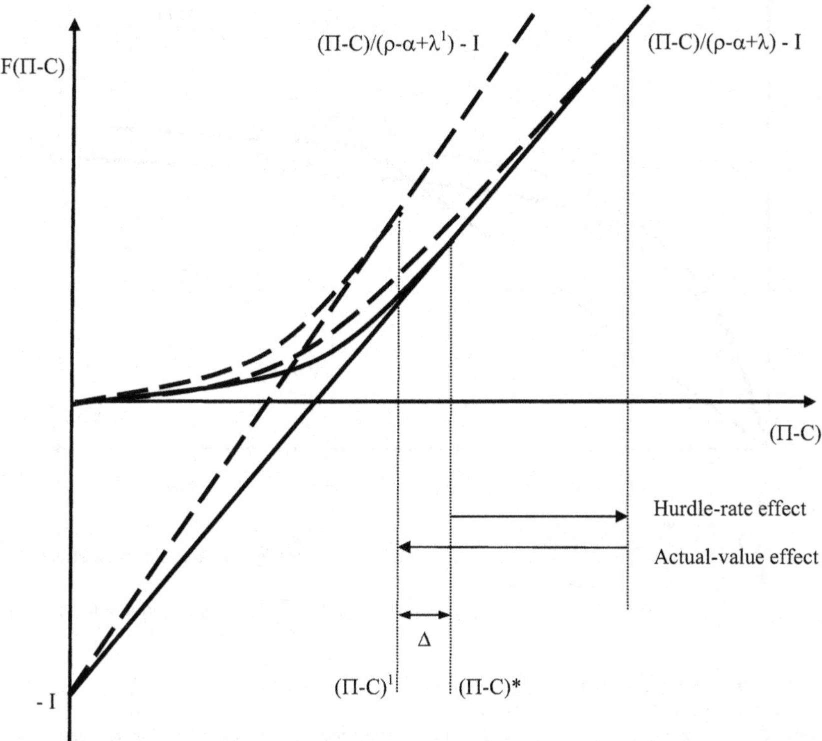

Legend:
Π = per hectare extra profit of the transgenic crop
C = per hectare costs of minimum distance
F (Π-C) = value of the investment opportunity
I = irreversible costs when adopting the technology
ρ = discount rate
α = drift of the Brownian motion
λ = mean arrival rate of the Poisson process
λ^1 = mean arrival rate after the increase in distance requirements
(Π-C)* = minimum net extra profit in order to adopt the technology at the initial state
$(\Pi\text{-}C)^1$ = minimum net extra profit in order to adopt the technology after the increase in distance
Δ = net effect on the minimum net extra profit given by the increase in distance

Figure 1. Effect of an increase in distance requirements on the threshold level of (Π-C) for the adopting farm

If what described above was the only force in action the effect of the policy would be a lower minimum adoption size. However, the increase in the minimum distance determines a raise in costs reducing the value of (Π-C). Hence, for the most common values of the parameters it is not possible, *a priori*, to conclude what is the effect on the minimum adoption size of the *ex-ante* regulation. This is shown in Figure 2.

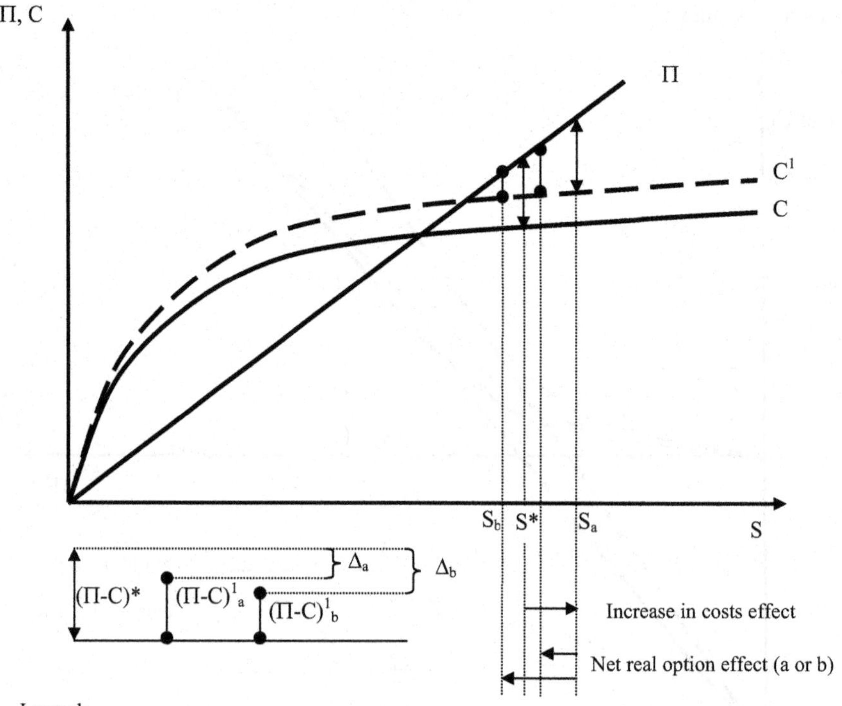

Legend:
Π = per hectare extra profit of the transgenic crop
C = per hectare costs of minimum distance
(Π-C)* = minimum net extra profit in order to adopt the technology at the initial state (see Figure 1)
$(\Pi\text{-}C)^{1}_{a,b}$ = minimum net extra profit in order to adopt the technology after the increase in distance
$\Delta_{a,b}$ = net effect on the minimum net extra profit given by the increase in distance (see figure 1)
S^{*} = minimum adoption size at the initial state
$S_{a,b}$ = minimum adoption size after the increase in distance

Figure 2. Effect of an increase in distance requirements on the minimum adoption size of the farm (S)

Only the case of very high drift rates α and low mean arrival rates λ, that is, low probability of *ex-post* liability, turns the sign of the square brackets positive. In this case the option value is prevailing and *ex-ante* regulations would be biased towards larger farms.

Conclusions

The release of transgenic crops in Europe will most likely be controlled by *ex-ante* regulations and *ex-post* liability rules. The regulations chosen and liability rules imposed, affect the expected benefits from adopting transgenic crops at farm level. In this paper we have shown that for the case of a minimum distance requirement this will not always be scale-neutral. However, for reasonable parameter ranges the net direction of the effect needs to be verified empirically. The presence of irreversible costs seems to play a role in determining *ex ante* the policy effect only in extreme cases. This observation holds for the case where *ex-post* liability costs, if a farmer is held liable, equal the *ex-ante* expected benefits. We expect the results to be similar, if the liability costs are linear in *ex-ante* expected benefits.

Appendix 1

From the definition of the value of the farmer's investment opportunity in GM crops

$$F(\Pi - C) = \max E\left[(V(\Pi - C) - I)e^{-\rho t}\right] \qquad \text{(A1)}$$

the investment problem can be solved by dynamic programming. The first step is to define the Bellman equation as

$$\rho F(\Pi - C)dt = E\left[dF(\Pi - C)\right]. \qquad \text{(A2)}$$

This equation equates the return over dt computed from a capital whose value is F using a discount rate ρ to the expected change in the value of the investment opportunity F. This means that the optimality condition for the farmer is when the value of the investment opportunity in GM crops changes over time in the same way as a normal capital investment.
Total liability is still considered the maximization problem and defined as a percentage ϕ of (Π-C) that follows a jump process with a mean arrival rate of λ. Hence, the following combined stochastic process for (Π-C) and TL is assumed:

$$d(\Pi - C) = \alpha(\Pi - C)dt + \sigma(\Pi - C)dz + (\Pi - C)dq \qquad \text{(A3)}$$

where dz is a Wiener process with the property $dz = \varepsilon_t\sqrt{dt}$, where ε_t has zero mean and unit standard deviation and the relative expected values are $E(dz)=0$ and $E(dz^2)=dt$, dq is the increment of the jump process and

$$dq = \begin{cases} 0 & \text{with probability } 1 - \lambda \text{dt} \\ -\phi & \text{with probability } \lambda \text{dt} \end{cases}.$$

Using Ito's lemma for the combined Brownian motion and Poisson process, the expected value of dF can be defined as

$$E[dF] = \frac{\partial F(\Pi - C)}{\partial(\Pi - C)}\alpha(\Pi - C)dt + \frac{1}{2}\frac{\partial^2 F(\Pi - C)}{\partial(\Pi - C)^2}\sigma^2(\Pi - C)^2 dt \\ + \lambda\{F[(\Pi - C) + (\Pi - C)(-\phi)] - F(\Pi - C)\}dt \tag{A4}$$

Substituting (A4) into (A2) and simplifying for dt gives the following second-order differential equation

$$\frac{1}{2}\left(\frac{\partial^2 F(\Pi - C)}{\partial(\Pi - C)^2}\sigma^2(\Pi - C)^2\right) + \frac{\partial F(\Pi - C)}{\partial(\Pi - C)}\alpha(\Pi - C) - \lambda\{F(\Pi - C) - F[(1-\phi)(\Pi - C)]\} = \rho F(\Pi - C)$$

which can be rearranged as

$$\frac{1}{2}\left(\frac{\partial^2 F(\Pi - C)}{\partial(\Pi - C)^2}\sigma^2(\Pi - C)^2\right) + \frac{\partial F(\Pi - C)}{\partial(\Pi - C)}\alpha(\Pi - C) \\ -(\rho + \lambda)F(\Pi - C) + \lambda F[(1-\phi)(\Pi - C)] = 0 \tag{A5}$$

Knowing that the value of the investment opportunity must also satisfy the following boundary conditions

$$F_{(\Pi - C)=0} = 0 \tag{A6.1}$$

$$F^*(\Pi - C) = \frac{\Pi - C}{(\rho - \alpha + \lambda)} - I \tag{A6.2}$$

$$F'^*(\Pi - C) = \frac{1}{(\rho - \alpha + \lambda)} \tag{A6.3}$$

a solution must take the form

$$F(\Pi - C) = A_1(\Pi - C)^{\beta_1} + A_2(\Pi - C)^{\beta_2} \text{ with } \beta_1 > 1 \text{ and } \beta_2 < 0 \tag{A7}$$

To ensure condition (A6.1) the coefficient A_2 must be assumed equal to zero. Hence, from equation (A7), (A5) simplifies to

$$\frac{1}{2}\beta_1(\beta_1 - 1)\sigma^2 + \beta_1\alpha - (\rho + \lambda) + \lambda(1-\phi)^{\beta_1} = 0. \tag{A8}$$

Assuming $\phi = 1$ this leads to the solution

$$\beta_1 = \frac{1}{2} - \frac{\alpha}{\sigma^2} + \sqrt{\left(\frac{\alpha}{\sigma^2} - \frac{1}{2}\right)^2 + \frac{2(\rho + \lambda)}{\sigma^2}} > 1. \tag{A9}$$

Boundary conditions (A6.2 – A6.3) can be used to get the value of the unknown variable A_1 and the optimal value of (Π-C) for the investment decision. This results in the following relation

$$(\Pi - C) = \left(\frac{\beta_1}{\beta_1 - 1}\right)(\rho - \alpha + \lambda)I \,. \tag{A10}$$

Appendix 2

Given $\frac{\partial \lambda(d)}{\partial d} < 0$ and $\rho > \alpha$ the sign of $\frac{\partial \beta_1}{\partial d} < 0$ while $\frac{\partial \left(\frac{\beta_1}{\beta_1 - 1}\right)}{\partial d} > 0$.

This can be easily seen from the following derivative:

$$\frac{\partial \left(\frac{\beta_1}{\beta_1 - 1}\right)}{\partial d} = \frac{\partial \beta_1}{\partial d}(\beta_1 - 1)^{-1} + \beta_1 \frac{\partial (\beta_1 - 1)^{-1}}{\partial d}$$

$$= \frac{1}{\sigma^2}\left[\left(\frac{\alpha}{\sigma^2} - \frac{1}{2}\right)^2 + \frac{2(\rho + \lambda)}{\sigma^2}\right]^{-\frac{1}{2}} \frac{\partial \lambda}{\partial d}(\beta_1 - 1)^{-1} - \beta_1(\beta_1 - 1)^{-2} \frac{1}{\sigma^2}\left[\left(\frac{\alpha}{\sigma^2} - \frac{1}{2}\right)^2 + \frac{2(\rho + \lambda)}{\sigma^2}\right]^{-\frac{1}{2}} \frac{\partial \lambda}{\partial d}$$

Setting $K = \frac{1}{\sigma^2}\left[\left(\frac{\alpha}{\sigma^2} - \frac{1}{2}\right)^2 + \frac{2(\rho + \lambda)}{\sigma^2}\right]^{-\frac{1}{2}}$

we can write

$$\frac{\partial \left(\frac{\beta_1}{\beta_1 - 1}\right)}{\partial d} = \frac{K \frac{\partial \lambda}{\partial d}}{\beta_1 - 1}\left(1 - \frac{\beta_1}{\beta_1 - 1}\right).$$

Given that both of the factors of the right-hand side are negative, the sign of the derivative is positive. Now, what is the prevailing sign of the following derivative?

$$\underbrace{\left(\frac{\beta_1}{\beta_1 - 1}\right)\frac{\partial(\rho - \alpha + \lambda)}{\partial d}}_{(-)} + \underbrace{\frac{\partial \left(\frac{\beta_1}{\beta_1 - 1}\right)}{\partial d}(\rho - \alpha + \lambda)}_{(+)}.$$

Substitute for the above result

$$\left(\frac{\beta_1}{\beta_1 - 1}\right)\frac{\partial(\rho - \alpha + \lambda)}{\partial d} + \frac{K \frac{\partial \lambda}{\partial d}}{\beta_1 - 1}\left(1 - \frac{\beta_1}{\beta_1 - 1}\right)(\rho - \alpha + \lambda)$$

and collect terms

$$\underbrace{\frac{\partial \lambda}{\partial d}}_{-}\left[\underbrace{\left(\frac{\beta_1}{\beta_1-1}\right)}_{+}+\underbrace{\frac{K}{\beta_1-1}}_{+}\underbrace{\left(1-\frac{\beta_1}{\beta_1-1}\right)}_{-}\underbrace{(\rho-\alpha+\lambda)}_{+}\right].$$

The prevailing sign is ambiguous and depends on the specific values of the parameters.

References

ACPC, 1999. *Canola's producers perspective on biotechnology*. Alberta Canola Producers Commission ACPC, Edmonton. [http://www.canola.ab.ca/tec/gmo.shtml]

Agnet, 2002. *GM free & the 6-mile (10km) exclusion zone*, September 19. [http://131.104.232.9/agnet/2002/9-2002/agnet_september_19.htm]

Bock, A.K., Lheureux, K., Libeau-Dulos, M., et al., 2002. *Scenarios for co-existence of genetically modified, conventional and organic crops in European agriculture*, Institute for Prospective Technological Studies and Joint Research Centre of the European Commission IPTS-JRC. Sevilla. [ftp://ftp.jrc.es/pub/EURdoc/eur20394en.pdf]

Clark, E.A., 2001. On the implication of the Schmeiser decision: the crime of Percy Schmeiser. *Genetics Society of Canada Bulletin,* June 2001.

Demont, M. and Tollens, E., 2001. *Uncertainties of estimating the welfare effects of agricultural biotechnology in the European Union.* Department of Agricultural and Environmental Economics, K.U. Leuven, Leuven. Working Paper no. 2001/58. [http://www.agr.kuleuven.ac.be/aee/clo/wp/Demont2001c.pdf]

Demont, M., Wesseler, J. and Tollens, E., 2002. *Biodiversity versus transgenic sugar beet: the one Euro question.* Department of Agricultural and Environmental Economics, K.U. Leuven, Leuven. Working Paper no. 2002/69. [http://www.agr.kuleuven.ac.be/aee/clo/wp/demont2002d.pdf]

European Commision, 2000. *Economic impacts of genetically modified crops on the agri-food sector: a first review*. European Commision, Brussels. Working Document Rev. 2, Directorate-General for Agriculture. [http://europa.eu.int/comm/agriculture/publi/gmo/fullrep/index.htm]

European Commision, 2002. *Frequently asked questions on the Commission's proposal on environmental liability*. European Commission, Brussels. Press Releases MEMO/02/10, 24 January 2002. [http://europa.eu.int/rapid/start/cgi/guesten.ksh?p_action.getfile=gf&doc=MEMO/02/10|0|AGED&lg=EN&type=PDF]

Gaisford, J.D., Hobbs, J.E., Kerr, W.A., et al., 2001. *The economics of biotechnology*. Edward Elgar, Cheltenham.

Germanò, A., 2002. *Voce "Biotecnologie in agricoltura". IV. Digesto discipline privatistiche*. Utet, Torino.

James, C., 2002. *Global status of commercialized transgenic crops*. International Service for the Acquisition of Agri-biotech Applications ISAAA, Ithaca. ISAAA Briefs no. 27. [http://www.isaaa.org/kc/Publications/pdfs/isaaabriefs/Briefs%2027.pdf]

Kershen, D.L., 2002. *Legal liability issues in agricultural biotechnology*. National Agricultural Law Center, University of Arkansas School of Law, Fayetteville. [http://www.nationalaglawcenter.org/assets/article_kershen_biotech.pdf]

Kolstad, C.D., Ulen, T.S. and Johnson, G.V., 1990. Ex post liability for harm vs. ex ante safety regulation: substitutes or complements? *American Economic Review,* 80 (4), 888-901.

Marra, M.C., 2001. Agricultural biotechnology: a critical review of the impact evidence to date. *In: IFPRI & AARES preconference workshop "Agricultural biotechnology: markets and policies in an international setting", Adelaide, January 22 2001.*

Monsanto Canada Inc. vs. Schmeiser, 2001. Decision by Judge MacKay, Federal Court of Canada FCT 256, March 29, 2001.

Morel, B., Farrow, S., Wu, F., et al., 2003. Pesticide resistance, the precautionary principle and the regulation of *Bt* corn: real and rational option approaches to decision making. *In:* Laxminarayan, R. ed. *Battling resistance to antibiotics and pesticides: an economic approach.* Resources for the Future, Washington, 184-213.

Network of Concerned Farmers NCF, 2003. *Issues & information.* [http://www.non-gm-farmers.com/issues_legal_litigation.asp]

Office of Science and Technology Policy OSTP, 1986. *Coordinated framework for regulation of biotechnology, 51 Fed. Reg. 23302, June 26, 1986.*

Schmidt, H., 2002. *Die heutigen rechtlichen Rahmenbedingungen für die Koexistenz der biologischen Landwirtschaft mit benachbarten transgenen Kulturen in Deutschland, Diskurs Grüne Gentechnik.* Bundesministerium für Verbraucherschutz, Ernährung und Landwirtschaft, Berlin.

Wesseler, J., 2003. Resistance economics of transgenic crops under uncertainty. *In:* Laxminarayan, R. ed. *Battling resistance to antibiotics and pesticides: an economic approach.* Resources for the Future, Washington, 214-237.

Winston, M.L., 2002. *Travels in the genetically modified zone.* Harvard University Press, Cambridge.

---

[1] For examples on short and long term impacts of GM contamination on conventional and organic farmers see Bock et al. (2002).

[2] For some authors the control of volunteers can be easily implemented by using other herbicides, such as 2,4-D or MCPA, in association with Roundup (ACPC 1999). Other scientists see this issue as much more problematic with important implications for farm practices (Clark 2001). Moreover, in the case of canola, the GM farmer willing to switch back to the conventional crop will face severe problems due to the control of volunteers with the risk of not being able to save seed for the following sowing season and the threat of being sued by the seed developer.

[3] If this is true, as will be clear in the model specification, *ex-ante* regulations are not the sole forces influencing the profitability of the GM crop and, consequently, the adoption decision.

[4]Council Regulation (EC) 1804/1999 of July 19, 1999 under article (10) states that products labelled as organic have to be free of genetically modified organisms or parts thereof. As the current analytical limit is at a level of about 0.1%, this has been interpreted as a *de facto* threshold (Bock et al. 2002).

11b

Comment on Soregaroli and Wesseler: Minimum distance requirements and liability: implications for co-existence

Volker Beckmann[#]

The future institutional environment for the co-existence of GM crops, conventional and organic crops in Europe is likely to combine measures of *ex ante* regulation and *ex post* liability rules. Recognizing Europe's heterogeneity in farm structures, crop patterns and legal environments, the European Commission decided to follow the principle of subsidiarity and states that "measures for coexistence should be developed and implemented by the Member States" (European Commission 2003). It could be expected that the Member States will develop a variety of different measures that will have a profound impact on the adoption rate of GM crops.

Against this background the paper by Soregaroli and Wesseler deals with an interesting question. How do *ex ante* regulation and *ex post* liability affect the adoption of transgenetic crops on different farm or plot sizes? The paper focuses its attention more specifically on the effect of different minimum distance requirements on the minimum field size of adoption. This question is without any doubt very relevant given the variance in size among Europe's farms. In Italy, Portugal and Greece, more that 70 % of all farms cultivate less than 5 hectares, compared to Denmark, Ireland and Sweden where the share of these small farms does not exceed 10 %. If the *ex ante* regulation and *ex post* liability for using GM crops induce additional costs, these costs may not be scale-neutral. It may be argued that minimum distance requirements in particular will disadvantage smaller farms compared to liability rules. This trade-off is the main theme of the paper.

Soregaroli and Wesseler push forward the modelling approaches to capture this trade-off and to my knowledge this is the fist paper that deals with these issues more systematically. The trade-off is modelled in a classical way using the approach by Kolstad, Ulen and Johnson (1990) and in a more advanced way considering irreversibility and uncertainty. With both models they arrive at the conclusion that it "... is not possible, *a priori*, to conclude what is the effect on the minimum adoption size of the *ex ante* regulation" (p. 176). Although I would not disagree, I would like to point out that some crucial assumption on the cost function may have an important impact on the results. I would like to elaborate on this point.

Soregaroli and Wesseler introduce some assumptions of the cost functions that may be very important and driving the results. The costs of regulations are defined as $C(s,d)$ with s as the size of the field and d as the minimum distance between the GM crop and the field's external limits. It is further assumed that the costs increase with distance and size but in both cases at a diminishing rate, $\partial C(s,d)/\partial s > 0$, $\partial^2 C(s,d)/\partial s^2 \leq 0$ and $\partial C(s,d)/\partial d > 0$, $\partial^2 C(s,d)/\partial d^2 \leq 0$. The costs of liability are defined as $\mu(s,d)$ with the assumption that $\mu(s,d)$ increases in s at a diminishing rate

[#] Dept. Agric. Econ. and Soc. Sci., Humboldt Univ. Berlin. E-mail: v.beckmann@agrar.hu-berlin.de

J. H. H. Wesseler (ed.), Environmental Costs and Benefits of Transgenic Crops, 183–184.

but decreases in d at an increasing rate, $\partial\mu(s,d)/\partial s > 0$, $\partial^2\mu(s,d)/\partial s^2 \leq 0$ and $\partial\mu(s,d)/\partial d < 0$, $\partial^2\mu(s,d)/\partial d^2 \geq 0$. Let us think about the field as a square. In this case we can define $s = a^2$ with a as the side length. The field size that is needed to keep the minimum distance d and, thus the costs C can be calculated as $C = a^2 - (a-2d)^2 = 4ad - 4d^2$ or alternatively as $C = 4\sqrt{s}d - 4d^2$. Now consider a different situation, where the farmer negotiates with his neighbours. In this case the cost function can be described as $C = (a+2d)^2 - a^2 = 4\sqrt{s}d + 4d^2$. This function has the following partial derivatives: $\partial C(s,d)/\partial s > 0$, $\partial^2 C(s,d)/\partial s^2 \geq 0$ and $\partial C(s,d)/\partial d > 0$, $\partial^2 C(s,d)/\partial d^2 > 0$. Thus, the costs steadily increase with the field size and constantly increase with the minimum distance requirements.

However, most important is that the minimum distance regulation creates some fixed costs since $\lim_{s \to 0} C = 4d^2$. This clearly would disadvantage very small farms. This disadvantage to small farms by minimum distance regulations is reinforced by a more realistic damage function. The damage function is not continuous, since damage is related to the threshold of food and feed being labelled as GM. In Europe economic damage will only appear if the fraction of GM products in non-GM crops exceeds the 1% threshold (see also Beckmann 2003). Suppose f_G defines the fraction of GM crops in non-GM crops and T defines the threshold. The damage function, then, could be redefined as

$\mu(s,d) = \begin{cases} 0 & f_G < T \\ > 0 & f_G \geq T \end{cases}$. If s is small compared to the size of the surrounding fields, the probability of being liable is zero perhaps over a larger range of s. If this holds true then minimum distance regulations will disadvantage and liability rules will advantage small farms. However, as the size of the field increases the effects are not that clear. This is what Soregaroli and Wesseler found. I would encourage them to calculate the model with different specifications of the cost function.

However, the Soregaroli and Wesseler paper offers a good starting point for further analysis of different minimum-distance regulations and liability rules on the adoption of GM crops in Europe. There is much more work to come along this line of research and it will be a rich field for empirical studies in the future.

References

Beckmann, V., 2003. *Governing coexistence: a case for coase?* Unpublished Manuscript.

European Commission, 2003. Commission Recommendation of 23 July 2003 on guidelines for the development of national strategies and best practices to ensure the coexistence of genetically modified crops with conventional and organic farming, 2003/556/EC. *Official Journal of the European Communities,* L 189/36, 29/07/2003, 1-12. [http://europa.eu.int/eur-lex/pri/en/oj/dat/2003/l_189/l_18920030729en00360047.pdf]

Kolstad, C.D., Ulen, T.S. and Johnson, G.V., 1990. Ex post liability for harm vs. ex ante safety regulation: substitutes or complements? *American Economic Review,* 80 (4), 888-901.

12a

Biotechnology, the US-EU dispute and the Precautionary Principle

Henk van den Belt[#]

Abstract

The international debate on biotechnology is extremely polarized. Opponents such as Greenpeace International, Friends of the Earth and other NGOs often invoke the Precautionary Principle to advance their cause. The principle is also at issue in trade disputes between the USA and the European Union. There are several versions of the Precautionary Principle in circulation. The strong version adopted by many environmentalist organizations is logically untenable, while the weaker versions espoused by the European Commission and enshrined in international treaties are rather vague and ill-defined. The contested role of the Precautionary Principle bears testimony to public ambivalence towards scientific expertise in modern risk societies. A more open and democratic decision-making process on biotechnology will not by itself resolve the underlying uncertainties. Shifting the burden of proof to the proponents of biotechnology makes sense only if the required standard of proof is also specified. The debate on the Precautionary Principle appears to be a proxy for a larger debate on the future of world agriculture.
Keywords: biotechnology; Precautionary Principle; burden of proof; public participation; scientific credibility

Introduction: French GMO controversies

In many parts of the world fierce debates on biotechnology have been raging without as yet showing any signs of abating. Recent controversies in France on 'les OGM' (*organismes génétiquement modifiés* or genetically modified organisms, GMOs) exemplify the many different issues that are involved in such debates.

Representatives of top-level institutions like the agricultural research establishment INRA (*Institut National de la Recherche Agronomique*), the Academy of Sciences and the Academy of Medicine and Pharmacy are showing increasing concern about the prevalent anti-biotech mood among the French population and its implications for the country's position in international research and industrial competition. As recently as April 2000 the charismatic activist José Bové and eight other anti-GMO militants destroyed experimental test fields planted with transgenic oilseed rape in the vicinity of Gaudiès, an act for which they were later convicted in court. In a letter to the daily newspaper *Libération* dated 23 September 2002, the president and the general director of INRA, Bertrand Hervieu and Marion Guillou, emphatically spoke out in favour of conducting field trials with transgenic crops as

[#] Applied Philosophy Group, Wageningen University and Research Centre, Hollandseweg 1, NL-6706 KN Wageningen. E-mail: Henk.vandenBelt@wur.nl

J. H. H. Wesseler (ed.), Environmental Costs and Benefits of Transgenic Crops, 185–197.

being indispensable for getting to know the real risks of 'gene flow' (Hervieu and Guillou 2002). They saw it as an obligation that a public research institute like INRA simply owed to France and its citizens: "Abandoning field tests would make France, in a certain sense, mute and blind". As a public agency, INRA also has an important part to play in any research supporting the innovative improvement of French agriculture, including biotechnology. "But we should take care that such progress is shared and approved by all the actors in the agricultural and food chain, from the farmer to the consumer". According to the president and the general director, INRA was already engaged in this 'dialogue'. They also claimed that "INRA had been the first to apply the precautionary principles [*principes de précaution*] to the GMO question and to put them into practice", without however explaining what this amounted to (or why the plural was being used). Finally, the two board members pointed out that INRA, together with other French research institutes, takes part in the research consortium *Génoplante* devoted to the study of plant genomics, in which private companies like Bayer and the seed business also participate. Thus the letter ended in upbeat fashion: "Constructive endeavour and dialogue with all the actors in the debate, sharing of orientations and results, a broad partnership [*partenariat large*]: these are the principles that guide our research. Let's use the opportunity offered by the debate on GMOs and on the field trials to renew the terms of the contract between science, industry and society and to finally put into place a veritable social management of innovation" (Hervieu and Guillou 2002).

Despite this invitation to co-operation, opponents of biotechnology did not grasp the outstretched hand. In October 2002 a number of critics wrote an open letter to INRA. They concluded that after the institute's leaders had spoken out and taken position in the current GMO debate, "INRA is from now on no longer a site of neutral expertise, but a partisan actor [*un acteur engagé*]" (Inf'OGM 2002). They maintained that the directorate's attitude even called the existence of a research institute in the service of the public good into question. The alleged need of field trials with transgenic crops to determine the true risks of gene flow was disputed; tests with non-modified crops would do just as well to follow the spread of pollen. Moreover, the large-scale commercial introduction of transgenic crops in North America and Argentina had abundantly demonstrated that the 'contamination' of non-GM crops and wild relatives already occurs to such an extent that the continued viability of organic farming is seriously threatened. The real reason for field tests in France, the critics suggested, is not to ascertain the exact risks of inadvertent spread of foreign genes but to assess new, transgenic crop varieties under actual conditions of culture in order to facilitate their subsequent commercial introduction. The 'broad partnership' with various business firms of which the INRA directorate prided itself so much, was in the eyes of the critics nothing less than a scandalous privatization of public research institutes. While public monies contribute 70 percent of the funds for the *Génoplante* consortium, public representatives occupy only 50 percent of the seats in the managing board. Private companies were said to have an inordinate influence on the direction of research. "Tell me who your business partners are and I will tell you what research you are going to undertake" (Inf'OGM 2002). The critics also accused the INRA directorate of paying no more than lip-service to the need for 'dialogue' with the larger public: in fact, the findings of a citizens' conference on GMOs held in 1998 and a debate on field trials organized in 2002 by a commission of wise men were completely ignored because their outcomes did not please the proponents of biotechnology (for comments on these abortive attempts at public participation, see also Testart 2000). Finally, as regards the claim that INRA took the lead in applying the Precautionary

Principle to the GMO question, the critics observed that while INRA had indeed instituted its own *Comité d'éthique et de précaution* (Comepra), the committee's report for the year 1999-2000 was almost kept as a secret and only released to the public after five months of insistent pressure. So much for INRA's self-declared policy of openness and dialogue.

After INRA, in December 2002 the Academy of Medicine and Pharmacy and the Academy of Sciences also published reports on the GMO question. While the report of the former institution counted only a few pages, the latter's report – prepared under the direction of Professor Roland Douce, director of the Institute of Structural Biology in Grenoble – was more extensive (Académie des Sciences 2002). Both reports pleaded, among other things, for lifting the existing European moratorium on the commercial introduction of new transgenic crops, in force since October 1998, and thereby aroused the ire of some environmentalist organizations. In response, the 'anti-globalist' group Attac and green watchdog organizations like OVALE (*Observatoire de vigilance et d'alerte écologique*) and CRII-GEN charged the authors of the two reports with having insulted science by debasing the two academies to the rank of mouthpieces for the big companies of the 'genetic-industrial and agrifood complex'. They were also accused of having exposed themselves as a fifth column in support of the Bush administration, which in January 2003 announced plans to lodge a complaint with the WTO against the European moratorium on GMOs (Attac France 2003b) – plans that were postponed at that time in view of the already strained relations between the USA and Europe due to the imminent war on Iraq.

The authors of the Académie des Sciences report were also criticized for their allegedly close personal relations with various business interests (Rhône-Poulenc / Aventis, Limagrain, the French seed industry confederation CFS, etc.), which presumably had severely compromised the independence of their judgment. Deputies of Attac at the National Assembly questioned the 'scientific objectivity' of the two reports and demanded the installation of a parliamentary committee of inquiry in order to investigate the 'conflicts of interest' that might have influenced their preparation (Attac France 2003a; Frat 2003). The authors of the reports, however, played down the importance of the adduced industrial connections and at any rate denied that these had biased their judgement. Professor Douce, for one, stated: "I am being attacked because I once led a research unit of CNRS / Rhône-Poulenc. That work ended in 1995 and since then I have not been involved in the development of GMOs. There is behind this entire affair an anti-GMO movement whose power I could not have imagined" (Frat 2003). He and his colleagues also claimed to have been harassed and threatened with violence: "[Douce] told *Nature* that such was the ferocity of the critical reaction that he would now think twice before giving public advice in the future" (Butler 2003). The president of the Academy of Sciences, Beaulieu, called on the French government to defend "the honour of scientists attacked in their mission of delivering independent and educated information to society" (Butler 2003). Indeed the ministers for research and technology and for national education, Claudie Haigneré and Luc Ferry, unreservedly condemned the personal attacks and the threats of physical intimidation as attempts to cast doubt on the moral and intellectual integrity of Academy members and to silence them (Agrisalon.com 2003).

It is indeed possible that the opponents of biotechnology have exaggerated the business connections of the authors of the Academy report and made too much of the ensuing conflicts of interest. Still this is a very sensitive issue. In order to maintain the public credibility of science sometimes even the mere semblance of relying on outside financial interests has to be avoided. Ironically, the report of the Academy of Sciences was thoroughly aware of this need for financial independence. It stated in the first chapter: "Maintaining a funding of research which guarantees a certain level of independence of the researchers with regard to economic imperatives is indispensable to preserve the credibility of risk assessment" (Académie des Sciences 2002, see Recommandations spécifiques aux chapitres).

Attac France distanced itself from the personal attacks and physical threats directed at Roland Douce and the other authors of the two reports. The group clarified its position as follows: "ATTAC reaffirms its wish to see a real social debate being conducted on GMOs. ATTAC is by no means hostile to research on GMOs in a confined environment, but demands that tests in the open field be prohibited. As long as there are no scientific proofs demonstrating a total absence of risk, the precautionary principle must be applied" (Attac France 2003a). Here, a strong version of the Precautionary Principle (in what follows to be abbreviated as PP) is invoked. This version is entirely at odds with the tenor of the official report on the PP which Philippe Kourilsky and Geneviève Viney issued in 1999 to Prime Minister Jospin: "Precaution may not in fact be equated, on pain of misunderstanding the sense of the principle, with the unrealistic demand of zero risk" (Kourilsky and Viney 2000, p. 12).

Our quick scan of recent GMO controversies in France has shown a few things. There are deep divisions of mutual distrust between established scientific institutes and large segments of the population, especially environmentally concerned citizens. While the scientific and political establishment preaches the virtues of 'dialogue' and involving the general public, the outcomes of concrete attempts at participatory decision-making like organized social debates and citizen's conferences are not taken seriously if they deviate too much from the official commitment to biotechnology. Environmentalist groups in their turn are deeply suspicious of 'expert opinion' on GMOs even if it comes from what were formerly prestigious scientific institutions. Advisers are seen to follow a political agenda rather than stick to the bounds of their scientific competence. Due to more or less close relationships with external business partners that seem to characterize the contemporary life sciences, charges of 'conflicts of interest' are easily raised and extremely hard to rebut (for similar concerns with regard to GMO field trials in the UK, see Myhr and Traavik 2003, p. 241). While both sides on the debate about GMOs appeal to the so-called PP, each party attaches its own favoured meaning to it.

Most problematic is the precarious credibility of scientific expertise. Even the official Kourilsky-Viney report recognizes the fundamental difficulty of the situation: "Scientific expertise surely provides insights [*des connaissances*] at the service of decision-making, but in situations of precaution where it operates on the limits of knowledge, *the expert does not know*. On the basis of what he knows, he expresses an enlightened opinion or conviction, which is however not entirely free from prejudice. He therefore inevitably transgresses the limits of his own knowledge and may for that reason be readily contradicted by his own peers" (Kourilsky and Viney 2000, p. 61, italics mine). This situation is responsible for a large part of the public ambivalence towards scientific expertise in modern risk societies, as laypersons cannot rely on their own unaided senses but need the mediation of science and technology to chart the hazards and risks that may threaten their existence (Beck 1992).

Different views of precaution: the US, the EU and NGOs

On 13 May 2003 the Bush administration announced that it would file a WTO case against the moratorium on genetically modified crops and foods that is *de facto* in force in the European Union since October 1998, when new approvals were frozen (Office of the United States Trade Representative USDA 2003). Earlier plans in that direction were stalled in the run-up to the war on Iraq. As many commentators had expected, the end of the military confrontation in Iraq signalled a new willingness on the part of the US government to initiate a major trade war: "If the United States and France continue to feud over Iraq, one place the Bush administration can be expected to seek revenge is the WTO" (Feffer 2003). Some commentators question the wisdom of this move, quite apart from the legal merits of the case: "[This] announcement is likely to play poorly in Europe, however, where many consumers already feel the U.S. government has been trying to shove gene-altered food down their throats" (Gillis 2003; see also Pew Initiative on Food and Biotechnology 2003). In the accompanying document to the announcement, US Trade Representative Robert Zoellick called the EU moratorium illegal, in violation of WTO rules and 'non-science-based': "Numerous organizations, researchers and scientists have determined that biotech foods pose no threat to humans or to the environment. Examples include the French Academy of Medicine and Pharmacy, and the French Academy of Sciences ..." (Office of the United States Trade Representative USDA 2003).

The US seems to have a very strong legal case to challenge the EU moratorium before the WTO, but European Commissioner David Byrne called the timing of the suit "a little eccentric" as the EU was already moving to lift the moratorium and replace it with new legislation on labelling and traceability (Reuters 2003). Of course, the US government would consider the new European rules on labelling and tracing of GM foods also to be unduly burdensome and costly to American farmers and exporters, and may in future be expected to challenge these in turn. However, such rules, if applied with consistency and without discrimination, would probably be much less vulnerable to a WTO challenge (Feffer 2003).

It is not likely that EU officials are going to take recourse to the PP in order to defend the moratorium; instead they will meet the WTO challenge with a lot of equanimity and move on to introducing legislation on labelling and tracing in order to ensure consumer choice. The real test case will be when the latter are also challenged; but in that event the dispute enters the ideologically charged domain of consumer choice where the US position is also vulnerable, because American consumers are *de facto* denied the choice of GM-free products (Pew Initiative on Food and Biotechnology 2003). According to an ABC News poll, 93 percent of Americans support labelling of genetically engineered foods (The Campaign 2003).

The US-EU disagreement over GMOs is just one episode in a longer series of disputes. Over the past decade or so, the European and American approaches to environmental, health, safety and consumer regulation have drifted further apart. Whereas during the 1970s and 1980s the US regulatory regime was generally much more strict and risk-averse than the European regime, the situation was reversed during the 1990s. In David Vogel's imagery, the hare and the tortoise changed places (Vogel 2003). After 1990, America started to move like a tortoise when a conservative pro-business majority in the Republican Party blocked further regulatory initiatives, helped by the fortunate absence in the US of major incidents such as the mad-cow disease, the HIV-contaminated blood scandal (especially in France) and a number of food scares which in Europe undermined the confidence and trust of citizens and consumers in their regulatory authorities. The creation of a single European market necessitated the strengthening of regulatory standards: a high level of health and environmental protection was critical to the legitimacy of a growing bureaucracy in Brussels. Simultaneously, the process of policy-making became more open and accessible to non-business constituencies. Already the Maastricht Treaty of 1992 declared the PP to be a key principle of Community environmental policy; in due course, this very principle would become the avowed cornerstone of EU regulation in the areas of food safety, environment, human health, animal health and plant health.

All the while the US government has been sceptical of the so-called PP, suspecting that it may be used too easily as an excuse for protectionism. The European Union had indeed invoked this principle to defend the ban on hormone-treated-beef imports from the United States, which the latter successfully challenged before the WTO (Charlier and Rainelli 2002). The US government is prone to counter any invocation of the PP with a mantra-like appeal to 'sound science' and to the fairly narrow provisions of the SPS agreement of the WTO (SPS stands for sanitary and phytosanitary measures). The implied suggestion is that the PP goes beyond sound science and is therefore arbitrary. In February 2000, the Commission of the European Communities issued a communication on the PP to strengthen its policy position in order to defend the EU better from future legal challenges by other WTO members (European Commision 2000).

The Commission argued that, regardless of divergences in the used terminology, the PP had already become a rule of customary international law in the areas of health and environmental protection. The Commission referred, *inter alia*, to the North Sea Declaration (1987), the Rio Declaration (1992), the preamble of the Convention on Biological Diversity (1992), the Convention of Climate Change (1992) and the Protocol on Biosafety (2000). A well-known formulation of the 'precautionary approach' is to be found in Principle 15 of the Rio Declaration: "*Where there are threats of serious or irreversible damage, lack of full scientific certainty shall not be used as a reason for postponing cost-effective measures to prevent environmental degradation*". Another well-known but unofficial definition of the PP, not quoted in the Commission's communication, was spelled out in a January 1998 meeting at Wingspread in Racine, Wisconsin. The *Wingspread Statement* summarized the principle thus: "*When an activity raises threats of harm to human health or the environment, precautionary measures should be taken even if some cause and effect relationships are not fully established scientifically*" (Raffensperger and Tickner 1999, p. 353-354). In the communication, the Commission declined to give a precise definition of the PP, arguing that the meaning of the concept will be fleshed out by decision-makers and courts of law. Still the Commission offered in rather convoluted prose what looked like the rudiments of a definition: "[The Precautionary Principle] *covers those specific circumstances where scientific evidence is insufficient, inconclusive or uncertain and there are indications through preliminary objective scientific evaluation that there are reasonable grounds for concern that the potentially dangerous effects on the environment, human, animal or plant health may be inconsistent with the chosen level of protection*" (European Commision 2000, p. 10). This formulation alludes to 'the chosen level of protection' because the Commission held that each WTO member has the independent right to determine the level of environmental or health protection they consider appropriate (p. 11). The Commission also introduced a sharp distinction between 'risk assessment' and 'risk management', that is, between science and politics. Whereas a 'prudential approach' may be part of (scientific) risk assessment (e.g. by taking into account a pre-defined safety margin in risk evaluation), the application of the PP is held to belong to (political) risk management. Risk management is the preserve of political decision-makers, according to the Commission: "Judging what is an 'acceptable' level of risk for society is an eminently *political* responsibility" (European Commision 2000, p. 4). It would appear that deciding on an acceptable level of risk is just the flip-side of choosing a certain level of protection. The Commission further held that application of the PP has to satisfy customary procedural criteria like proportionality, non-discrimination, consistency, examination of potential costs and benefits, review and assignment of responsibility. Through the thicket of added qualifications it is rather difficult to discern what the PP basically stands for.

Existing definitions and formulations of the PP – and there are many – all tend to beg the crucial questions. Is there ever full scientific certainty or sufficient, conclusive and fully certain scientific evidence? In other words, wouldn't it be possible to invoke the PP on each and every occasion? Do we need a minimal threshold of scientific certainty or plausibility before we may (or should) undertake preventative action? How strong must the 'indications' and the 'reasonable grounds' be before we should do something about the presumed threat (but remember that we have by definition only preliminary evidence!)? And do we really know how to prevent harm if we are so much ignorant about the underlying cause–effect relationships? The definitions that are currently on offer fail to spell out the precise conditions that have to be fulfilled before the PP may be invoked or the nature of the preventative action that has to be taken. The types of action suggested range from implementing a ban, imposing a moratorium while further research is conducted, allowing the potentially harmful activity to proceed while closely monitoring its effects, to just conducting more research. The PP does not have a very precise meaning as long as such crucial aspects are left largely unanswered.

A specific problem with the communication of the European Commission is that it does not offer us any guidance on how we should go about bridging the gap between the scientific risk assessment and the policy objectives of risk management, i.e., the effective realization of a *chosen* level of protection (or an *acceptable* risk level). Does it make sense at all, if there is so much scientific uncertainty, to think in terms of realizing specific levels of protection?

Unlike the European Commission, many environmentalist NGOs do not subject the application of the PP to a series of additional procedural criteria. In practice, this often means that the PP is given a more definite meaning by effectively reducing it to an absurdity. Normally no minimal threshold of plausibility is specified as a 'triggering' condition, so that even the slightest indication that a particular product or activity might possibly produce some harm to human health or the environment will suffice to invoke the principle. And just as often no other preventative action is contemplated than an outright ban on the incriminated product or activity. Greenpeace provided a clear example of this approach when it rang alarm bells after a team at Cornell University showed in 1999 that monarch butterfly larvae exhibited increasing morbidity and mortality when fed in the laboratory with milkweed leaves dusted with pollen from transgenic Bt-maize (for a more extensive analysis of this case, see Van den Belt 2003). For Greenpeace it was immediately clear that monarch butterfly populations in the wild, and one hundred other species of butterflies as well, were gravely endangered by the growing of genetically engineered maize and of all other transgenic crops. The NGO therefore demanded an immediate stop to these practices. It was later found that monarch butterfly populations in the American Mid-West, where transgenic maize is cultivated on a large scale, were not affected at all (Ortman et al. 2001).

Closely linked to various versions of the PP is the idea of *reversing the onus of proof*. Thus the adherents of the Wingspread Statement declare that "the applicant or proponent of an activity or process or chemical needs to demonstrate that the environment and public health will be safe. The proof must shift to the party or entity that will benefit from the activity and that is most likely to have the information" (Raffensperger and Tickner 1999, p. 345-346). Greenpeace also holds that effective implementation of the PP requires a shift in the burden of proof (Greenpeace International 2001). In its communication the European Commission maintains that action under the head of the PP sometimes implies reversing the onus of proof, but that such reversal cannot be the general rule (European Commision 2000, p. 21). Shifting the burden of proof seems a fairly straightforward way to ensure, as the German philosopher Hans Jonas demanded, that greater weight will be given to the 'prognosis of doom' than to the 'prognosis of bliss' (Jonas 1984, p. 34).

The PP is sometimes also associated with the idea that the introduction of a new technology like genetic modification needs the 'informed consent' of the population and therefore requires open, transparent and democratic processes of decision-making *before* commitments have been made to research, development and marketing (Barrett 2000). The European Commission also gestures towards the need to ensure a wider and more active participation by the public, and even a biotech company like Monsanto nowadays makes a solemn 'pledge' to engage in dialogue and move in step with a large array of 'stakeholders'. To me the question of democratic decision-making about technology is a special issue that is perhaps better kept apart from the PP. It raises questions of its own, such as: How seriously do we take the idea of 'informed consent'? Are we willing to grant veto power to minorities who remain tenaciously opposed to a technology that the majority wants to embrace? Is this compatible with the existence of a 'free' market economy? One way to link the PP with public participation is to reason that we need to broaden the process of decision-making because, due to the fundamental scientific uncertainties with which we are confronted, 'the expert does not know' (cf. the earlier quotation from Kourilsky and Viney 2000; – this line of reasoning is followed by Testart 2000). Myhr and Traavik (2003, p. 242) also argue for expanded peer-review processes and extended peer communities as a mechanism to balance scientific advice with the involvement of other parties. If the experts do not know, then everybody supposedly becomes an expert. But will a more open and democratic decision-making process on biotechnology by itself resolve the underlying uncertainties?

The strong version of the PP

According to the strong version of the PP the mere prospect of potentially harmful effects of a new technology is enough to stop its introduction and deployment. But why should the prospect of harmful effects take precedence over the prospect of beneficial effects, quite apart from the inherent likelihood of each of these possibilities? The obvious answer seems to be that such a priority is defensible only when the harmful effects are of such magnitude that they carry catastrophic (or, as Jonas would say, 'apocalyptic') potential. The infinite costs of a possible catastrophic outcome necessarily outweigh even the slightest probability of its occurrence.

It is, however, not difficult to see that the strong version of the PP – also dubbed the 'catastrophe principle' – is logically untenable (for the analogy of this version with

Pascal's famous but equally untenable 'wager argument' for believing in the existence of God, see Van den Belt 2003). Take the application of this principle to the problem of global warming. Environmentalists often argue that even if it is not conclusively established that the emission of carbon dioxide and other gases causes an enhanced greenhouse effect, the mere prospect of an ecological catastrophe due to such a scenario should lead us to curb our emissions of greenhouse gases drastically now. By the same logic, however, one could conjure up the possibility of a coming ice age. The mere prospect of this equally catastrophic scenario should then induce us to avert this outcome by stepping up the emission of greenhouse gases. The strong version of the PP would thus lead to contradictory recommendations. In a similar way, it could be argued that this principle commits us to each of two contradictory policies: (1) we must not develop GM crops, and (2) we must develop GM crops. The first alternative is argued vehemently by many environmentalists who appeal to the PP. To support the second possibility, Gary Comstock conjures up a dramatic scenario in which people are forced to seize upon the remaining reserves of nature in a desperate effort to overcome food shortages resulting from global warming. He then argues, in the style of the environmentalists, that "lack of full scientific certainty that GM crops will prevent environmental degradation shall not be used as a reason for postponing this potentially cost-effective measure" (Comstock 2000).

Reversing the burden of proof?

So the strong version of the PP is untenable. But what about the proposed shifting of the onus of proof towards those who advocate a new technology or activity? Reversing the burden of proof would amount to substituting the maxim 'guilty until proven innocent' for the age-old legal principle 'innocent until proven guilty'. Biotech enthusiasts and anti-regulationists resent this departure from what they consider time-honoured legal sanity (Miller and Conko 2000). They are prone to counter the frequent invocation of the PP with an equally insistent demand of 'sound science'. While one side claims the moral high ground, the other side attempts to seize the scientific high ground.

The critics of the PP assert that the burden that environmentalists and regulators want to impose on the proponents of new technologies tends to be unbearable (Miller and Conko 2000). In the name of absolute safety the latter are asked nothing less than to demonstrate conclusively that the new technologies they advocate offer no possible harm. This is a formidable, perhaps even logically impossible task. You cannot prove a negative (cf. Wildavsky 1995, p. 430). Moreover, a risk-free world is not a real option. Thus a consistent application of the PP would in the final analysis stifle all innovation.

Should we therefore follow the adage 'innocent until proven guilty'? Even in the area of criminal justice we do not use this principle in an absolutist way. We may try to reduce the risk of condemning an innocent person by demanding ever more exacting standards of proof, but only at the expense of increasing the risk of acquitting culpable offenders. So we must recognize that there is an inevitable *trade-off* involved in the design of our system of criminal justice. We may attempt to set our standards as high as we can, but somewhere a balance must be struck, lest the system will become unworkable by making it too difficult to pass sentence on the majority of wrongful offenders. (In statistical testing there is a similar trade-off to be made between the chances of committing a type-I or a type-II error, i.e. rejecting the null hypothesis of 'no effect' when it is in fact true or failing to reject the null hypothesis when in fact it is false. By selecting a significance level we implicitly strike a particular balance. Ideally, this balance should depend on our estimation of the – economic and other – costs associated with either of the two types of error.)

The above analysis shows that the matter at issue is not just where to place 'the' burden of proof. As soon as we allow for more or less exacting *standards* of proof, an extra dimension of variation immediately becomes visible. In other words, the burden we want to put on the shoulders of one or the other party becomes more or less heavy, depending on whether we set our standards of proof more or less high. This consideration may help us to escape from the unduly polarized opposition of PP versus sound science.

In most countries, companies aiming to commercialize GM crops have to submit their products to scrutiny for health effects and environmental impacts. This scrutiny can be more or less searching. The ideal of those who swear by 'sound science' is a fully quantified risk assessment. However, it is only possible to meet this objective in more limited contexts, where direct and short-term hazards such as toxicity or pathogenicity are at issue. Even then the expression 'sound science' is disingenuous, for it obscures the extra-scientific value judgments that necessarily enter into the whole exercise, e.g. identification of hazard types, pathways of exposure, baselines of acceptability, trade-offs between type-I and type-II errors (see also Thompson 2003). In other contexts, where indirect, cumulative or more subtle ecological effects are at issue, the format of the fully quantified risk assessment is unattainable. Adherents of 'sound science' are tempted to downplay such less straightforward hazards as purely hypothetical risks that can safely be ignored. However, as the proponents of the PP are never tired in pointing out, lack of evidence of harm is not evidence of lack of harm. If we are really concerned about such hazards, we can put in additional investigative effort to learn more about their plausibility or likelihood. It would be absurd to halt our inquiries with an appeal to 'sound science'.

The wider context

A recent European directive on the deliberate release of GMOs into the environment lays down that any company that wants to introduce or commercialize a transgenic crop should carry out a 'full' environmental risk assessment taking into account 'direct, indirect, immediate and delayed effects' (European Commission 2001). This new regulation of GM crops goes much further than current US registration requirements, although some American biologists also argue for a more comprehensive approach (Obrycki et al. 2001).

The new European Directive surely places a heavy burden of proof on biotech companies intending to introduce GMOs. Whether or not they are able to take that burden on their shoulders will partly depend on the definition of a standard protocol or methodology for conducting environmental risk assessments. The danger to be avoided is that the obligations imposed on these companies will become 'open-ended', putting them entirely at the mercy of regulatory agencies and NGOs asking for ever-escalating assurances of environmental safety. This suspicion will be enhanced by the fact that the drafting of the Directive has avowedly been informed by the PP and that regulatory authorities may give consent to the introduction of GMOs only after they have been satisfied that the release will be safe for human health and the environment.

The fairly comprehensive scope of the required environmental-risk assessment need not be offensive in itself, if rules of fair play for the regulation of GM crops can be developed. More clarity is also needed about the societal values that have to be taken into account in evaluating risks. The outcome of the assessment is clearly contingent, for instance, on whether or not chemical-intensive methods in agriculture are taken as a normative baseline or whether or not a strong commitment to organic agriculture as a viable option is maintained (Levidow 2001; Myhr and Traavik 2003). The pros and cons of a *Bt* maize hybrid or any other transgenic crop in Europe might be quite different from those in the USA. Europeans are usually strongly attached to farmland, as their countries lack vast tracts of national parks and other 'wilderness' areas (Hails 2002). Indeed, Willy de Greef, head of regulatory affairs at Syngenta Seeds, holds that the debate in Europe on GM foods is not fundamentally about safety, but is in fact a proxy for a larger debate on how farming should be done (Hileman 2001). GM crops have become a symbol for all that Europeans do not like in modern agriculture. John Feffer sees an even deeper 'clash of civilizations' behind the US-EU dispute on GMOs, formulaically described as '*terroir* versus McWorld'. '*Terroir*' stands for the (French) belief that conditions such as soil and weather produce distinctive tastes. "In Europe, people want to know how their food was raised and made. For quality control, they generally trust farmers over biotechnicians" (Feffer 2003).

While the European Union may eventually opt for a combination of organic and conventional farming and a rather limited acceptance of agricultural biotechnology – only insofar as the latter is compatible with a 'co-existence' regime –, the critical question is whether this strategy also makes sense on a global scale. It is not surprising that Greenpeace International offers us the option of organic farming as a worldwide solution. However, several agronomic experts argue that we shall badly need all the various tools of modern biotechnology to feed a growing world population and sustain natural biodiversity (Trewavas 1999; Conway and Toenniessen 1999). The ecologist's response is that thermodynamic considerations make it unlikely that GM plants can actually increase food production and, at the same time, repel pests, resist herbicides and compete with weeds for water and nutrients for any prolonged period of time (Jordan 2002). Although the scientific debate is somewhat technical and esoteric, the stakes are clearly high.

It thus appears that the polarized debate on the PP is just a proxy for a larger debate on the future of world agriculture.

References

Académie des Sciences, 2002. *Les plantes génétiquement modifiées*, Paris. Rapport sur la Science et la Technologie no. 13.

Agrisalon.com, 2003. *Claudie Haigneré et Luc Ferry dénoncent "des attaques inadmissibles contre les scientifiques" (10/02/2003).* [http://www.agrisalon.com/06-actu/article-9352.php]

Attac France, 2003a. *OGM: la nostalgie de la cohabitation. Deux ministres du gouvernement Raffarin et un député socialiste font, au même moment, des déclarations de soutien au lobby pro-OGM (12/02/2003).* [http://france.attac.org/a1804]

Attac France, 2003b. *Quand l'Académie des Sciences vole au secours des industriels et de l'administration Bush (15/01/2003).* [http://france.attac.org/a1701]

Barrett, K., 2000. *Applying the Precautionary Principle to agricultural biotechnology.* Science and Environmental Health Network. [http://www.biotech-info.net/PP_Barrett.pdf]

Beck, U., 1992. *Risk society: towards a new modernity.* Sage, London.

Butler, D., 2003. Ministers back gene-crop advisers. *Nature,* 421 (6925), 775.

Charlier, C. and Rainelli, M., 2002. Hormones, risk management, precaution and protectionism: an analysis of the dispute on hormone-treated beef between the European Union and the United States. *European Journal of Law and Economics,* 14 (2), 83-97.

Comstock, G., 2000. *Are the policy implications of the Precautionary Principle coherent? : viewpoints.* [http://www.cid.harvard.edu/cidbiotech/comments/comments72.htm]

Conway, G. and Toenniessen, G., 1999. Feeding the world in the twenty-first century. *Nature,* 402 (suppl.), C55-C58.

European Commision, 2000. *Communication from the Commission on the Precautionary Principle, Brussels, 02.02.2000, COM (2000) 1,* Brussels. [http://europa.eu.int/comm/dgs/health_consumer/library/pub/pub07_en.pdf]

European Commission, 2001. Directive 2001/18/EC of the European Parliament and of the Council of 12 March 2001 on the deliberate release into the environment of genetically modified organisms and repealing Council Directive 90/220/EEC. *Official Journal of the European Communities,* L106/1 E, 17/4/2001, 1-38. [http://europa.eu.int/eur-lex/pri/en/oj/dat/2001/l_106/l_10620010417en00010038.pdf]

Feffer, J., 2003. Trans-Atlantic food fight: the stakes in the U.S.-Europe battle over genetically engineered crops. *American Prospect,* 14 (5). [http://www.prospect.org/print/V14/5/feffer-j.html]

Frat, M., 2003. Les scientifiques soupçonnés de conflit d'intérêt contre-attaquent. *Le Figaro* (12 February 2003).

Gillis, J., 2003. Suit to seek end of biotech crop ban. *Washington Post* (May 13, 2003), A13. [http://www.washingtonpost.com/wp-dyn/articles/A47197-2003May12.html]

Greenpeace International, 2001. *Safe trade in the 21st century: the Doha edition.* Greenpeace International, Amsterdam. [http://www.greenpeace.org/politics/wto/doha_report.pdf]

Hails, R.S., 2002. Assessing the risks associated with new agricultural practices. *Nature,* 418 (6898), 685-688.

Hervieu, B. and Guillou, M., 2002. Oui aux OGM aux champs. *Libération,* 23 September 2002.

Hileman, B., 2001. Polarization over biotech food. *Chemical and Engineering News,* 79 (21), May 21.

Inf'OGM, 2002. *OGM: Opinion Grossièrement Manipulée?* October 2002. [www.infogm.org]

Jonas, H., 1984. *The imperative of responsibility: in search of an ethics for the technological age.* The University of Chicago Press, Chicago.

Jordan, C.F., 2002. Genetic engineering, the farm crisis, and world hunger. *BioScience,* 52 (6), 523-529.

Kourilsky, P. and Viney, G., 2000. *Le principe de précaution.* Odile Jacob, Paris.

Levidow, L., 2001. Precautionary uncertainty: regulating GM crops in Europe. *Social Studies of Science,* 31 (6), 842-874.

Miller, H.I. and Conko, G., 2000. The science of biotechnology meets the politics of global regulation. *Issues in Science and Technology,* 17 (1), 47-54.

Myhr, A.I. and Traavik, T., 2003. Genetically modified (GM) crops: precautionary science and conflicts of interest. *Journal of Agricultural and Environmental Ethics,* 16 (3), 227-247.

Obrycki, J.J., Losey, J.E., Taylor, O.R., et al., 2001. Transgenic insecticidal corn: beyond insecticidal toxicity to ecological complexity. *BioScience,* 51 (5), 353-361.

Office of the United States Trade Representative USDA, 2003. *U.S. and cooperating countries file WTO case against EU moratorium on biotech foods and crops: EU's illegal, non-science based moratorium harmful to agriculture and the developing world.* 13 May 2003. [http://www.ustr.gov/releases/2003/05/03-31.pdf]

Ortman, E.E., Barry, B.D., Buschman, L.L., et al., 2001. Transgenic insecticidal corn: the agronomic and ecological rationale for its use. *BioScience,* 51 (11), 900-903. [http://www.bioone.org/pdfserv/i0006-3568-051-11-0900.pdf]

Pew Initiative on Food and Biotechnology, 2003. *Should the U.S. press a WTO case against Europe's genetically modified food policies? (Originally broadcast live on February 13, 2003).* [http://www.mindfully.org/GE/2003/US-Press-WTO-GMOs13feb03.htm]

Raffensperger, C. and Tickner, J.A. (eds.), 1999. *Protecting public health and the environment: implementing the Precautionary Principle.* Island Press, Washington DC.

Reuters, 2003. *EU says a US attack over GMOs would be 'eccentric'.* 13 May 2003. [http://www.planetark.com/dailynewsstory.cfm/newsid/20770/story.htm]

Testart, J., 2000. How to let ordinary people in on the future: be careful, take precautions. *Le Monde Diplomatique (English version)* (September 2000).

The Campaign, 2003. *GE foods tutorial: why labeling?* [http://www.thecampaign.org/education/brochurelabels.htm]

Thompson, P.B., 2003. Value judgements and risk comparisons: the case of genetically engineered crops. *Plant Physiology,* 132 (1), 10-16.

Trewavas, A., 1999. Much food, many problems: a new agriculture, combining genetic modification technology with sustainable farming, is our best hope for the future. *Nature,* 402 (6759), 231-232.

Van den Belt, H., 2003. Debating the Precautionary Principle: "guilty until proven innocent" or "innocent until proven guilty"? *Plant Physiology,* 132 (3), 1122-1126.

Vogel, D., 2003. The hare and the tortoise revisited: the new politics of consumer and environmental regulation in Europe. *British Journal of Political Science,* 33 (4), 557-580.

Wildavsky, A., 1995. *But is it true? A citizen's guide to environmental health and safety issues.* Harvard University Press, Cambridge.

12b

Comment on Van den Belt: Biotechnology, the US-EU dispute and the Precautionary Principle

William A. Kerr[#]

The 'Precautionary Principle' has become the focal point of the debate over how genetically modified organisms (GMOs) should be licensed for commercial use and the rules under which they should be traded internationally. The reason for the debate is that while exercising precaution in the face of scientific uncertainty is a generally accepted principle, turning that principle into an operational decision-making mechanism has proved to be extremely difficult and controversial. The result is that what was meant to be an innocuous part of science-based decision making within the Risk Analysis Framework has become politicized. The politicization of the Precautionary Principle has meant, among other things, that a great deal of what has been written about the Precautionary Principle is either advocacy or purposely opaque. The paper by Henk van den Belt is a refreshing departure from advocacy and obfuscation as it lays out in clear, finely crafted arguments the central questions pertaining to the Precautionary Principle as a mechanism for deciding policy. I highly recommend it to anyone interested in biotechnology, the environment, technology management and/or science policy.

The debate over the Precautionary Principle is often couched in terms of a conflict between the European Union's approach to science policy – and biotechnology in particular – and that of the United States. Van den Belt is careful to point out this fallacy. He chooses to examine the acrimonious debate between anti-GMO activists and the French scientific establishment over conducting field trials of genetically modified crops by way of illustration. France is often portrayed as the major advocate of the Precautionary Principle in the EU, yet even within France the scientific establishment's interpretation of the Precautionary Principle is greatly at odds with that of anti-GMO advocates. The contention arises because the Precautionary Principle was enshrined in a number of multilateral environmental agreements and the World Trade Organization (WTO) as well as EU legislation on food safety and the environment before its operational procedures had been decided. While the views on how the Precautionary Principle should be operationalized are extremely diverse, two have become the focal point of the debate – decisions by 'scientific experts' versus 'zero risk' (Van den Belt's strong version of the Precautionary Principle). Anti-GMO activists have grasped the latter and made it their mantra because they believe that it can be used to prevent any further use and development of agricultural biotechnology – zero risk cannot ever be scientifically proved.

One major contribution of Van den Belt's paper is to show why the 'zero risk' interpretation of the precautionary principle is logically untenable. This is done by showing that the idea of zero risk always cuts both ways, meaning that there is always a 'zero risk' argument that would support the further development of biotechnology –

[#] Van Vliet Professor, University of Saskatchewan, Saskatoon, SK, Canada

J. H. H. Wesseler (ed.), Environmental Costs and Benefits of Transgenic Crops, 199–201.

including a wonderful analogy to Pascal's untenable wager argument regarding the worship of God. Equally important as a contribution is his pointing out that 'reversing the burden of proof' arguments often put forward by those wishing to introduce bias into the use of the Precautionary Principle is a 'red herring' or non-argument because it is the setting of standards that is the central element in decision-making, not who bears the burden of proof.

One area that Van den Belt does not delve into is what is at the heart of the trade dispute between the EU and the US over the EU's invocation of the Precautionary Principle. If the EU were to allow deference to scientific expertise to form the basis of its decision making under the Precautionary Principle, the US would likely accept it. The EU, however, has chosen to allow other, non-scientific, considerations to inform its choices explicitly – what Isaac (2002) calls a 'social rationality' basis for operationalizing the Risk Analysis Framework. In particular, the EU Commission's communication on the Precautionary Principle (European Commision 2000) specifically states that when the Precautionary Principle is invoked, subsequent decisions will have a political element and that socio-economic considerations can be part of the decision process. For a trade economist, allowing these two criteria raises 'red flags'.

The reason why the scientific-principle approach was included in the WTO's Agreement on Sanitary and Phytosanitary Measures (SPS) was to prevent these regulations from being used as unfair barriers to trade. While the intent of having the decision being a political one may be the belief that making difficult trade-offs under scientific uncertainty is a politicians' role, explicitly allowing for it also allows politicians to use it to extend protection when asked for it by political constituents – negating the intent of the SPS. The same can be said of socio-economic considerations – which could be interpreted in a protectionist way as, for example, "some EU farmers losing from import of genetically modified foods". Suspicion in the US runs high due to the willingness of the EU to ignore its scientific experts in the case of banning the import of beef produced using growth hormones (Kerr and Hobbs 2002). In the case of beef produced using growth hormones and GMOs, however, there is also an institutional failure at the WTO. Unlike the case of traditional producer protectionism, there is no way in the WTO to accommodate the politicians' need, at times, to extend protection when it is demanded by consumers or other groups in civil society such as environmentalists (Perdikis, Kerr and Hobbs 2001). Until this question is addressed directly at the WTO, politicians and trade policy makers will be tempted to try and find ways to extend protection within the SPS, including how the Precautionary Principle is operationalized.

References

European Commision, 2000. *Communication from the Commission on the Precautionary Principle, Brussels, 02.02.2000, COM (2000) 1*, Brussels. [http://europa.eu.int/comm/dgs/health_consumer/library/pub/pub07_en.pdf]

Isaac, G.E., 2002. *Agricultural biotechnology and transatlantic trade: regulatory barriers to GM crops*. CABI Publishing, Wallingford.

Kerr, W.A. and Hobbs, J.E., 2002. The North American-European Union dispute over beef produced using growth hormones: a major test for the new international trade regime. *The World Economy,* 25 (2), 283-296.

Perdikis, N., Kerr, W.A. and Hobbs, J.E., 2001. Reforming the WTO to defuse potential trade conflicts in genetically modified goods. *The World Economy,* 24 (3), 379-398.

13a

Do patent-style intellectual property rights on transgenic crops harm the environment?

Timo Goeschl[#]

Abstract

This paper examines the linkages between the system that society uses to incentivize R&D by private innovators in the area of crop improvement on the one hand and the environment on the other. This examination is an important addition to the technology-assessment exercise conducted in the context of transgenic crops since it focuses on the organization of the R&D process rather than on the outputs. The paper first demonstrates that design choices with respect to the system of rewards under which crop improvement is carried out determine important characteristics of R&D outputs. In particular, it shows that choosing a patent-style system of intellectual property rights (IPR) will impact on the rate, direction, pace and mode of technological change in the agricultural system. This is relevant in an environmental context because the R&D outputs thus generated interact with biological systems. Specific production and adoption characteristics of these outputs therefore matter in environmental terms. While the presence of these environmental impacts is a generic characteristic of carrying out crop R&D under patent-style IPRs, the extent of these deviations differs between conventional and transgenic crops and is determined by a number of biological, technological and legal key determinants. A comparison of the differences in these key determinants between conventional and transgenic crops shows that there are some areas in which there is no difference between conventional and transgenic crops, in particular with respect to the mode and direction of technological progress. In those areas where we find differences, the differential environmental impact of moving from conventional to transgenic crops is ambiguous.
Keywords: R&D; intellectual property rights; patents; transgenic crops; environmental impacts

Introduction

There is a sizeable and expanding literature that examines the direct effects of releasing genetically modified crops into the environment (for a survey, see Conner, Glare and Nap 2003). This literature concerns the possibility, probability and consequential environmental harm of such a release and forms an integral part of the process of technology assessment in agriculture. In this assessment process, the products of the societal research and development (R&D) process are thus to be subjected to due scrutiny.

[#] Department of Agricultural and Applied Economics, University of Wisconsin-Madison, Madison WI 53706, USA. E-mail: goeschl@aae.wisc.edu

J. H. H. Wesseler (ed.), Environmental Costs and Benefits of Transgenic Crops, 203–217.

The starting point for this paper is that the literature on the environmental impacts of transgenic crops may answer the question of the impact of specific R&D *outputs*, but not the more fundamental question of the impact of the R&D *process*. While the former is a question of how to manage a given technology, the latter is a question of how society should organize the biotechnological research process of which the specific technologies are an outcome. The organization of the R&D process that delivers crop improvements can be analysed from a number of perspectives such as sociology (Buttel 1999; Busch et al. 1991) and history (Ruttan 2001; Palladino 1996). For economists, an essential determinant of this organization is effected through the assignment of property rights to different actors at different stages of the R&D process (Swanson and Goeschl 2000). These property rights take the form of residual ownership over various inputs and outputs of the R&D process. The focus of this paper is a specific subset of property rights, namely those defined over the intangible asset of information that are the essential inputs and outputs of the crop improvement process. These property rights are commonly referred to as 'intellectual property rights' (IPR).

Why is the specific nature of IPR and structure of their ownership of relevance to the environment? There are two principal reasons. The first is that biotechnologies are endogenous in the sense that the organization of the biotechnological research process determines important aspects of their nature and shape. As we will show, organizing an R&D process under the specific reward system of IPR affects the volume and nature of the R&D outputs pursued. Since the R&D outputs at the centre of this paper are crop plants, these aspects are of direct environmental relevance. The ecological characteristics of crops determine their interaction with the biological environment of the agricultural system, while their production characteristics determine – at the margin – the returns to alternative uses of land and hence the relative allocation of intensive production, extensive production, and land outside agricultural usage. Systems of ownership over R&D outputs are therefore of direct environmental relevance. However, the organization of the R&D process in society determines ownership not only over R&D outputs, but also over the inputs into the R&D process. In the case of crops, one essential R&D input is genetic resources. These genetic resources need to enter into the R&D process on a continuous basis emanating from active agro-ecological system (Swanson 1999; Holden, Peacock and Williams 1993). The nature of how property rights are assigned over these R&D inputs is the second reason why society's choice of how to organize the R&D process has environmental relevance. In an abstract sense therefore, this paper examines the nature and impact of the linkages between the institutions that society chooses to incentivize agricultural R&D and the natural environment. The specific angle from which we will examine the environmental impacts will be land use patterns as the main determinant of environmental quality in agro-ecological systems.

To illustrate the environmental relevance of the organization of the R&D process, Figure 1 below gives a simple schematic representation of the R&D process and its linkage with land-use patterns. Essential informational inputs relevant for the biotechnological R&D process are generated in biodiverse systems that we will repeatedly refer to as 'reserves' to emphasize their conservation function. These systems are rich in genetic resources and are generally found in non-converted areas or areas of low-intensity production. The main form through which informational inputs arise in biodiverse systems is through the interaction of a diverse set of genetic resources with specific ecological conditions that vary through time and that generate information about currently successful ecological strategies such as resistance to a current pest that is used to produce more productive R&D outputs such as a virus-resistant crop variety (Swanson 1999). The information so produced is used in the R&D process to enhance existing cultivars. These cultivars form the output of the R&D process and are then applied in intensive production systems. The relationship between R&D inputs and outputs is therefore characterized by both complementarity and competition: in the R&D process, there is a reliance of new cultivars on available genetic resources while, in terms of land use, genetic resources and new cultivars compete for land resources. The challenge is to design and assign property rights over these inputs and outputs in a way that leads the R&D process to effect the correct land-use structure. The question regarding the environmental impact of intellectual property rights on crops can then be framed as one regarding the impact of assigning such rights over R&D outputs on key characteristics of biodiverse and intensive systems (such as relative size), and this is the angle from which we will approach the question in this paper.

The paper has two parts. The first examines to what extent IPR – as the preferred incentive regime for agricultural R&D – are a causal factor in inducing land-use patterns that deviate from what would be first-best from society's point of view. The conclusions of this first part are generic rather than technology-specific, and highlight the welfare loss relative to a putative (i.e. non-existent) first-best system. They point to quite fundamental questions about the adequacy of a conventional IPR system in managing biotechnologies that are fully developed elsewhere (Goeschl and Swanson 2003c).

The second part analyses to what extent the trend towards genetic modification as the preferred technology in agricultural R&D exacerbates or reduces the deviation from the optimum. In examining the second question, it is important to declare the baseline against which the comparison is made, and for this purpose the most appropriate choice must be conventional forms of breeding. The analysis shows that the net environmental impact of a trend from conventional towards transgenic crops is ambiguous since it is determined by complex interactions between technological, biological and economic parameters pulling in different directions.

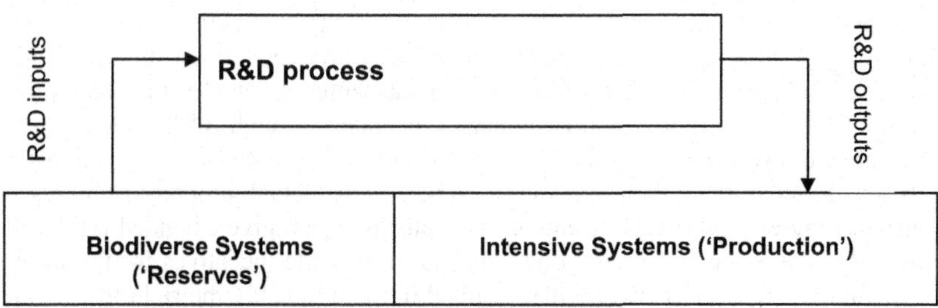

Figure 1. Schematic view of R&D process in land-use terms

To develop the argument, this paper first describes the relationship between intellectual property rights and the organization of the agricultural R&D process, and how IPR impact on the various dimensions of technological progress such as rate, pace, direction and mode. In the next section, the paper summarizes the results of some recent work that sheds light on how an R&D process conducted under IPR impacts on the environment. The following section then examines how a shift from traditional to transgenic forms of plant breeding modifies these impacts. The last section concludes.

IPR and the R&D process in plant breeding

Intellectual property rights

In order to examine the relationship between IPR and the R&D process, some background in the economics of property rights in general is required. Economists understand their paramount function in the economy to consist of directing economic activities and the allocation of goods and services. They do so by assigning ownership in certain assets in the economy to individual economic agents. These rights are usually assigned through a legislative process that is coupled with a judicial process in which disputes over rights can be settled. Ownership of property rights, thus secured, is generally regarded as "the most common and effective institution for providing people with incentives to create, maintain, and improve assets" (Milgrom and Roberts 1992, p. 288).

Property rights can be defined over both tangible and intangible assets. Arguably the most important type of property rights over intangible assets are intellectual property rights. This regime is often used to allocate ownership in industries that are focused on the production of useful information through a process of R&D. When R&D is a significant part of the production process within an industry, it is not always possible to obtain a reasonable rate of return on the product without an extended right of control over its subsequent use and marketing. This is because the end result of the R&D process is an idea, and this idea is then embodied in the products in which it is sold, and potentially lost on first sale. The software industry is a typical example: a computer program that balances a bank statement is first an idea, and then a specific list of computer instructions created to effect that idea. If there is no exclusive right to control the subsequent marketing of the good (or close facsimiles thereof), then the first purchaser of that good would have the right to produce competitive products without expending all of the R&D resources required to produce it initially. The first sale of the computer code would enable the purchaser to make a similar program and set up in competition with the first. This is problematic if the first seller invested years in the construction of the program while the second only invested the few minutes (and dollars) required to copy it. In industries in which a substantial amount of the value produced is attributable to the information it contains (generated through R&D), there would be no incentive to invest in this R&D in the absence of the capacity to control the marketing of its goods even after their transfer to others. Intellectual property-right regimes are analysed by economists as incentive mechanisms which give extended rights of control over the marketing of certain goods in order to provide incentives for the information-generating investments (R&D) that resulted in them, notably the right to exclude others from their use (Arrow 1962; Swanson 1995).

There is a great variety of rights denominated 'intellectual property': trade marks, copyrights, patents, plant variety rights etc., which differ in the strength of the right to exclude. The most important thing that all of these rights have in common is that they allow the holder to control some of the uses of the good subject to these rights *even after the good has left the rightholder's possession.* Thus, a person with a copyright on a book is able to sell the book but retains the exclusive right to copy it. A person with a patent on a machine is able to sell the machine while retaining the exclusive right to manufacture it. A person with a registered plant variety certificate is able to sell that plant while retaining the exclusive right to reproduce it for re-sale.

The function of this extended right of control is to vest the holder with an exclusive marketing right in the particular good, usually for a limited period of years. This allows the holder to obtain a reasonable rate of return on the book, machine, plant variety or other good that is subject to the recognized right. Note that this rate of return is only available to the extent to which users recognize and enforce this right after the good has already left the possession of the rightholder. To the extent that the other users are willing to purchase from prior purchasers, the rightholder's exclusive marketing right will be of little value. There is a substantial increase in the rate of return afforded by allowing rightholders to control the uses of their rights outside of their possession. Various studies have substantiated the investment-stimulating effect the introduction of IPR regimes has had on R&D intensive industries, in particular plant breeding (for a survey, see Fuglie et al. 1996). In the following section, we examine how IPR are assigned in the R&D process that generates crop improvements.

The vertical industry of plant breeding

In Figure 1, the R&D process appears as a 'black box' into which R&D inputs enter and from which R&D outputs leave. Here we 'open' this 'box' to understand the various actors involved in the various stages of the R&D process and how IPR are assigned along this vertical chain. Swanson and Goeschl (2000) present a schematic view of the agricultural R&D process that highlights the vertical structure of the biotechnological R&D industry and is reproduced in figure 2.

At its base, effective characteristics for new plant varieties develop naturally through the process of 'natural selection': only those which are able to survive existing threats (pests and environmental changes) remain and reproduce. Since the set of threats is constantly changing, the natural environment continuously produces new information on the characteristics that are relatively fit under current conditions. The maintenance of a relatively greater diversity of genetic resources and the dedication of greater amounts of lands to the retention of that diversity are the investment choices that determine the amount of information flowing out of this stage of the industry on the nature of the plants that work effectively in the prevailing environments.

The next stage of the industry consists of the individuals who observe the natural process of selection and aid in the dissemination of its information. 'Traditional farmers' have themselves survived by means of a process of observing this naturally produced information and the disproportionate use and transport of those plant characteristics which have aided survivability. They invest in the production of this information both by means of their land-use decisions (as mentioned above) and by dedicating their time and resources to the observation and discriminatory use of those genetic resources which are revealed by nature to be of greater fitness. Their choices each year result in the capture of some of the flow of information on what was successful in the environment prevailing in the current year. This information also accumulates as a 'stock': traditional plant varieties (landraces) encapsulate the accumulated history of the information that nature has generated and that farmers have observed and used disproportionately (Swanson 1999).

At the end of this process, the 'plant-breeding industry' has collected the set of varieties that farmers have created over millennia and hence the stock of naturally-produced information that is encapsulated within them. By investing in laboratory equipment and scientists, the breeding process becomes focused on the use of this set of information for the preparation of the best possible variety for current environmental conditions. The modern plant breeder has then used its investments to create a variety that is an amalgam of some subset of the traditional varieties. We now proceed to characterize the current structure of assignment of IPR across this vertical industry.

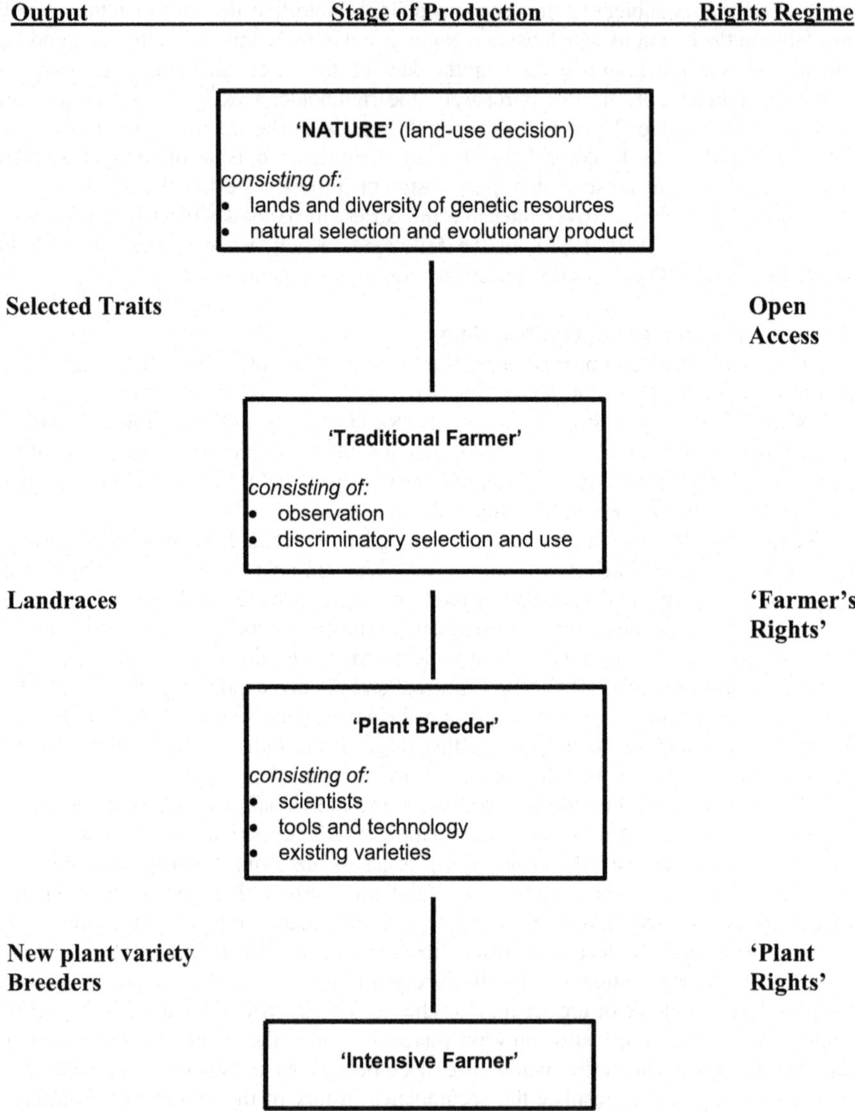

Figure 2. The R&D process of plant breeding (source: Swanson and Goeschl 2000)

Intellectual property rights in crop R&D

One important characteristic of the vertical industry of plant breeding is that it involves the flow of information between several stages of the R&D process. Between each stage, information is exchanged in a particular form between different agents: land-use decisions of landowners allow the appropriation of information about successful ecological strategies by any observer. Farmers engaged in extensive production generate sets of information embodied in landraces. These are subsequently used by plant breeders to incorporate value-adding traits into cultivars. Finally, farmers use the information embodied in new cultivars in the final production process. From an economic point of view, the question is how to ensure that the exchange of information along this vertical industry is carried out efficiently by adequately distributing rents to the different agents across the various R&D stages (Swanson and Goeschl 2000).

IPR can be thought of as the institution created to enable the voluntary exchange of information to occur efficiently just in the same way as property rights in general enable the exchange of conventional goods from agent to agent. What IPR govern the exchange of information along the sequence of R&D stages in plant breeding? Swanson and Goeschl (2000) provide a detailed analysis of the IPR structure. Their key observation is that IPR in the crop R&D process are instituted in an asymmetric fashion. Only one stage of the industry is invested with IPR, namely the retail end of the vertical industry where R&D outputs are marketed. Even though there are some recent policy initiatives to create so called 'farmers' rights' and there are some nascent national IPR systems governing biodiversity (for example in the Philippines), so far prior stages of the R&D process, in particular those involved in producing R&D inputs, do not benefit from a formally instituted and enforceable system of IPR protection.

Asymmetry in property-rights allocation along a vertical industry is not in itself an indicator of inefficient institutional design as Coase (1952) has demonstrated. In the presence of transaction costs, however, the nature of the asymmetry is of critical importance for the overall efficiency of the choice of institutions. Here we merely note the presence of this asymmetry and refer the reader to Swanson and Goeschl (2000) for a full discussion of whether the implicit differentiation of property-rights protection to different forms of land use are a potential source of inefficiency.

Impact of an intellectual property-rights regime: the nature of R&D

To complete our analysis of IPR and crop R&D, we need to examine in greater detail the impact the use of an IPR at the retail end of the vertical industry has on the R&D activity at this stage. This IPR hands the plant breeder the right to exclude others for a specific time period from the use of the information embodied in the cultivar either for the purpose of further R&D (unless allowed through a so-called 'plant breeders' exemption) or for the purpose of use in final production (unless allowed through a so-called 'farmer's privilege'). The intended consequence of IPR is to allow the creation of a temporary monopoly for the innovator and to reward the innovative step through the collection of monopoly rents.

What does the presence of such a reward system for plant breeders imply in terms of R&D pursued? Even though there is an extensive literature on the effects of IPR on R&D in general (e.g. Jaffe and Trajtenberg 2002; Merges and Nelson 1990; Nordhaus 1969), in order to answer this question in the context of crops, it is necessary to consider a fundamental distinction between the agricultural industry and other sectors (Goeschl and Swanson 2003c). This is caused by the fact that agricultural innovations such as new cultivars are threatened with obsolescence not only on account of better innovations entering the market, but also on account of pathogens present in the agricultural system adapting to new cultivars and causing a breakdown of their resistance. The significance of the pathogen problem in crop improvement can be gleaned from the observation that traits conferring virus (40%) and insect resistance (37%) account for 77 percent of the area sown with genetically modified crops (Nap et al. 2003). Pathogen evolution and consequential crop loss is therefore a major challenge to agricultural R&D (Oerke et al. 1994). How do firms operating under a patent system respond to the challenge implied by the simultaneous presence of two contests, a commercial one against other firms and a technological one against biological competitors?

We have developed the analysis of the impact of IPR in life-science R&D in a number of papers (Goeschl and Swanson 2002; 2003a; 2003c; 2003b). Lack of space permits little more than a summary of the four key conclusions from these papers to the extent that they reflect on the environmental impacts of organizing the R&D process under patent-style IPR. These key conclusions concern the rate of R&D, the direction of technological change, the pace of technological change, and the mode of technological change in the agro-environmental system.

Impact 1: IPR as determinant of the rate of technological change

The socially optimal scale of investment in R&D in the biotechnology sector balances the increased benefits from increased innovation (from increased R&D) and from reduced adaptation (from reduced scale of production using the current technology) against the cost of foregone production. Private firms invest in R&D in order to increase the output of private innovation (Goeschl and Swanson 2003a). Private R&D therefore decreases with the severity of the adaptation problem while the socially optimal amount of R&D increases.

Impact 2: IPR as determinant of the direction of technological change

The social optimum implies investment in technologies that decrease the rate of biological adaptation. A typical example is R&D into the optimal design and scale of ecological buffer zones (such as refuge areas). Analytical results indicate that IPR-incentivized firms have little incentive to invest in R&D for the purpose of reducing the rate of adaptations since the benefits of investing in mitigation technologies dissipate across the industry (Goeschl and Swanson 2003c).

Impact 3: IPR as determinant of the pace of technological change

Biological systems respond to variations in the 'step size' of innovations, even though the direction and extent of this response is not very well understood so far. While society will vary the size of innovations and hence the pace of technological change in accordance with the response of the biological system to the size of innovations, industry choice of the pace of technological change will be invariant to the nature of biological response (Goeschl and Swanson 2003b).

Impact 4: IPR as determinant of the mode of technological change

While society would prefer a cumulative adoption of new technologies, private industry operating under an IPR regime prefers sequential adoption of new technologies. The IPR-incentivized firm does not consider the positive impacts of its R&D with regard to a) the social gains that would be received by reason of its own innovations that occur within an existing patent's life; b) the social gains that would be received by reason of reduced levels of adaptations from the introduction of an additional technology by itself or another firm (reducing the scale of application of other technologies) (Goeschl and Swanson 2002).

These four key conclusions highlight that the application of patent-style IPR to incentivize research and development in crop improvement generates a number of deviations from what economists refer to as the 'social optimum'. In part, the presence of such deviations is not surprising. It is well known that patents are strictly second-best instruments to resolving the problem of knowledge production. However, the four impacts noted above point to additional problems that arise in the use of patent-style IPR in the specific domain of crop improvement.

The environmental consequences of IPR in crop R&D

What is the environmental relevance of the problems of using IPR in the agricultural R&D process? This question requires an analysis of how the use of IPR to incentivise agricultural R&D impacts on land-use patterns and the management of the environment.

Environmental consequences of IPR problems

Impact 1 states that agricultural R&D in crops conducted under a patent-type system of rewards will result in insufficient investment in R&D on account of the negative impact that the presence of evolving pests and pathogens has on the expected returns on R&D investment. As a result of the suboptimal R&D effort by industry, demand for inputs into the R&D process will not reach the level expected under first-best. The market failure on the R&D output market therefore spills over into the input market. Brown and Swierzbinski (1988) show in a static model that this type of spill-over leads to an undervaluation of biodiverse resources on the market and consequently to lower returns on forms of land use that promote the conservation of such resources. Goeschl and Swanson (2003a) show that this effect holds to an even greater extent under dynamic considerations and that incentives to convert land to intensive production above what would be optimal persist over time. These deviations from optimal land-use patterns on account of insufficient demand for genetic resources generated by the private R&D sector therefore pose the first set of environmental problems associated with the use of IPR in crop development.

The second conclusion from the formal analysis of the R&D problem points to the direction of technological change. The models indicate that firms have no incentive to invest in technological trajectories that lead to lower rates of adaptation in pests and pathogens. The reason is that the benefits of pursuing these trajectories have strong positive externalities to other firms in the industry (on account of increasing their expected patent rent) and a first-order negative effect on the innovating company since its own expected patent rent is reduced on account of higher R&D investments by its competitors. From society's perspective, however, such technological trajectories are highly desirable[1]. The failure by private firms to pursue them leads – again – to lower R&D investment than would be optimal, further reducing the demand for R&D inputs and driving the management of the agro-ecological system away from technologies that manage evolutionary dynamics.

Not only does R&D conducted under patent-style IPRs lead to the pursuit of sub-optimal R&D trajectories, it also impacts on the pace of technological change and hence on the evolutionary pressure that is exercised on the agro-ecological system. The pace of technological change in dynamic models of agricultural R&D is captured in the step size of innovations, in other words the degree of novelty introduced by subsequent technological vintages into the agro-ecological system. Although the ecology of host plants and their pathogens does not give a clear indication about the impact of varying the pace of technological progress on the evolutionary response of the biological system (for an example, see the paper by Schubert et al. in this volume), the economic analysis of the problem demonstrates that whatever the response is, society will prefer to vary the pace inversely to the dynamics thereby induced. If a higher pace of technological progress increases the speed of the evolutionary response of the system, a reduction in the pace is the socially desired response and vice versa. Under some fairly general conditions it can be shown that an industry operating under an IPR system will – by contrast – be invariant to the evolutionary response of the agro-ecological system to the chosen pace of technological change (Goeschl and Swanson 2003b). This implies a suboptimal management of the ecology of agricultural systems and a consequential welfare loss.

The last linkage posited by the analysis of the process of technological change under IPR is the mode of technological change. Against the background of the evolving nature of the agro-ecological system of hosts and pathogens, there are social gains from a coexistence of various technologies at any point in time in order to limit the evolutionary pressure on the system. The degree to which this coexistence is desirable depends on a number of factors, most importantly the instantaneous productivity loss from not using the first-best technology uniformly. On the other hand, the gains from a diversified portfolio of production technologies are not realized in an IPR system. The reason is that the rewards for successful innovation are mediated through a particular type of market structure, namely that of a monopoly. As previous research on the economics of industrial organization has demonstrated, deviations from this market structure towards those involving market sharing cannot be reconciled with the reward system's incentive function (Gilbert and Newbery 1982). Technological progress is therefore incentivized through the award of sequential monopolies. This implies that the predominant mode of technological progress is one of a non-diversified application of a single technology rather than an accumulation and simultaneous use of different technologies (Goeschl and Swanson 2002). Land-use patterns will reflect this mode of technological progress through a prevalence of monocultural applications in the productive sector.

In sum, recent research on the linkages between patent-style reward systems and the environment points towards a number of problem areas. Firstly, the volume and type of R&D outputs generated will usually deviate from the R&D outputs society would want to generate under first-best conditions. Environmentally problematic are the effects that the specific rate, direction, pace and mode of technological progress embodied in new crop plants have on the agro-ecological system with which these crops interact. Patent-style IPR systems imply a tendency on the one hand to carry out less R&D than would be optimal and on the other hand to apply R&D outputs in a way that does not optimally manage the evolutionary dynamics of the agro-ecological system. Secondly, the organization of the R&D process through patent-style IPR impacts negatively on the demand for biodiversity from outside the intensive-production sector. One would expect this to lead to lower returns on conservation of R&D inputs and hence to a reduction in preservation activity. The R&D outputs pursued under patent-style IPR and their impact on R&D inputs are therefore the key areas of concern over the environmental impact of patent-style IPR.

Determinants of the deviation from the social optimum

The previous section attempted to demonstrate the presence of deviations from a social optimum in the case of biotechnological R&D under patent-style IPR. This discussion has left open the factors that determine the extent of environmental problems thus generated. The literature examines a number of biological, technological and legal determinants that impact on the extent to which R&D activities will deviate from the social optimum. The biological determinants focus on the response of the agro-ecological system of hosts and pathogens. The first is the exogenous adaptation rate of pathogens. This measures the extent to which pathogens have solutions available to the innovations contained in new cultivars. The source of these solutions is either the presence of successful counterstrategies in the genetic pool of the pathogens (Munro 1997) or the generation of new counterstrategies through mutative processes (Weitzman 2000). The second determinant is the induced evolution function, namely the extent to which evolving adaptations spread through the pathogen population as a result of the scale of application of a new crop and as a result of the pace of technological change embodied. While the positive impact of the scale of application on the rate of adaptation is a well-established fact in the ecological literature, the impact of the pace of technological change on the rate of response is less well understood and more speculative.

The technological determinants focus on the characteristics of the R&D process. There, the first feature is the so-called 'hit rate' in the R&D process. This is a measure of the probability of a research success such as a lead for a new molecular entity that actually results in a final product approved for sale. In some areas, this hit rate has a very precise interpretation such as the ratio of successful leads to screens in pharmaceutical research (Artuso 1994), but in general, it denotes the quality of the search process for new solutions and is thus an indicator of the underlying knowledge base in the R&D process (Rausser and Small 2000). The second feature is the innovation function, which is a traditional knowledge-production function that transforms measures of R&D inputs (such as the volume of genetic resources processed, the amount of labour involved, etc.) into a probability of generating a new product. Various studies have analysed the marginal factor productivity of different inputs into this knowledge-production function in the plant-breeding sector (for example Evenson 1995). The innovation function governing this process can therefore be based on well-defined and quantifiable characteristics.

The legal determinant is the extent to which the IPR system allows the innovator to exclude others from using the innovation, once in place. This exclusion translates into the rent appropriability of the R&D process and is determined by design choices in the IPR systems such as the breadth or length of time over which exclusion can be exercised. The extent to which these design choices impact on the long-run incentives for R&D depends critically on the time horizon of the analysis (see O'Donoghue 1998), but at least in the short run, stronger rights to exclude increase R&D incentives. In the next section we analyse the differential impacts of transgenic crops by reference to the changes in the determinants of the environmental effects of R&D outputs that a trend towards transgenic technologies entails.

The differential impact of patents on transgenic crops

The environmental impacts described in the previous section are generic in the sense that any agricultural R&D process directed at crops and conducted under a set of conventional IPR will generate these types of impacts. These impacts therefore highlight the linkages that exists between the choices of an incentive system to reward R&D activity, but they also underline that these linkages are by no means specific to transgenic crops. At the same time, differences in the productivity of transgenic R&D, the productivity of R&D outputs embodying transgenes, their treatment in IPR law and various other changes in the technological characteristics of crop development will interact with the particular system of R&D rewards implied by IPR. What shape this interaction between the biotechnological trend to transgenic crops and the IPR regime under which these technologies are regulated will take remains to be seen in full, but a number of plausible developments will be charted below.

Focussing on the land-use impacts of patents on transgenic crops, the crucial question is how differences in the determinants of R&D between conventional crops and transgenic crops impact on the marginal returns to land in agricultural production and the changes on returns to land in conservation. This is critical since the marginal returns to different forms of land use are a major determinant of land-use patterns. Table 1 sets out the plausible impacts of the expected changes in key determinants of the R&D process on the rents generated in the intensive and the reserve sector and on the welfare loss associated with using an IPR regime (relative to a first-best reward system under perfect information).

Table 1 illustrates that in terms of land use, the expected differential impact of transgenic crops is the net impact implied by the expected differences in the key determinants of the R&D process. This net impact is highly ambiguous: while improvements in rent appropriability are expected to give additional incentives for preservation and to decrease the net rent to land in intensive use (thus increasing the optimal reserve size), other changes can be expected to shift the optimal combination of intensive and reserve use in the opposite direction by virtue of increasing returns on intensive use (such as increased productivity) and decreasing returns on preservation. This means that in terms of land-use impacts, the question of whether harm will increase is mostly an empirical one and cannot be answered merely on the basis of analytical deduction.

Table 1. Plausible differential impact of transgenic technologies in agricultural crops

Determinant	*Expected difference of transgenic to conventional crop*	*Economic impact*	*Land-use rent impact*	*Differential welfare loss*
Adaptation rate	None	None	None	None
Induced evolution function	Ambiguous	Lower or higher returns to R&D at industry level possible	Lower returns on intensive use	Increased
Hit rate in R&D	Higher	Higher demand for R&D inputs	Higher return on conservation	Reduced
Innovation function	Higher marginal rate of innovation	*First order*: higher demand for R&D inputs *Second order*: greater innovation size	*First order:* Higher return on conservation *Second order:* Higher return on intensive use	*First order*: Reduced *Second order*: Increased
Rent appropriability	Higher by virtue of better legal and technological opportunities to exclude	Higher returns on R&D output and higher demand for R&D inputs	Lower return on intensive use and higher return on conservation	Reduced

In terms of the management of the agro-ecological system, the trend towards transgenic technologies will not impact on the direction of technological change since genetic modification does not alter the fundamental lack of incentives to firms to generate technologies that address the adaptation dynamics of pathogens. Likewise, we would not expect an impact of the availability of transgenic manipulation on the mode of technological change: the fundamental incentives remain to phase technologies in the form of sequentially uniform applications rather than the accumulation of a diversified portfolio of differentiated technologies.

In sum, therefore, the expected impact of moving towards transgenic crops in terms of the environmental impact of patent-style IPR is ambiguous. There are a number of factors that point to transgenic crops actually decreasing some of the environmental impacts experienced under the combination of IPR and conventional breeding techniques such as the new technology's ability to increase rent appropriability and higher productivity of R&D. However, these effects are modulated through an asymmetric system of property rights (see section Intellectual property rights in crop R&D) and the net effect of the differential impacts is not obvious. Negative impacts can be postulated in terms of managing the adaptive dynamics of pathogen population and if transgenic crops induce a shift in the balance of the extensive and intensive margins in land use towards additional conversion of lands to agriculture. However, there are also significant areas such as the direction and mode of technological change that are unlikely to be affected by a trend towards transgenic crops.

Conclusions

This paper has examined the linkages between the system that society uses to incentivize R&D by private innovators in the particular area of crop improvement on the one hand and the environment on the other. This examination is an important addition to the technology-assessment exercise conducted in the context of transgenic crops since it focuses on the organization of the R&D process rather than on its outputs. The paper first demonstrates that design choices with respect to the system of rewards under which crop improvement is carried out determine important characteristics of the outputs of the R&D process, namely both volume and type of R&D outputs pursued. The paper shows that choosing a particular type of reward system has implications for the rate, direction, pace and mode of technological change in the agricultural system. It draws on ongoing research that has shown that, in general, applying such a system of rewards to R&D processes the outputs of which interact with evolving ecological systems, implies deviations from the social optimum along several dimensions of technological change.

While the presence of these deviations from the social optimum is a generic characteristic of carrying out R&D under patent-style IPR, the extent of these deviations differs between conventional and transgenic crops and is determined by the nature of some key determinants of a biological, technological and legal nature. A comparison of the differences in these key determinants between conventional and transgenic crops shows that there are some areas in which there is no difference between conventional and transgenic crops, in particular with respect to the mode and direction of technological progress. In those areas where we do find differences, the differential environmental impact of moving from conventional to transgenic crops is ambiguous.

These results have two major implications: The generic results highlight the need to re-evaluate the use of patent-style IPR in the area of crop development in general, and not just in the context of transgenic crops. The public perception in Europe that transgenic crops represent a discontinuity in the technological trajectory may offer a political opportunity to carry out such a re-assessment and to consider whether alternative forms of rewarding R&D activity in the domain of crop improvement might be preferable. Whether superior systems are available is not obvious, but certainly merits additional investigation. The specific results of the paper offer guidance in the differential evaluation of transgenic crops relative to conventional outputs of the R&D process. It will have to depend on the empirical results of this evaluation whether it is concluded that the combination of patent-style IPR and transgenic crops results in outcomes that are clearly undesirable from a social perspective. On the basis of theoretical analysis alone, this evaluation appears ambiguous and therefore ultimately inconclusive.

References

Arrow, K., 1962. Economic welfare and the allocation of resources for invention. *In:* Nelson, R. ed. *The rate and direction of inventive activity*. Harvard Business School Press, Cambridge.

Artuso, A., 1994. *An economic and policy analysis of biochemical prospecting*. PhD dissertation, Department of Economics, Cornell University, Ithaca.

Brown, G. and Swierzbinski, J., 1988. Optimal genetic resources in the context of asymmetric public goods. *In:* Smith, V.K. ed. *Environmental resources and applied welfare economics*. Resources for the Future RFF, Washington, 293-312.

Busch, L., Lacy, W.B., Burkhardt, J., et al., 1991. *Plants, power, and profit*. Blackwell, Oxford.

Buttel, F., 1999. Agricultural biotechnology: its recent evolution and implications for agrofood political economy. *Sociological Research Online,* 4 (3), 292-309.

Coase, R., 1952. The nature of the firm. *In:* Stigler, G. and Boulding, K. eds. *Readings in price theory*. RD Irwin, Chicago, 331-351.

Conner, A.J., Glare, T.R. and Nap, J.P., 2003. The release of genetically modified crops into the environment. Part II. Overview of ecological risk assessment. *The Plant Journal,* 33 (1), 19-46.

Evenson, R., 1995. *The valuation of crop genetic resource preservation, conservation and use*, Paper prepared for the Commission on Plant Genetic Resources, Rome.

Fuglie, K., Ballenger, N., Day, K., et al., 1996. *Agricultural research and development: public and private investments under alternative markets and institutions*. Economic Research Service, United States Department of Agriculture, Washington DC. Agricultural Economics Report no. 735. [http://www.ers.usda.gov/publications/aer735/]

Gilbert, R.J. and Newbery, D.M.G., 1982. Preemptive patenting and the persistence of monopoly. *American Economic Review,* 72 (3), 514-526.

Goeschl, T. and Swanson, T.M., 2002. *Lost horizons: the noncooperative management of an evolutionary biological system*. Fondazione Eni Enrico Mattei, Milano. FEEM Working Paper no. 89-2002. [http://ssrn.com/abstract=344322]

Goeschl, T. and Swanson, T.M., 2003a. The interaction of dynamic problems and dynamic policies: some economics of biotechnology. *In:* Laxminarayan, R. ed. *Battling resistance to antibiotics and pesticides: an economic approach*. Resources for the Future, Washington, 293-329.

Goeschl, T. and Swanson, T.M., 2003b. *On biology and technology: the economics of managing biotechnologies*. Fondazione Eni Enrico Mattei, Milano. FEEM Working Paper no. 42-2003. [http://ssrn.com/abstract=419080]

Goeschl, T. and Swanson, T.M., 2003c. Pests, plagues, and patents. *Journal of the European Economic Association,* 1 (2/3), 561-575.

Holden, J., Peacock, J. and Williams, T., 1993. *Genes, crops and the environment.* Cambridge University Press, Cambridge.

Jaffe, A.B. and Trajtenberg, M., 2002. *Patents, citations, and innovations: a window on the knowledge economy.* MIT Press, Cambridge.

Merges, R.P. and Nelson, R.R., 1990. On the complex economics of patent scope. *Columbia Law Review,* 90 (4), 839-916.

Milgrom, P. and Roberts, J., 1992. *Economics, organization, and management.* Prentice-Hall International, Englewood Cliffs.

Munro, A., 1997. Economics and biological evolution. *Environmental and Resource Economics,* 9 (4), 429-449.

Nap, J.P., Metz, P.L.J., Escaler, M., et al., 2003. The release of genetically modified crops into the environment. Part I: Overview of current status and regulations. *The Plant Journal,* 33 (1), 1-18.

Nordhaus, W.D., 1969. *Invention, growth, and welfare: a theoretical treatment of technological change.* MIT Press, Cambridge.

O'Donoghue, T., 1998. A patentability requirement for sequential innovation. *Rand Journal of Economics,* 29 (4), 654-679.

Oerke, E.C., Dehne, H.W., Schönbeck, F., et al., 1994. *Crop production and crop protection: estimated losses in major food and cash crops.* Elsevier, Amsterdam.

Palladino, P., 1996. *Entomology, ecology, and agriculture: the making of scientific careers in North America 1885-1985.* Taylor and Francis, London.

Rausser, G.C. and Small, A.A., 2000. Valuing research leads: bioprospecting and the conservation of genetic resources. *Journal of Political Economy,* 108 (1), 173-206.

Ruttan, V.W., 2001. *Technology, growth and development: an induced innovation perspective.* Oxford University Press, New York.

Swanson, T.M., 1995. Uniformity in development and the decline of biological diversity. *In:* Swanson, T.M. ed. *The economics and ecology of biodiversity decline: the forces driving global change.* Cambridge University Press, Cambridge, 41-54.

Swanson, T.M., 1999. The management of genetic resources for agriculture: ecology and information, externalities and policies. *In:* Peters, G.H. and Von Braun, J. eds. *Food security, diversification and resource management: refocusing the role of agriculture?* Ashgate, Aldershot, 212-231.

Swanson, T.M. and Goeschl, T., 2000. Property rights issues involving plant genetic resources: implications of ownership for economic efficiency. *Ecological Economics,* 32 (1), 75-92.

Weitzman, M.L., 2000. Economic profitability versus ecological entropy. *Quarterly Journal of Economics,* 115 (1), 237-263.

[1] One pertinent illustration of the conflict between societal and industry interests regarding the direction of technology implementation exists in the form of U.S. Environmental Protection Agency regulations on refuge requirements. See Hurley elsewhere in this volume for results derived against the background of a static market structure.

13b

Comment on Goeschl: Do patent-style intellectual property rights on transgenic crops harm the environment?

Henk Hogeveen[#] and Tassos Michalopoulos [#, ##]

Introduction

Intellectual property rights (IPR) are known to be an effective institution for providing people with incentives to create, maintain and improve assets. However, in the breeding of live organisms (plants as well as animals) IPR did not play an important role. In plant breeding there was an institution known as breeder's rights to prevent copying. Following the introduction of biotechnological research into breeding, the role of IPR became more important, for both traditional and biotechnological breeding technologies. Two subsequent GATT/WTO agreements (Uruguay 1993 and TRIPS agreement) brought agriculture and trade-related intellectual property rights to fall under the mandate of the free-market regulations. Consequentially, governments tend to withdraw from the R&D process and transfer the R&D-stimulating role to the private sector. This means that IPR become even more important. As Goeschl has stated, in the discussions with regard to biotechnological developments (including transgenic crops) IPR play an important role. Part of this role is based upon the fundamental question whether life can be patented. In this comment, however, the relation between IPR in plant breeding and effects of plant breeding on the environment will be discussed. In order to do this, the role and incentives of the stakeholders in this field should be examined. Using these roles, the effects of IPR in research and development (R&D) with regard to plant breeding can be discussed.

Roles of stakeholders

Around the discussion on IPR, genetic modification, environment and R&D, the following stakeholders can be distinguished: government, consumer, breeding company, farmer and research institute. The role and incentives for these stakeholders will be discussed in the line of a western, capitalist society.

Government

The main goal of the government (ideally) is to safeguard the interest of the whole society. To achieve this the government will (try to) establish a minimum level of welfare for all inhabitants of a country. This means that business activity is necessary to generate income. The environment (natural resources) provides part of

---

[#] Farm Management Group, Wageningen University, Hollandseweg 1, 6706 KN Wageningen, The Netherlands. E-mail: henk.hogeveen@wur.nl

[##] Applied Philosophy Group, Wageningen University, Hollandseweg 1, 6706 KN Wageningen, The Netherlands. E-mail: tassos.michalopoulos@wur.nl

J. H. H. Wesseler (ed.), Environmental Costs and Benefits of Transgenic Crops, 219–223.

the welfare, and therefore maintaining or even improving the environment is considered part of the responsibility of the government. The government can stimulate a proper direction of society by making regulations or by stimulating the appropriate development by subsidies, R&D investments etc.

Breeding company

The main goal of each private company is to create shareholder value. Those who invest in a company want a return on their investment, either by an increasing value of their shares or by dividend. A company can do so by creating successful competitive products (price increase), adopting strategies that will increase its market share, and by means of reduction of production costs. Improving products or production processes by R&D can help to reach the goals, especially when the rights to use these products or production processes are protected by IPR. However, a (breeding) company will not directly care about negative side effects (for instance on the environment). Only if their products harm the environment in such a way that it will negatively influence the image and in the long run the market share of a company, will the environment become of interest.

Farmers

From a business-economics point of view, almost all farms are losing money. Therefore, the goal of a farmer (most farms are family farms) can be seen as to continue being a farmer and to make enough money to support the family. From that point of view, they want their costs, including costs for their crops, as low as possible, with a price for their own products as high as possible. When they spend more money on new breeds, they expect to earn that money back, either by lower costs in crop production or by higher prices because of the higher quality of their products. The decision which crop to use, is taken by each individual farmer, given the place of the farmer in the production chain. However, because of their number, farmers can directly influence neither the price structure, nor the selection of the technologies to be adopted. In that way, wishing to maintain an income adequate to cover his financial needs and subsequently to be able to remain within the farming practice, the farmer is feeling pressure to keep himself as close to the frontier as possible. The tendency of an IPR-regulated system for adopting a subsequent rather than a parallel mode of technology adaptation as remarked in the commented paper, enforces this process. That is, as the subsequent mode of technological adaptation is chosen due to its better financial perspectives for the breeding companies, while it might fail to cope with the response of the biological system (e.g. widespread development of resistance).

Consumers

The interest of the consumer is to pay the lowest price for a product with a certain quality. Breeding-technology progress often has a direct impact on lowering product prices. Concerning quality, there are three quality aspects involved: sensory aspects (taste, size, colour etc.), safety aspects (possible chemical or bacteriological contamination) and intrinsic aspects (the way a product is produced, i.e., harm for the environment). For each of these quality aspects there is a production function: how much is a consumer willing to pay for a better taste, for a safer product or for a product produced in an environmentally friendly way? However, this function is not merely an object of observation for the producers, as effective marketing strategies often aim precisely at increasing this margin.

Research institutes

Research institutes need money in order to carry out research. The incentives of researchers are not only financial, but also the 'honour' to publish (scientific) papers and the status it gives in the international research community. This is especially the case with fundamental, curiosity-driven research, which has the role to clarify our knowledge of the 'world'.This fundamental research may never be applied, but the knowledge is shared and publicly available. With more applied research, the goal is either to develop new directions for new technologies or products, or to study the effects of these new directions on economy, welfare, environment etc. The IPR on research results may be used to generate money for new developments. An asymmetry that has to be noted, however, is that while applied research is usually based on (some of) the outputs of previously done, publicly available (and usually publicly funded) fundamental research, under the notion of IPR the knowledge it produces is private.

Flow of products and money in various scenarios

In this section two (extreme) scenarios with regard to IPR and R&D will be described:

1. A public system; there are no IPR possible on breeds.
2. A private system; IPR are intensively used and because of that the government withdraws from R&D activities.

These two extreme situations are described under the assumption that a constant amount of R&D is done in order to establish progress.

Public system

In a situation where private R&D expenditures cannot be covered by IPR (or something like a breeder's right) there will be very little incentive for breeding companies to invest in R&D. The only advantage from investments will be during a short period after introduction. After that, other breeding companies will soon be able to introduce the same variety. This means that the government will pay for almost all R&D. Results of this R&D will be publicly available and all breeding companies are able to utilize the new knowledge. Competition between companies will be the result and breeding material will be relatively cheap for the farmer. Consequently, the price of the end product for the farmer will be low. Prices for the consumer can also stay low. On the other hand, the money the government spends on R&D must also be acquired. That means that the taxpayers (often also consumers) have to pay higher taxes.

In an ideal situation, the government will not only pay for the development of improved breeds, but also for possible side-effects of new technologies, for instance effects on the environment. This information can lead to new regulations to protect the environment.

With the current globally operating breeding companies, there is one major disadvantage, which is the flow of money from country to country. If the government of one country spends money on R&D, the companies, farmers and consumers of another country also benefit, which means that different countries should share R&D expenditures. In this sense, a 'National Breeder's Right' may prove to be required.

Furthermore, questions can be raised as to whether the public sector can show the necessary management and marketing skills to be regarded equal to those in the private sector in terms of production efficiency.

Private system

When the government does not pay for R&D in plant breeding, investments of private companies should be made worthwhile by some kind of IPR system. In such a system the breeding company pays for the R&D, either by its own research or by financing a research institute to conduct research on its behalf. The company will earn that investment back by means of an increased price of the breeding material. The farmer will adopt the new product in order to enjoy its increased advantages, the more so because if he chooses to do otherwise he may have a production disadvantage in comparison with other farmers who do adopt it. When the advantages are great enough, the farmer will pay this higher price. In case of the improvement of production, the benefits of cheaper production should be enough to make up for the higher price of breeding material. However, the profit increase for the farmer will be lower than in the public system. On the other hand, if the quality of the products increases, the consumer has to make up for the increase in price of the breeding material. Hence, the prices for the consumer will be higher than in the public system. This can be explained by the fact that the private sector aims at profit optimization, while for the public sector societal welfare is expected to be the main mandate, requiring a sustainable R&D sector, but moderating profit expectations.

As a consequence of this private system, there will hardly be any R&D into side-effects of new technologies, unless demanded by the consumers/market. There is no incentive for private companies to do otherwise. It is possible that breeding material might harm the environment. This lack of R&D into side-effects might be made up for by regulations such as for the pharmaceutical industry (cf. the FDA in the US), where in order to get a registration (and thus the right to sell a product) the company must show that the product is safe. This type of regulation builds on known types of side-effects. New side-effects are found by application of the new product. This means that damage could already be done before regulation is adjusted. More than this, while the private sector is the one that enjoys the profit of marketing its products, it is the public sector that carries the burden of restoring potential – unexpected – harmful side-effects. A possible way out of this perspective, would be to make the private sector responsible for restoring *any* harm produced as a side-effect of its production, whether it was unexpected or not. Such an approach would result in increased testing prior to marketing, less side effects, but higher product prices.

Concluding remarks

The two extreme and very simple scenarios described above are very simple examples of the real situation. Neither of these scenarios exists. The current situation is a mixture between both extremes. However, there is a tendency towards the private system, which is supported by the WTO agreements on agriculture and IPR. The two examples demonstrate that R&D in progress in plant breeds, whether it is financed by the government (public sector) or by breeding companies (private sector) is in the end going to be paid for by the society (consumers and/or farmers). There are possible reasons that the private system is more efficient in the R&D process and in thinking of the demands of consumers and farmers than the public system. The private system can prove to have harmful impacts on aspects such as the environment and societal structure as long as these aspects keep on being regarded as side-effects of the supply chain (indirectly linked to its scope via consumers' demands), and keep on not being included directly in its targets, next to profit optimization. Possible ways for avoiding potential harmful impacts resulting from this situation would be the increase of public-sector involvement in plant breeding, or the creation of adequate incentives for the private sector to be more focused on possible side effects.

Finally we should be aware that the discussion with regard to IPR in breeding is much wider than the scope of this comment. The ethical question on the possibility to patent life, the difficulties for developing countries with regard to IPR, the societal implications that arise due to dismissing traditional farming practices (seed saving, versus dependency on breeding companies), and the inefficiency involved when patents from various companies have to be added together form another aspect of the discussion.

14a

Agricultural biotechnology and globalization: U.S. experience with public and private sector research

Greg Graff, David Roland-Holst and David Zilberman[#]

Abstract

This paper examines the challenges and opportunities facing the agricultural biotechnology sector as it works to sustain innovations and further propagate its benefits into the new century. Drawing on US experience, we survey the milestones of technological, legal and economic precedence and discuss institutional mechanisms for public and private partnership that can help agricultural biotechnology fulfil its immense promise. In particular, we emphasize the importance of the public sector in facilitating private agency, via promotion of basic research, dissemination of innovations, and as a guarantor of property rights.
Keywords: public and private R&D; intellectual property rights

Introduction

We enter a new century equipped with remarkable tools for improving human circumstances; the most promising of these arising from biotechnology. Over the four decades since the discovery of DNA, extraordinary innovations have taken us to the brink of creation itself, conferring the ability to directly modify and even originate new organisms. There are inevitably controversial aspects of this technical revolution, and considerable research remains to be done on issues such as environmental risk and public health. Despite this, however, the economic potential of biotechnology is now firmly established in thriving life-science industries that are contributing to hundreds of aspects of medical therapy and also to the original life science, agriculture.

In establishing and sustaining this remarkable process of innovation, one of the most important factors was public–private partnership. In this paper, we examine the salient historical and present-day features of this partnership, with particular attention to the incentive properties needed to facilitate research and product development. In leading examples, such as the Green Revolution and medical biotechnology, we see how the right combination of economic institutions and legal precedence can deliver rapid and sustained innovation to the marketplace. Drawing upon these cases, we indicate how agricultural biotechnology (ag-biotech) can overcome a new set of challenges, posed mainly by globalization, in propagating its benefits across the greater part of humanity.

[#] Department of Agricultural and Resource Economics, University of California, Berkeley. E-mail: zilber@are.berkeley.edu.
Opinions expressed here are those of the authors and should not be attributed to their affiliated institutions

J. H. H. Wesseler (ed.), Environmental Costs and Benefits of Transgenic Crops, 225–245.

The history of marketable scientific innovation, of which biotechnology is only one example, has special economic characteristics that merit restatement for the present discussion. Acting independently, both public and private agents tend to underinvest in research, albeit for different reasons. We give a brief overview of these conceptual issues in the next section. This is followed by a survey of recent history for indications of how public–private partnerships have evolved successfully, drawing lessons from the Green Revolution, medical biotechnology and ag-biotech itself. Then we discuss the special characteristics of biotechnology that relate to public–private partnership. The next section reviews the main institutional challenges to biotech proliferation, including imperfections in legal, educational and economic institutions, and discusses new initiatives for overcoming these. Most prominent among these is the Intellectual Property Rights (IPR) Clearinghouse; a new institutional mechanism that we believe can play an essential role in propagating the benefits of science in the service of mankind. The next section is devoted to a forward-looking discussion of the future of IPR in biotechnology, with implications for the wider agenda of technology and globalization. The a section reviews the special challenges and opportunities posed by globalization, including essential disparities in strategies for human-capital development. The final section of the paper presents concluding remarks.

The economics and role of public and private research: some theoretical considerations

There is a large body of literature on the economics of public and private research (Huffman and Evenson 1993; Alston, Pardey and Taylor 2001), including several lines of arguments that elucidate the role of research in the different sectors. In particular, we can distinguish between several major approaches to the problem. From the outset, we distinguish between innovations that are embodied in new products and those that are disembodied, including cultural practices etc. The private sector is more likely to conduct research and pursue development of embodied innovation. Our emphasis here is on this type of innovation, even though capturing the rents associated with improved seeds may be difficult and in some cases the private sector may underinvest.

Economic welfare analysis of research funding

Welfare economics attempts to identify resource allocations that improve overall societal economic welfare and compares them with choices made by the policymakers and the private sector. In particular, the public sector allocates resource funding with the putative aim of maximizing welfare for all sectors of society. Doing so, they strive to maximize the benefits to firms producing and using the technology and to consumers who utilize final products produced with this technology, minus the cost to generate and use the technology. On the other hand, the objective of firms generating the technology is to maximize their profits, i.e. revenues they obtain less the cost of the technology. Private-sector firms that engage in research and development (R&D) to generate new technologies take into account consumer interest only as it affects demand and revenues. However, consumers and users of publicly subsidized technologies are assumed to enjoy a surplus above what they pay for the technologies. Without this embodied surplus, their observed willingness to pay for new technology would be diminished.

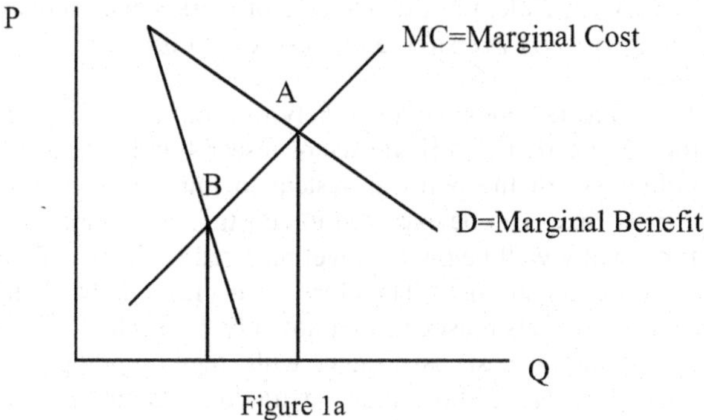

Figure 1a

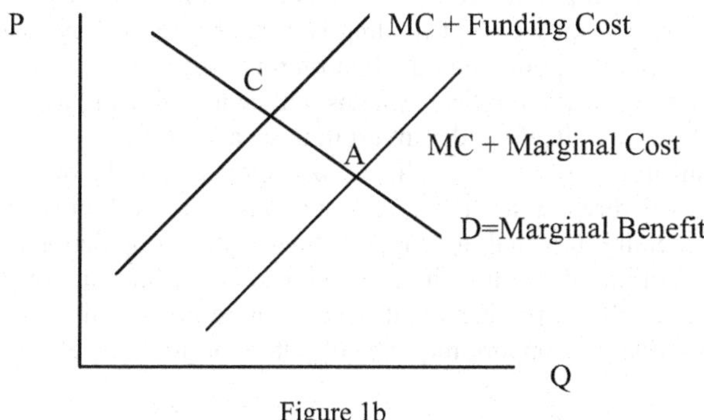

Figure 1b

Figure 1. (a) The private-sector under-investment in R&D due to monopoly power; (b) The public-sector under-investment in R&D due to resource constraint

However, these surpluses are not taken into account by producers, and this can lead to underinvestment by private agents who conduct R&D to generate technology products. Figure 1a illustrates this outcome. The private equilibrium is at point B (where marginal revenue is equal to the marginal cost of the technology), while the public optimum is at point A (where demand that is equal to the marginal benefit and combined consumer surplus is equal to the marginal cost of the technology).

Funding considerations

The private sector is assumed to finance investment and earn revenue only through the sale of products and services. Thus, they will invest in R&D activities seeking innovations that can be embodied in new products to be sold in the market. The best examples are mechanical innovations such as vehicles, chemical innovations such as new drugs or pesticides, etc. Private companies may also finance research that leads to patents with greater market potential than originally envisioned or than can be realized by the originating firm. The basic idea behind start-ups, for example, is to engage in high-risk research that may result in high-value patents that can then be sold to other companies that benefit from this type of research. On the other hand, private-sector companies are less likely to invest in research activities that result in knowledge and that cannot be protected by patents. Thus, imperfections in the patent system, in terms of both specificity and scope of protected intellectual content, are reflected by gaps (Black Holes) in the universe of marketable innovation because of what might be termed R&D aversion, unwillingness to invest because of incomplete IPR.

Public-funding considerations are also likely to limit the scope and absolute level of private-sector R&D activities. Private-sector research is financed for the most part by income. Willingness of the political system to impose taxes and the ability of governments to raise taxes can be expected to constrain investable funds available to the public sector, usually well below the level required to finance all activities whose social benefits exceed social costs. Therefore, in deciding how to invest the funds, public-sector decision makers must consider not only the explicit cost of each activity, but also the opportunity cost associated with their funding constraints. This opportunity cost arises because allocating an extra dollar to public research may entail reduction of public investment in other activities that are also socially beneficial.

We need to take into account that the opportunity cost of public funds results in significant reduction in public-sector investment in R&D (see Figure 1b). In an ideal case, where the opportunity cost of funding is zero, the optimal level of research will be determined at point A, but when the opportunity cost is high, it will be at point C. Thus, societies with weaker political systems will be less able to raise funds to support public research and will likely under invest in these activities.

One mechanism that will help public sectors to overcome the constraint associated with tax monies is direct assistance (e.g. donations). In developing countries, where research is generally targeted to support low-income populations and economic development, sources of funding include both taxation and aid. Again, aid is not sufficient to cover all the products that have positive net social benefits, and higher aid levels will reduce the opportunity cost of public project funding but not eliminate it.

Political economy and institutional considerations

There has been a growing realization among economists that public decisions are not necessarily made with the sole objective of maximizing social welfare, and political processes can result in choices that are less than optimal. To a large extent, these choices reflect the relative political power of social interests with varying degrees of influence over agencies that manage public allocation of resources. De Gorter and Zilberman (1990), for example, suggest that if farmers control the Department of Agriculture, which is the primary arbiter of funding of agricultural research, this may lead to underinvestment in research, regardless of the opportunity cost of public funds. By this reasoning, farmers only consider the impact of extra research on their own profit and not on consumer demand, except when it affects demand for agricultural products. The demand for food products is inelastic (namely, an increase in supply induces significant reduction in commodity prices). Thus, farmers who control research may not invest too heavily in supply-enhancing research, fearing drastic reductions in commodity prices that would harm producer profit, yet benefit consumers.

Having said this, agricultural research policies appear to differ across countries and even regions, but this can probably be accounted for by variations in the constituencies with salient influence over agriculture ministries and related policy institutions. In areas where consumers have a say about public research funding, we can expect to observe higher public (and subsidized private) investment because it may lead to increased supply and reduction in commodity prices. When agribusiness has significant influence on research funding, one can expect to see research with an emphasis on the supply and (especially) marketing and trade side. Because they can operate further downstream than farmers, agribusinesses are more tolerant of supply expansion if it can be translated into larger market capture.

As farmers remain the primary constituency in OECD agricultural policy, we can expect some degree of underinvestment relative to the optimum. One way to alleviate farmers' concerns about public research resulting in lower farm income is to subsidize farmers and increase their income. Agricultural policy in developed countries can be viewed as a package where public sectors have invested in research to increase the supply, and in return subsidize farmers to attenuate their induced losses. However, even this mechanism may not result in an optimal level of research because of the funding constraints faced by the public sector.

The organizational characteristics of public-sector research have important influences on the evolution of its agenda and priorities. Public-sector research is conducted in universities and research institutes. University professors produce joint products, teaching and research, and much day-to-day research at universities is conducted by PhD students. Both faculty and students are judged by their scholarship, which give primary recognition to originality. Thus, the nature of university research often leads to projects that are small-scale and self-contained, oriented toward conceptual breakthrough and relatively narrow intellectual outcomes. This line of research is essential for new innovations, but it is only part of a bigger picture.

In addition to conceptual origination, there is also an essential role for derivative or applied research that meets specific customer needs, e.g. testing for safety or conformity to other regulatory requirements. These types of research may be considered 'mundane' from a scientific perspective but they are essential for product development, i.e., to realize tangible social benefits from research investment. Partly for this reason, scientists and government research institutes are more oriented toward derivative research, and there is a division of labour within the public sector between the research conducted at universities and that at government research institutes. There are also significant and systematic differences between R&D done by universities and companies, as companies are much more interested in product development and revenues than enhancing general knowledge.

Systematic differences in institutional research emphasis lead to sorting of individual researchers based on their preferences and characteristics. They might be seen to pursue three ends: *fame, fortune and freedom.* Graff and Zilberman (2001) argue that universities may offer more fame and freedom, while the private sector may give more emphasis to the pursuit of fortune. Even within industries, there are differences in terms of risk and earning. Individuals who have both the temperament and opportunity to take more risk are more likely to be associated with start-ups financed by venture capitalists than to work in the secure and predictable environment of a major corporate research lab.

A thorough examination of the research environment, including organizational characteristics, economic incentives and political economy, can reveal much about the current landscape of research, how we arrived here, and what might be expected to appear on the near horizon of marketable innovation. To summarize for the present discussion, the results of research on research have several important implications.

Underinvestment both in public and private sectors

Research in both sectors is insufficient from a social perspective, and thus mechanisms to enhance investment in R&D and increase its productivity are likely to be valuable. These would include efforts to leverage complementarities between public and private research. The problem of underinvestment in research is much more severe in developing countries than in developed countries. The developed countries of the North have a larger capacity to finance public research, larger markets for agricultural producers, and more extensive commitments to human-capital development, all of which contribute to a good foundation for public and private agricultural R&D. In developing countries, aid can supplement public-sector research resources, but other challenges remain.

Complementarity of research characteristics conducted by the public and private sectors

The private sector will engage in R&D projects that will likely result in profitable products. Thus, R&D efforts should be directed at areas that are socially beneficial but are not profitable and not likely pursued by the private sector. Two categories for such research include research that has public-goods properties, namely, it results in outcomes that can be shared by others without the capacity to claim ownership, and research that may lead to 'orphan products'. This is research that ends up in products that have insufficient revenues and the potential to cover the costs. However, if one adds the consumer and social surplus to this revenue base, then the net social benefit exceeds social costs. Two typical 'orphan' markets are specialty crops in developed countries and subsistence crops in developing countries. The definition of orphan markets is evolving over time as the cost of R&D of new products declines and the private sector becomes more unlikely to engage in these research activities.

Importance of intellectual property rights protection

Since public investment in R&D depends on the earning capacity, mechanisms that enhance, for example, functions of well-defined property-rights systems with effective enforcement mechanisms and low transaction cost, are likely to induce the private sector to engage in research activities. However, since patents, in essence, generate monopoly rights, there may be excessive IPR protection, and patents and IPR may be abused. Patent rights owned by an incumbent firm may prevent new firms from entering the industry and producing new products that may require the use of innovations covered by the patents. Thus, design of the patent system has to balance social benefits associated with inducement of R&D by private companies with the social cost associated with the generation of extra monopoly power. In some cases, the patent system that aims for better utilization of knowledge 'commons' may result in constraints and transaction costs that will curtail future investment in research. This consequence of the patenting system is referred to by Heller and Eisenberg (1998) as the 'anticommons' problem.

Research and development as a dynamic phenomenon

The anticommons problem is one manifestation that R&D activities are part of the dynamic processes associated with build-up and utilization of the stocks of knowledge. While most of our theory is built on static assumptions, research is evolving. Within each research line, there is a transition from the laboratory to the factory, to the shop and to the field. The baton is handed down from researchers in the public sector who do conceptual work to private-sector actors who may do product-oriented R&D and engage in production and marketing efforts. Once the problem is solved, much of the research line that aims to address it may be redundant. But new research issues may rise. That suggests that we have cycles in research. Yesterday's solved puzzles are today's practices, and today's practices create tomorrow's problems.

The evolution of the research problems may cause continuous changes in the research system, and the allocation of public and private money to research efforts within fields is evolving over time. Productivity of the research line is changing and public-funding priorities must reflect it. As a problem, say, prevention of a disease, is solved, research in this area becomes a lower priority and thus funding should be reallocated to other areas where the benefit is higher. Similarly, introduction of new bodies of knowledge or techniques may open new fields of research and may lead to diversion of public money and corporate funding toward these new areas. Computer science did not exist 50 years ago, and now it is a major area of research investment (Mowery and Rosenberg 1989). We suggest that not only should allocation of public and private resources between private and public sectors be examined constantly and evaluated over time but, more importantly, allocation of resources within sectors and between fields and areas of research should also be constantly examined and re-evaluated.

Of course reallocation of resources between research lines is problematic. The skills of scientists are not malleable, and changes in research productivities and resulting research priorities of different research lines may be challenged and met with many objections. The basic discoveries in molecular and cell biology, which gave rise to genetically modified organisms, opened new avenues of research that may lead to reallocation of public research funding from areas that have been well established in the past. Furthermore, the use of genetically modified organisms and other technologies based on molecular and cell biology may also affect the nature of agricultural products and markets. Thus, to understand the challenge of the new technology, we have to view their place within a historical context.

Historical perspectives

Historically, most of the important innovations we all benefit from were generated by practitioners who identified a need and met it by exploiting knowledge resources that have accumulated in society. Many of the most important mechanical innovations in agriculture (the plough, tractor etc.) were such applications. Over time, as research in engineering has become more formulized and there is more reliance on university and other institutional research, yet most mechanical innovations are still developed by private sector companies. The most important chemical innovations (e.g., insecticides, artificial fertilizers) may have originated in scientific discoveries by people such as Haber and others. However, the rights to develop many of these technologies have moved to the private sector, and most research on new pesticides and improved agrochemical management (including fertilizer) have been done by the private sector. Of course, in many cases private agents relied on and took advantage of new knowledge developed in the public sector and incorporated it in their activities, sustaining a continuous public–private transfer of knowledge and (de facto or explicit) partnership.

While research may have been an indispensable starting point, from the user perspective (especially in the developed world), the source of chemical products has been the private sector. In developing countries, governments in some cases act as buyers and intermediary suppliers of fertilizers because of market failures, mainly lack of profitability and earning potential for the private sector. Such public-sector engagement in education and especially provision of inputs may reflect so-called orphan markets, where social benefits may justify intervention where private profit is not sufficient.

One area where the public sector was the dominant generator of technology and supplier of inputs has been genetic material. In the past, most of the research in seed development was conducted primarily by the public sector. Since it was difficult to appropriate benefits from development of new seeds, public-sector plant breeders, both in government and international research centres, developed the superior varieties. These were then disseminated to national research centres which, through selective breeding, adapted these improved breeds to local conditions. Seeds thus distributed to farmers became, through expanded agricultural productivity, one of the most important private contributions from public research (Evenson and Gollin 2003). What became known as the Green Revolution was an important example of how public-sector commitment to research led to immense improvements in global welfare.

As part of this process, public-sector provision of seeds led to an open system where breeders the world over could exchange gene plasm and other genetic materials, leading to increased efficiency. Within this system, however, elements of privatization already began to emerge. Once hybrids were established, private companies started making investments to differentiate this genetic material, adding some productivity/quality attributes and, more importantly, making it non-reproducible. Sterile hybrids conferred property rights that were sustainable, overcoming one of the essential incentive problems for private-sector investment in this area. Furthermore, Green Revolution varieties had been intensive users of fertilizers, and the introduction of these public-sector-generated varieties was co-ordinated with more intensive provision of private-sector-generated fertilizers and other inputs. Nevertheless, the public sector, through its direct and indirect control of genetic materials, retains significant responsibility for the evolution of crop systems.

Biotechnology: lessons from U.S. experience with rapid and sustained genetic innovation

Scientific innovation has a very long history, but for most of that time its relationship with the marketplace has been an occasional and less than faithful one. The ultimate economic potential of early scientific discoveries was rarely foreseen and material rewards usually eluded the discoverers. Beginning with Edison and Bell a century ago, however, the laboratory and marketplace established a mutually beneficial relationship that has now produced undreamed-of technological assets and enriched some of the world's largest private profit and non-profit institutions. This dramatic success was due not only to the inspiration of inventors and entrepreneurs, but also to an evolving relationship between public and private agencies in research and product development. In this section, we examine how this partnership accelerated and sustained technological innovation and product development.

In the context of agriculture, research and product development with biotechnology are different from those with selective breeding, and many of the principles and institutional systems that are applied to traditional plant breeding are of limited relevance to biotechnology. To some extent, the evolution of ag-biotech is more analogous to that of chemical innovation than of seed breeding with traditional methods. Some of the breakthrough innovations leading to the development of biotechnology tools and products have been made in public-sector institutions, but the private sector has dominated the effort to commercialize the technology. Several factors contribute to the privatization of ag-biotech:

1. The Bayh-Dole Act (which allows U. S. universities to sell the rights to patents generated with public funding).
2. The establishment of IPR on living organisms (the Chakrabarty case, Diamond vs. Chakrabarty 1980) and utility patents for seeds.
3. The proliferation of university offices of technology transfer.
4. The availability of venture capital funds for start-ups in biotechnology.

The main cause for the high degree of privatization of ag-biotech is the nature of innovation. Patents are statements of concepts that are novel and useful. Most patents are not applied and do not earn income. Commercialization of most patents requires significant extra investment. The patent system provides not only incentives for research but mostly for development and commercialization. Patent ownership is an asset essential for obtaining finance for further technology development. The offices of technology transfer served as a mechanism to reduce transaction costs in moving IPR from universities to the public sector. Biotechnology is typified by partnerships between university researchers, start-up companies and major corporations that allow sharing of risk and division of labour so each organization concentrates on tasks where they have relative advantage. Universities emphasize conceptual research, start-ups pursue early stages of risky commercialization, and major companies while having their own programmes in basic research and investing in risky development activities are dominant in product commercialization.

In the United States, the public–private partnership and privatization of ag-biotech have had several important consequences. The most salient characteristic is the emergence of an educational/industrial complex around biotechnology generally. Major companies are locating near research centres. The Silicon Valley is the most obvious example of a symbiotic co-evolution of an industrial hub and educational institutions. Industrial centres near research universities in the San Francisco Bay area (among others), Davis and San Diego in California, and in Saskatoon, Canada, have been crucial for the evolution of agbiotech.

Academics are joining the entrepreneurial community, providing deeper insight for capital markets into the process of innovation. A key to a successful commercialization of university innovation is sustained involvement of the innovator (Graff et al. 2003). Thus, successful technology transfer entails not only a transfer of IPR, but of innovation-specific human capital.

Universities are discovering that research can lead to substantial, diversified and sustained new income streams. The public sector gains from royalties, contracts and grants, as well as donations resulting from commercialization of its technology. As Graff et al. note, in the United States royalties hardly cover 1 percent of the research expenditures of the US universities, but they were concentrated in specific areas, in particular medicine. Some agricultural innovations generated a large stream of income. Overall, though, royalties are not likely to be a major source of income for universities, and intensive public support of research will be needed to maintain a viable public research system.

University research has emerged as an important mechanism of industrial competition. With technology transfer, established firms not only face competition from other firms but also from new innovations originated at universities and financed by venture capitalists.

Companies such as Monsanto and Novartis have spent billions of dollars to develop a viable technology package to generate new seeds and hybrids through genetic modification. We now have a foundation that can be advanced and improved to transfer genes to plants and to produce a diversified portfolio of seeds commercially at a reasonable cost.

It needs to be noted, however, that transferring control of enabling technologies to private hands may restrict access to innovations and impede new innovation. An important risk of privatization for ag-biotech is the emergence of an 'anticommons' problem. Wright (1998) provides some anecdotal evidence that lack of access to process innovation is preventing commercialization of genetically modified varieties of specialty crops. Even when access to technology is available eventually, it becomes very costly and the transaction cost associated with obtaining access may prevent undertaking many worthy projects. The case of Golden Rice, where access to more than 70 patents was needed in order to obtain 'freedom to operate' and develop the technology, is illustrative of this problem.

It is useful to distinguish between process innovations that are also referred to as enabling technologies (e.g., tools of genetic manipulation) and product innovation (e.g., functional genomic knowledge about the functions that certain genes may serve) in assessing the impact of privatizing IPR that was originated in the public sector. The anticommons problems may be especially severe with process innovations. When these technologies are patented and their use is restricted to research or banned altogether, the capacity to develop applications relying on this technology is limited. In some cases, Wright suggests that commercialization of public-sector innovations might have been blocked because of lack of access to patents that originated by public-sector researchers, but were transferred to the private sector. The extent to which access to public-sector innovations will be available in the future depends on licensing arrangements with the private sector.

The impact of licensing on the evolution of industries is illustrated by comparing the impact of the Cohen Boyer vs. *Agrobacterium* patents. The Cohen-Boyer patent (for the basic process of medical genetic engineering) has not been licensed exclusively. Its use has been licensed for a relatively small fee per application. It generated immense revenues to the University of California, Stanford, and the innovators. The affordable non-exclusive licenses enable fast diffusion of the technology and did not hamper their growth of medical biotechnology.

On the other hand, the rights for the *Agrobacterium* (a crucial process for planting genes in plants) were transferred exclusively to Monsanto from Washington University. Monsanto's restriction of use of the technology by researchers in other companies has been a source of much resentment (especially in Europe). Lack of access to this technology presumably thwarted commercialization of some innovations. Parker, Zilberman and Castillo (1998) argued, based on interviews, that innovators and offices of technology transfer wait for patents of enabling technologies to expire to issue dependent technologies (to avoid hold-up).

Privatization of ag-biotech has led to emphasis on the development of genetically modified technologies that served the needs of the North and targeted major crops. As theory predicts, companies like Monsanto have targeted the most profitable crops for genetic modifications. Monsanto and the other companies launched Roundup Ready and *Bt* varieties for major crops (soybeans, corn, cotton). Yet, these technologies are not applied for small crops or subsistence crops in developing countries. The major companies are not likely to apply these technologies for the orphan crops, and the public research institutions that will develop such technologies may be concerned about IPR availability.

Overcoming IPR constraints to spread the benefits of biotechnology

Thus far, ag-biotech innovation and product development have been confined largely to research systems in OECD countries, yet the economic and social potential of these technologies is global in scope. For example, *Bt* cotton has been widely adopted in the United States and conferred significant gains there in terms of reduced pesticide dependence and lower consumer costs. Recent studies of India (Qaim and Zilberman 2003) show even more dramatic per hectare gains, however, and research in China (Pray et al. 2002) also associates its adoption with improved worker health and reduced environmental side-effects. More generally, higher pest intensity in developing countries and more limited alternatives for pest control further amplify the relative benefits of pest-mitigating biotechnology, including collateral gains in terms of reduced chemical loading of soil, water and other resources.

Despite this emergent evidence, the world remains sharply divided when it comes to technological research, innovation and assimilation. Instinctive resistance to innovation might seem prosaic for everyday consumer technologies, but it has graver implications in the context of human health and nutrition. In the developing world, especially in some of the poorest countries, there has been precious little basic or applied research of the kind we are discussing, either of the public or private-sector variety[1]. Even in China and India, which have strong scientific traditions and many public and private laboratories, the trends we delineated in the above are only beginning to be established.

We believe that the potential of biotechnology is also underutilized in applications for minor crops that include fruits and vegetables in developed countries. Application of biotechnology generally in developing countries and to specialty crops in developed nations probably requires more intensive investment of the public sector in research, development and commercial licensing because such investment may not be profitable from a private perspective but may be desirable from a social perspective given benefits to consumers and users of the technology.

Indeed, the international research centres and public and private aid agencies are funding or considering investment to enhance biotech research and development capacity in developing countries. Commodity groups in the United States and developed countries are funding research and development activities to enhance the application of biotechnology for specialty crops[2]. In both cases, lack of access to intellectual property is one of the primary obstacles. One way to overcome these obstacles is the establishment of an Intellectual Property Rights Clearinghouse (IPRC); a new institution that can serve several purposes.

To understand some of the potential benefits of the IPRC, it is important to compare the way intellectual-property management differs between the private and public sectors. The private sector recognizes IPR constraints as part of the cost of doing business. New projects are not introduced without 'freedom to operate", i.e., the potential to capture rents from embodying a given technology in new products. In the course of their own research agendas, public-sector researchers do not have the information needed to foresee such downstream IPR constraints. This information gap can seriously limit the potential for future commercialization of their innovations. One objective of the clearinghouse is to provide researchers in the public sector with greater visibility on the freedom to operate issue, harmonizing their information set with that of their colleagues in the private sector.

Private-sector organizations use their IPR holdings to secure access to other needed components of intellectual property. One reason, for example, for merger arrangements and strategic alliances between firms is to enlarge and diversify IPR portfolios, thereby increasing their flexibility in research, development and commercialization. Graff et al. (2003) found that public-sector institutions actually have a significant share of the ag-biotech patents. These are concentrated in research universities in the United States and in the OECD countries. In 2000, private-sector entities owned 22 percent of value-weighted US agricultural biotechnology patents, and 44 percent of these private patents were owned by the "Big 5" (Monsanto, 19 percent; DuPont, 10 percent; Dow, 7 percent; Syngenta, 5 percent; and Aventis, 3 percent). The rest of the private sector, mostly start-ups and smaller companies, owned 34 percent of all ag-biotech patents. Similar proportions are observed in other OECD patent systems (EU, Japan and Patent Cooperation Treaty). Using cluster analysis in case studies, Graff et al. (2003) documented that private-sector organizations have patented broadly across the various technology classes necessary for most applications of ag-biotech. The range of research projects that can be supported by public-sector-owned IPR is also significantly enhanced by a wide range of unpatented innovations that are accumulating in the private-sector institutions.

While the public sector has a significant IPR ownership, it is diffused among many institutions. No individual institution has more than 2% of total patents, and the diffused ownership of IPR by public-sector institutions weakens the sector's power to negotiate and leverage greater public interest into biotech applications. The clearinghouse will provide mechanisms to combine public-sector IPR and, thus, make it a stronger block in possible technology negotiations.

While patent ownership is divided, the rights of use have largely been transferred to the private sector. To achieve more effective collective action among public-sector organizations, it is essential to know the actual portfolio of technologies controlled by the public sector. Information on the actual control of technologies is quite sensitive and, thus far, not available in one central location. This lack of transparency increases risk and transaction cost for potential entrants in both research and market development, seriously hindering the innovation and the realization of its benefits. Another role of the clearinghouse is to collect updated information about technology ownership and to advise individuals where to obtain technologies they need.

Private ownership of patents by corporations is perceived to be a major constraint of technology use in developing countries and for orphan crops. However, in some cases, obtaining the access to patents that are owned by universities may be as difficult or even more difficult. Some researchers in developing countries actually maintain that they have a harder time obtaining rights to utilize technologies from public offices of technology transfer than from private companies. Companies provide technologies to orphan markets simply for the sake of public-relations gains. Such goodwill motives may induce them to give away the rights to use the technology, especially in developing products that do not threaten established markets or other financial interests. For some university inventors, the income from use of technology is of major importance, and they may be reluctant to waive their rights to the revenues generated by their technology. One possible role of the clearinghouse is thus to establish arrangements for facilitating access to public-sector and especially university patents for orphan markets.

Greater transparency can also facilitate clear delineation of market scope, reducing risks of spill-overs to competing interests. In this sense, some barriers to technologies that originate in the public sector may be the result of imprecise marketing. Companies may obtain the rights of a patent for all markets, while in reality they may be interested in applying the patent to a small number of crops in OECD countries.

Once they own the rights, liability considerations, transaction costs and other factors may limit the capacity of researchers to utilize technologies for orphan markets. One possible role of the clearinghouse is to share knowledge and research cost to develop precise technology-transfer procedures that will lead to more efficient and socially beneficial IPR management. The above analysis suggests several objectives for an IPR clearinghouse for agricultural biotechnology:

1. Reduce transaction costs for the commercialization of innovations (Shapiro 2000).
2. Expand the universe of accessible technologies (for research and product development).
3. Improve efficiency of technology-transfer mechanisms and practices in public-sector institutions.
4. Increase transparency of IPR ownership.
5. Provide mechanisms to expedite IPR negotiation and access.
6. Consolidate the public interest in technology origination and development (Graff and Zilberman 2001).

There have been two recent attempts to develop IPRCs. The Rockefeller and McKnight Foundations are collaborating with 13 major universities in the United States to establish Public-Sector Intellectual Property Resource for Agriculture (PIPRA). This initiative aims to increase public-sector scientists' freedom to operate and provide access to IPR to develop technologies for orphan crops. The new organization will have two elements:

1. A database of IPR ownership and rights in ag-biotech (a team of experts will advise researchers, administrators and managers about practical intellectual-property management strategies and IPR ownership and access).
2. A mechanism such as a technology pool (the clearinghouse will consist of supporting public-sector institutions that will share technologies and users, i.e., researchers in developing countries). Institutional members who contribute to the pool will have access to it. Namely, the universities basically combine all technologies that they control into a pool available to subscribers. Actually, the technology can be sorted and arranged according to its function to ease access within the IPR maze. The pool will aim to provide a set of technologies that will allow pursuing a broad range of ag-biotech applications. The pool can also be used strategically to trade access to technologies from the private sector.

Another clearinghouse is the African Agricultural Technology Foundation (AATF). Also supported by the Rockefeller Foundation, it aims to facilitate research and the introduction of new sophisticated crop varieties (including biotechnology). It emphasizes technology transfer when using public research and will help scientists to overcome IPR and regulatory requirements. AATF aims to negotiate with the public sector directly to obtain licenses for technologies in Africa used for humanitarian causes. This organization will go beyond technology transfer, providing some funding for research, particularly to overcome IPR regulatory constraints. Its main emphasis, however, is to work with technology owners and project partners (including donors to negotiate overall licenses). The AATF will be the licensee, and then sublicensed to research teams and product developers.

In the medical arena, an interesting institution is the Management of Intellectual Property in Health (MIHR) R&D clearinghouse. Its motivation is to gain access to IPR and develop therapies for diseases (tuberculosis, AIDS, malaria) afflicting the poor. Its main areas of work include: (1) identification and codification of best practices for licensing to achieve the goals of the public sector; (2) provision of training to scientists, universities and research institutes in managing intellectual property to benefit the public sector in both developed and developing countries; and (3) consulting services to developed and developing country groups concerned with research and product development.

Since it embodies both the promise of sustained innovation and the risk of exploitation, private ownership of technology will remain a controversial subject for the foreseeable future. The responsibility of public entities is clearly to facilitate the former and mitigate the latter, and effective policies toward biotechnology will necessarily reflect this. Facilitating access to IPR is the primary impetus for the initiatives discussed above, but the following considerations are also important:

Clear delineation of patents

Designing optimal patents is a challenge. If patents are too broad, they may hamper future research and may undermine access to the commons of intellectual and scientific discovery. If they are too narrow, they will undermine incentives for private discovery and incentives to develop and commercialize discovery. The latter incentive effect may be even more important because it applies to both private and public discovery. Research to develop methodologies for precise patenting and licensing, as well as implementation of its findings as new knowledge accumulates, is of paramount importance.

Currently, genomic knowledge is patentable, and the discovery of a gene sequence and its use can confer monopoly power. However, genomic discoveries are no longer novel and, therefore, do not justify patents in most cases. Companies deciphering genetic codes now earn their primary income via information services (i.e. selling databases). On the other hand, functional-genomic discoveries, which identify the function of genes and their potential applications, are more logical candidates for patenting. The evolving distinction between genomic and functional-genomic innovations illustrates the importance of adjusting patent criteria as the state of knowledge advances.

Biodiversity and biotechnology: a two-way street

The relationship between biotechnology and biodiversity is a contentious one, and is generally not well understood. On the one hand, there is a public perception that biotechnology reduces biodiversity. On the other, there is a widely held sentiment that agricultural technology institutions (public and private) seek to appropriate biodiversity resources in developing countries.

On the first point, biotechnology actually has the potential to contribute to crop biodiversity (Qaim, Yarkin and Zilberman 2004). They argue that while classical breeding has narrowed crop diversity significantly, biotechnology could make possible retention of a large proportion of today's crop varieties. *Bt* technology, for example, enables local varieties to be made pest resistant, obviating the need to adopt and adapt more homogeneous 'global' varieties, as was the norm during the Green Revolution. As a result, the US now has more than 1000 varieties of *Bt* soybean, most of which are single-gene variants of local legacy varieties. Far from homogenizing the gene pool, the introduction of ag-biotech in OECD markets has acted to protect and even increase biodiversity.

The issue of biodiversity and (implicitly) North-South property rights might seem more ambiguous. Genetic material from the developing world has certainly contributed to science and practical technology in OECD economies, but the productivity gains of technology transfer in the opposite direction have been enormous. There is a growing literature on the economics of biodiversity that shows it is, say, valuable at 'hot spots' with plants of apparent value (Rausser and Small 2000). However, for most locations, it is very low (Simpson, Sedjo and Reid 1996). The likelihood of discovery of new wonder drugs is a result of bioprospecting, which limits the capacity to change the access to biodiversity. Our discussion on technology transfer shows that universities' royalties are very low and cannot support their research. Similarly, biotechnology compensation will be low and should not be counted as a major source of income for developing countries.

Like the Green Revolution, public and private agencies will accomplish their primary objective (public interest and profit, respectively) only if they achieve their secondary mission, increasing agricultural productivity and food security in the developing world. From an economist's perspective, land is an immobile factor of production, and for this reason globalization of ag-biotech cannot succeed without local assimilation. Some observers see the advent of ag-biotech as a process of global consolidation, but the evidence on *Bt* reveals the opposite, a process of technology dispersal and localization. Instead of adapting innumerable farmers to a few varieties, ag-biotech appears to be adapting a few technologies to innumerable local varieties. This suggests not exploitation, but a partnership to overcome barriers to increased production for the world's majority enterprise, small farming, building upon the global legacy of biodiversity.

Having said this, the evolution and eventual success of such a partnership will depend critically on innovation and technology sharing, where the latter encompasses both man-made and natural technology (e.g. biodiversity). This in turn will depend upon clear delineation, ownership and market articulation of property rights, and much remains to be done in these areas. The IPRC can perform an essential service here, by increasing transparency and reducing transaction costs, but public institutions will have to fill many gaps in global standards for more complete markets to develop in this area.

Education - A North-South partnership and human-capital development

Biotechnology is in its infancy. This technology, using the tools of molecular biology, promises a future where biological solutions for many industrial problems will become more efficient and environmentally friendly. While most of the technology has been developed in the North, most of the world's genetic resources are in the South. At present, much of the research is developing tools to utilize genetic materials, but many of the opportunities in the future will arise from better understanding of functional genomics. Much of this research can be facilitated by North-South partnerships, and it is important for the South to participate more fully in this. Better intellectual capacity to take advantage of biodiversity will allow the South to take a better bargaining position to negotiate its role in partnerships.

Biotechnology is a modern technology that, to a large extent, was originated and sustained by university research, and many of the centres of this industry have been built in proximity of universities. It has thereby become apparent that, to succeed in biotechnology, a country needs to develop and maintain superior higher education, developing its own educational–industrial complex to generate human capital and marketable intellectual property. This observation alone defines an agenda for education-oriented development assistance, whether it be private or public, bilateral or multilateral[3]. Perhaps the greatest challenge, but ultimately the greatest opportunity, for fuller North-South partnership in biotech innovation is education.

Technology generally, and biotechnology in particular, are strong complements for human capital, and research and development are especially human-capital-intensive. The geographic and institutional symbiosis between modern universities and the technology sectors is an important example of this, and it is an example that developing countries have difficulty emulating for many reasons. A combination of underinvestment in education, private-capital insufficiency, and (in many cases) small market size have prevented the emergence of significant research capacity in most developing countries. Even those with large and long-established scientific traditions, such as China and India, are in the earliest stages of building the public–private research alliances that are hallmarks of dynamic technology sectors in OECD countries.

These facts reveal the need for expanded international partnership, both public and private, to develop capacity for biotech innovation and commercialization in the South. On the public side, aid agencies should reaffirm their commitments to human-capital development generally, and scientific capacity in particular, recognizing this as the key to sustained productivity growth and higher living standards. Private interests, for their part, should take new initiatives to leverage human resources in developing countries, transferring technology and capital into new markets and thereby gaining first-mover advantage in these emerging biotech markets. China, India and some other large and populous developing countries are already attractive candidates for this kind of entrepreneurship, while smaller and less advanced countries should be seen in a regionalized perspective.

The future of biotech

The majority of private investment in ag-biotech R&D has been accompanied by activities aimed at protection of IPR and development of proprietary applications of original scientific ideas. By its nature, scientific discovery is uncertain and requires significant initial financial commitments, followed by (often larger) investments for product development, testing and marketing. Generally speaking, neither public nor private agents would invest in development of a downstream research product without secure IPR at each stage. The decisive institutional reforms in the 1980s were legal precedents defining and protecting IPR, including the Bayh-Dole Act and decisions establishing the patentability of living organisms (the Chakrabarty case).

In this section, we give an overview of the salient issues that will influence the future evolution of biotech. After discussing necessary conditions for public–private research partnership, we examine emerging research priorities, and then close with a more speculative discussion about future trends in this dynamic sector.

Necessary conditions for public–private partnership in innovation

The inherent division of labour between public and private research reflects the substantive (and constructive) differences between the two. Universities and the scientific community place a high premium on originality and creativity, and in this way are more likely to come up with new paradigms and new research agendas. Within private companies, especially larger and more established ones, emphasis is on research that will enhance the bottom line. Their research is thus driven by a larger universe of criteria, including process efficiency, regulatory conformity and demand-driven design standards.

In principle, these differing objectives and priorities all represent socially (and individually) desirable product characteristics. Thus, the parallel agency of public and private research could yield significant complementarity, but of course this depends critically on the prevailing regime of incentives. The incentives, in turn, are strictly disciplined by the legal regime governing property rights and economic conditions of the destination market. As we have seen, these conditions have been quite virtuous for life-science research in OECD countries over recent decades, but much less so in poor countries.

In ag-biotech, new basic discoveries have emerged from public research and this new knowledge can be transferred to private agents because the legal regime allows them to acquire (appropriate) and retain partial or complete property rights. With ownership comes the incentive to invest in capacity to utilize this new technology and commercialize it. Thus the legal system enables faster technology transfer and commercialization because it recruits new (and arguably more appropriate private) resources to this task. Conversely, partial retention of property rights may sharpen the research incentives of the public partners. Meanwhile, other public dissemination institutions might further accelerate the process of technology transfer and, ultimately, promote more competitive innovation and commercialization.

As the research by Graff et al. (2003) suggests, it often happens that original innovators in public-sector research migrate (partially or completely) downstream to the start-up or industry stages, engaging directly in the development and application that allowed for commercialization of their technologies. These forays, while incurring some risks of conflict of interest, can have many benefits. The expertise of these individuals can facilitate much more effective transfer of the technology, significantly reducing the moral hazards associated with new technology acquisition. Considering the financial incentives sometimes offered to these 'technology couriers', risk reduction must be quite valuable to venture-capital entrepreneurs. Of course, these people also represent the human-capital component of the intellectual property in question, and thus are an investable asset in themselves. Whether or not they migrate back to basic-research institutions, their experience and continued contact will also influence the agendas of their former labs and those of their colleagues. In this way, migration of this kind influences both the basic research and the product-development environments.

As to incentives for public partners, this kind of technology transfer has certainly generated royalties for universities. As Graff et al. suggest, however, the order of magnitude of these royalties is relatively small and their relative impact on budgets even less significant. Nevertheless, new monies have been allocated to certain avenues of research that have not been highly funded before, such as medical biotechnology which has benefited significantly from these new resources. However, the main contribution of commercialization to the university, at least in the case of the United States, was through private-sector donations and contracts and grants. For example, Monsanto invested tens of millions of dollars into the University of St. Louis. Some of the new facilities in universities in California have been financed by contributions from companies and, especially, their owners. In addition to the formal transfer of funds, which were modest, there were also informal transfers that were more significant.

Despite largesse from private sources, earned or contributed, public-sector contributions far exceeded royalties and donations and have been the most important source of financing for ag-biotech. The data also suggest that, while investment in ag-biotech has increased in the public sector, most of it is being committed from the private sector. Certainly, there are important cases of public sector investment in ag-biotech, but this source is dwarfed by its private-sector counterpart.

The major concern about the increase in the use of genetically modified organisms for production of genetically modified materials in agriculture is fear that an important agricultural input, seeds, will transfer from the public to private sector. As we argued earlier, control of other important agricultural inputs as well as provision of medicines is the offering of private ends. In both cases, even though the private sector is the producer of most of the products, a large investment in research leading to the development of public policy has emerged to assure social optimality, and it will protect each individual and group that may be neglected if the private sector is left uncontrolled. Thus, the policy challenge is to develop similar institutions in the case of ag-biotech.

Emergent public research priorities

One perennial feature of research and its agenda, whether public or private, is continuous evolution. The primary drivers in the latter case are market-related, but the forces that animate public interest are more diverse. New societal problems and needs may instigate new priorities for investigation, in turn instigating new research agendas and even new disciplines. At the same time, once a research agenda has reached maturity, resources tend to shift from the frontier scientists towards practitioners.

Agricultural technology has gone through the same processes again and again. Much of the research in the last 100 years in the United States has focused on developing new varieties etc. This agenda resulted in improved high-yield varieties, chemical fertilizers, synthetic pesticides etc., but also in the discovery of new issues, such as the negative side-effects of DDT. In this case, the publication in 1962 of Rachel Carson's book *Silent Spring* gave rise to an environmental agenda that has become a major area of emphasis in both public and private sectors. For example, the recent NRC reports point out that integrated pest management, biological control and other biology-based technologies have become areas of emphasis in the private and public sector. Research in soil science and water has evolved an emphasis on issues of water quality and other environmental dimensions.

While the private sector has taken a leading role in the development of improved systems of production of livestock, university research has shifted its focus to environmental side-effects of these systems. To some extent, the increase in private funding of research has reflected maturity of technologies, and the emphasis on improvement of proprietary knowledge in established fields. At the same time, public research has emphasized some new avenues of research; some ultimately transferred and further developed by the private sector, or some addressing issues that have public-good properties. An example of the latter is measurement of the environmental impact and side-effects of new technologies that are not embodied in different products, and these type of technologies will continue to be emphasized by the public sector, since they do not promise significant return to private investment.

Salient future trends in agricultural biotechnology

As in any technological sector, the future ag-biotech is highly uncertain. Despite this, however, we feel it worthwhile to close the present discussion by highlighting a few salient issues that are sure to influence the course of future events. Each of these topics is worthy of its own research agenda, but we mention them only in passing to evoke deeper thinking and discussion about how to facilitate best the realization of biotech's enormous economic and human potential.

Resolution of IPR issues will reduce the costs and accelerate the introduction of new technologies when the economic conditions are ripe. The legal, political and economic universe of IPR is undergoing very rapid evolution, particularly with the impetus or public and private initiatives to promote globalization. With some exceptions, it is fair to say that most of these trends will lead to the goals embodied in the clearinghouse concept: greater transparency, lower transactions costs etc. This process will contribute decisively to the global proliferation and acceleration of innovation.

More stringent patent requirement will reduce IPR pressures. Like IPR issues generally, international policy toward patent law is changing fast, stimulated by the same public and private initiatives to establish economically rational international standards for tradable entitlement to innovation. Success in this area will be measured by how rapidly innovation progresses in both scope and depth.

As patent rights for basic technology expire, some tensions will be reduced. As we have seen in information technology, telecoms, and many other sectors, the usefulness of innovative technology generally outlives the right to own it, and in any case it usually makes contributions far beyond the boundaries of original property rights. For this reason, the natural process of rights expiration can be seen to contribute to a technology commons, enlarging mankind's stock of shared intellectual capital and broadening the basis for future innovation.

Public-sector institutions will develop technology pools and arrangements for swapping technologies with private-sector players. In terms of policy and market strategy, biotech can still be seen as a relatively new game, and the pubic sector is no more experienced than any of the other players. For this reason, one can expect to see public-sector strategy in this sector to evolve dramatically over the coming decade, much as it did with the advent of other path-breaking technologies (atomic energy, space travel etc.). Looking ahead, it is reasonable to expect more linkage between public policies in this area as governments strive for greater domestic and international policy coherence. It is also reasonable to expect progress in public–private partnerships, extending beyond basic regulatory duties to facilitating practices such as sponsorship of technology pools, clearinghouses and more complex incentive arrangements.

Private players will use their IPR assets to earn income and promote their technologies. From an investment and innovation perspective, the success of biotechnology is to an important extent self-fulfilling. Most biotech companies have very high rates of profit retention and reinvestment and, as the market success of today's innovations are consolidated, private resources dedicated to innovation will steadily expand. The semi-conductor industry provides a useful role model here, where multi-billion dollar R&D budgets have emerged from firms that were non-existent two decades ago.

Conclusions

Ag-biotech enters a new century with a remarkable set of accomplishments. The innovations of the four decades since DNA was decoded are revolutionary, and now we look to globalization to consolidate and expand the economic benefits which have until now been enjoyed primarily by the wealthy countries. If the promise of ag-biotech is to be fulfilled, the successes of public and private research partnerships in the OECD must be repeated around the world. Between the present and a bright horizon of opportunity, however, is a chasm of technological inequality between North and South. Bridging this gap will be one of the greatest challenges to lasting improvement of the human condition.

In this paper, we have briefly reviewed the lessons of biotech's successes in the United States, with particular reference to the way in which institutional factors facilitated partnership between public and private research interests. With these experiences in mind, we then discussed a series of ideas about how this success can be replicated elsewhere, both within the South itself and in partnership between North and South. Generally speaking, we believe the paramount considerations are appropriate institutions and incentives. If public interests can facilitate the development of these, we believe the private sector will identify society's unmet needs and provide solutions.

References

Alston, J.M., Pardey, P.G. and Taylor, M.J. (eds.), 2001. *Agricultural science policy: changing global agendas*. Johns Hopkins University Press, Baltimore.

De Gorter, H. and Zilberman, D., 1990. On the political economy of public good inputs in agriculture. *American Journal of Agricultural Economics,* 72 (1), 131-137.

Diamond vs. Chakrabarty, 1980. 447 U.S., 303.

Evenson, R.E. and Gollin, D. (eds.), 2003. *Crop variety improvement and its effect on productivity: the impact of international agricultural research.* CABI, Wallingford.

Graff, G. and Zilberman, D., 2001. An intellectual property clearinghouse for agricultural biotechnology. *Nature Biotechnology,* 19 (12), 1179-1180.

Graff, G.D., Cullen, S.E., Bradford, K.J., et al., 2003. The public-private structure of intellectual property ownership in agricultural biotechnology. *Nature Biotechnology,* 21 (9), 989-995.

Heller, M.A. and Eisenberg, R.S., 1998. Can patents deter innovation? The anticommons in biomedical research. *Science,* 280 (5364), 698-701.

Huffman, W.E. and Evenson, R.E., 1993. *Science for agriculture: a long-term perspective.* Iowa State University Press, Ames.

Mowery, D.C. and Rosenberg, N., 1989. New developments in U.S. technology policy: implications for competitiveness and international trade policy. *California Management Review,* 21 (1), 107-124.

Parker, D., Zilberman, D. and Castillo, F., 1998. Offices of technology transfer: privatizing university innovations for agriculture. *Choices,* 13 (1), 19-25.

Pray, C.E., Huang, J.K., Hu, R.F., et al., 2002. Five years of Bt cotton in China: the benefits continue. *Plant Journal,* 31 (4), 423-430.

Qaim, M., Yarkin, C. and Zilberman, D., 2004. *The impact of biotechnology on crop genetic diversity.* Department of Agricultural and Resource Economics, UC Berkeley. Discussion Paper.

Qaim, M. and Zilberman, D., 2003. Yield effects of genetically modified crops in developing countries. *Science,* 299 (5608), 900-902.

Rausser, G.C. and Small, A.A., 2000. Valuing research leads: bioprospecting and the conservation of genetic resources. *Journal of Political Economy,* 108 (1), 173-206.

Sadoulet, E. and De Janvry, A., 1992. Agricultural trade liberalization and low income countries: a general equilibrium-multimarket approach. *American Journal of Agricultural Economics,* 74 (2), 268-280.

Shapiro, C., 2000. Navigating the patent thicket: cross-licenses, patent pools, and standard-setting. *In: Paper prepared for the conference "Innovation policy and the economy," National Bureau of Economics Research, Washington, D.C., April 11, 2000.*

Simpson, R.D., Sedjo, R.A. and Reid, J.W., 1996. Valuing biodiversity for use in pharmaceutical research. *Journal of Political Economy,* 104 (1), 163-185.

Wright, B.D., 1998. Public germplasm development at a crossroads: biotechnology and intellectual property. *California Agriculture,* 52 (6), 8-13.

---

[1] For more general discussion of market mechanisms in developing countries, see Sadoulet and De Janvry (1992)

[2] Alan Bennett, Director of Office of Technology Transfer, University of California, personal communication

[3] Development assistance to overcome the North-South technology gap is simply an example of the old 'giving a fish versus teaching to fish' adage, but with more profound growth implications because of endogenous growth externalities

14b

Comment on Graff, Roland-Holst and Zilberman: Agricultural biotechnology and globalization: how will public-private partnership evolve?

P.J.G.M. de Wit[#]

The paper takes two examples of public–private-partnership successes in agriculture and medical science: the green revolution and medical biotechnology. The green revolution was initiated by international research centres such as CYMMIT, CIP and IRRI: (i) the accessions and varieties were distributed to national institutes and breeding companies, (ii) the local extension services provided the farmers with advice on using well-adapted varieties, (iii) these activities were followed by the appearance of hybrids developed by breeding companies. Nowadays extension services have become privatized and farmers often lack information especially in the developing world; In contrast to the green-revolution period, the present extension services in many countries have disappeared or become privatized.

The difference between the green revolution and agro-biotechnology (ag-biotech):

1. Ag-biotech is an industrial issue, while variety improvement was an issue for farmers or small breeding companies.
2. Small breeding companies cannot afford the costs; in The Netherlands biotech companies have built a consortium of small biotech companies to perform contract research for them on a non-competitive basis; the costs are just too high for one company.
3. Small independent biotech companies often do not exist for a long time; they are bought by established companies; scientists from some private biotech companies in the US and Europe return to the university labs (Fame, Freedom, Fortune).
4. There is a crisis in developing countries as multinationals merge and buy small biotech companies: overall there is a decrease in private investment in ag-biotech.
5. After these mergers there are fewer private partners left to support research at universities, which counteracts the flourishing relations between public and private sector.
6. In addition, due to the GMO debate big companies are reluctant to invest in GMOs; the profits are often marginal; too many costly patents.
   This is in contrast with the situation in medical biotech.

The solution:
1. Make genome sequences of crops publicly available as soon as possible.
2. Let small start-up companies work on functional genomics.

[#] Wageningen University, Wageningen, The Netherlands. E-mail: paul.dewit@wur.nl

J. H. H. Wesseler (ed.), Environmental Costs and Benefits of Transgenic Crops, 247–248.

There is still much to be done; only 5% of the genome variation is exploited in breeding lines; a patent does not mean anything if it is not further developed into a product.

The problem of the North–South dialogue:

Incomplete delineation and protection of property rights.
1. Low per capita income.
2. Capital insufficiency.

Solutions:
1. Agricultural Clearing House; release constraints for commodity crops.
2. Found small public DNA labs in developing countries.
3. The (international) public sector should invest in genomics and functional genomics; little of the genetic variation in plants has yet been exploited in breeding programmes. Make genome sequences publicly available as will be done with the genome sequence of rice, a commodity crop; there are still so many biological questions to be solved that could be commercialized by the private companies globally.
4. Biotechnology is still in its infancy; there are big differences in consumer technology, medical research and Ag-biotech (food and nutrition; emotional factors).

The dilemma:

Does the (local) farmer benefit from more industrial genomics activities? The prices for his products will not increase; he has to produce more to keep the same standard of living! (a chicken egg in the Netherlands has hardly increased in price over the last 30 years!).

15a

Will consumers lose or gain from the environmental impacts of transgenic crops?

Jill E. Hobbs[#] *and William A. Kerr*[#]

Abstract

Opposition to genetically modified food encompasses environmental concerns, food-safety concerns and ethical objections. Potential environmental benefits from transgenic crops are not well accepted. Genetic modification is a credence attribute that cannot be detected by consumers without labelling. In the absence of labelling a pooling equilibrium results in an adverse quality effect for those consumers who prefer not to consume genetically modified (GM) food for environmental or food-safety reasons, but may result in a beneficial price effect for all consumers if the innovation is drastic. Labelling enables consumers to express their environmental preferences through the marketplace and can mitigate the adverse quality effect, but only in the absence of cheating. Both mandatory GM and voluntary non-GM labelling will impose segregation costs on the non-GM sector, leading to an increase in prices. The challenge will be to allow technological advances in agriculture that increase yields, reduce costs and improve product quality, while respecting consumer preferences. Future research could assist by improving our understanding of the consumer decision-making process, including how consumers react to new information and how consumers would respond to future GM products with direct consumption or environmental benefits.
Keywords: consumer welfare; labelling; transgenic crops

Introduction

Transgenic crops have been adopted rapidly in some countries in response to their perceived agronomic and economic benefits to producers. Herbicide-tolerant varieties and crops genetically engineered to be resistant to targeted pests offer the potential for reduced agricultural-chemical use, lower input costs and higher yields. Where approved for release, the relatively rapid adoption of transgenic varieties of soybean, corn and canola means that food derived from genetically modified crops quickly found its way onto retail shelves. Many processed foods contain soy protein or soy derivatives, and the majority of soybean crops grown in the US and Canada are now transgenic varieties. Consumer reaction to food derived from transgenic crops has been decidedly mixed, ranging from vocal opposition among some consumers and interest groups, to confusion or puzzled indifference among others. This paper examines the consumer reaction to genetically modified (GM)[1] food, focusing on the

[#] Department of Agricultural Economics, University of Saskatchewan, 51 Campus Drive, Saskatoon, Saskatchewan, S7N 5A8, Canada. E-mail: jill.hobbs@usask.ca and E-mail: william.kerr@usask.ca

J. H. H. Wesseler (ed.), Environmental Costs and Benefits of Transgenic Crops, 249–262.

potential consumer benefits and costs from the perceived environmental impacts of transgenic crops.

Consumer reaction to GM food is complex and multi-faceted. Broadly speaking, consumer concerns can be split into four groups: specific food-safety concerns, fear of the 'unknown' consequences of consumption, ethical concerns and environmental concerns. Although the primary focus of this paper is on environmental concerns, it is useful to outline the other issues briefly. Specific food-safety concerns relate to allergens and the use of antibiotic-resistant marker genes. Some consumer groups have expressed concerns that transgenics could result in the transference of allergens into foods in which they were not previously present. This risk is explicitly recognized in the European and North-American regulatory systems and the use of potentially allergenic genes is strictly regulated. The use of antibiotic-resistance marker genes to track the presence of a modified gene also raised concerns that this practice could contribute to the growth of antibiotic resistance in humans and animals (Gaisford et al. 2001).

Specific food-safety concerns relate to known risks that can be addressed through the regulatory system. In contrast, food-safety concerns over the potential long-run, unknown consequences of ingesting GM food present a more difficult problem. These concerns are not based on clear scientific evidence; instead they reflect a lack of confidence in the ability of scientific analysis to identify long-run risks. An element of uncertainty is introduced into the consumer decision-making process. For the most part, consumer unease over the long-run consequences of consuming GM food has been exacerbated by a series of high-profile food-safety scares, including Bovine Spongiform Encephalopathy (BSE) in beef. Lack of trust in the regulatory system and weakened confidence in the ability of science to determine long-run food-safety risks were consequences of high-profile food-safety problems.

The third group of concerns can be classed as ethical objections to transgenic technology and the patenting of genetic material. For some consumers, mixed with this is often a suspicion of or hostility towards the large multinational firms that are seen to dominate the agricultural biotechnology sector. Finally, some consumers and environmental groups have expressed concerns over the potential environmental consequences of introducing transgenic crops into cropping systems. While the purpose of this paper is to examine the environmental aspect of consumer responses, it is important to recognize that consumer unease over transgenics is multi-faceted. As a result, 'genetic modification' of food has become a lightening rod for a powerful coalition of interest groups.

Examining the potential environmental impacts of transgenic crops

Concerns over the potential environmental impacts of transgenic crops include fears about evolutionary resistance in target organisms, potential outcrossing with weedy relatives, the use of 'terminator' genes, reduction in biodiversity and damage to non-target organisms[2] (Gaisford et al. 2001). Evolutionary resistance in target organisms as a result of continuous use of specific herbicides or pesticides is an ongoing concern in cropping systems and is not limited to transgenic crops. The possibility of outcrossing between the transgenic variety and wild weed relatives through cross-pollination is a paramount concern for many environmentalists who fear the transference of herbicide resistance to weeds could make them harder to control, disrupting ecosystem balances. The use of so-called 'terminator' genes to prevent seed germination in transgenic crops has raised fears that this gene could contaminate other crops; however, sterile seeds cannot contaminate the gene pool as the trait cannot be passed on. Environmental groups have raised concerns that transgenic crops with a clear agronomic advantage could increase monoculture production and reduce biodiversity. As with the development of resistance, however, this concern is not unique to transgenic crops and could occur with the development of superior conventional varieties. Finally, potential negative effects on non-target organisms such as butterflies or beneficial insects and fungi from transgenic crops designed to target a different specific pest have been raised. Intense media attention and conflicting scientific studies, for example, related to Monarch butterflies[3], have fuelled these fears (Gaisford et al. 2001).

While the potential negative environmental impacts of transgenic crops have garnered most of the media and public attention, there is also a range of potential environmental benefits, which have a lower profile in terms of public awareness. The most important environmental benefit is the potential to reduce the application of chemical pesticides and herbicides. The transfer of genes from *Bacillus thuringiensis (Bt)* into crops such as corn and cotton produces toxins that kill target pests, reducing the need to use chemical pesticides. Through transgenics, herbicide-tolerant crops such as corn, soybeans and canola have been developed that are resistant to broad-spectrum herbicides. In theory this should enable farmers to kill weeds with fewer chemical applications and facilitate the use of 'zero-till' production systems. Transgenic technology offers the promise of future plants engineered to be resistant to major diseases and tolerant of environmental stresses; in turn this could reduce the use of fungicides and chemical fertilizers or reduce the need to irrigate (Gaisford et al. 2001). Reduced soil erosion is another potential agronomic benefit if fewer pre-emergent chemical applications are required, since this practice can result in higher moisture conservation and can reduce impacted soil problems from heavy machinery.

Clearly, there are a range of potential environmental problems and environmental benefits from transgenic crops. The extent to which consumers perceive the costs to outweigh the benefits, or vice versa, will influence their reaction to food identified as genetically modified. The next section explores the potential impacts on consumers.

Impacts on consumers

Consumers are affected in two ways by transgenic crops: direct (tangible) consumption effects and indirect (intangible) existence-value effects. Direct consumption benefits result from the physical purchase and consumption of GM food. Reduced use of agricultural chemicals may lead to lower pesticide-residue levels in food, with health benefits (or reduced health risks) for consumers. This is a positive quality effect. There may also be beneficial price effects for consumers. If transgenic technology reduces production costs, assuming competitive markets, these cost savings should be passed through to consumers in the form of lower food prices. While competitive markets is a strong assumption with respect to downstream food processing and retailing markets in most developed countries, technological advances in agriculture have tended to result in falling real food prices over the long run.

Counteracting these direct consumption benefits, are negative consumption effects for some consumers. For those consumers with specific food-safety concerns, or uncertainty over the long-run health effects of consuming GM food, there is an adverse quality effect, as these consumers will perceive GM food as lower quality.

Indirect existence benefits and costs also arise from transgenic crops. A positive existence value implies that some consumers value the potential environmental benefits from transgenic crops. A negative existence value represents the opposite situation; consumers who believe that transgenic crops are harmful to the environment perceive that there is a market failure resulting in the over-production of a negative externality. Consumers receive disutility or utility from the consumption of GM food derived from transgenic crops believed to be harmful or beneficial to the environment, respectively. These are additional adverse and beneficial quality effects. Thus, an 'eco-unfriendly/-friendly' discount or premium may apply to foods identifiable as GM. Whether consumers gain or lose from transgenic crops depends on the extent to which the market accurately reflects their preferences and if information asymmetry is present. If GM foods are labelled, consumers can express their environmental preferences through the market.

Credence attributes and the role of quality signals

With the exception of beneficial price effects, the impacts of GM food on consumers are credence attributes. Unlike search attributes, which consumers can evaluate prior to purchase – for example the colour of an apple – or experience attributes, which consumers can evaluate after consumption – for example the juiciness of an orange – consumers cannot detect the presence of credence attributes even after consumption (Nelson 1970). The presence of a genetically modified organism is a credence attribute.

Credence attributes create an information-asymmetry problem for the consumer unless the attribute is transformed into a search attribute through labelling. In the absence of labelling, adverse selection leads to over-supply of 'low-quality' (as perceived by consumers) goods in the marketplace in the classic 'lemons' effect (Akerlof 1970). Consumer uncertainty over product quality reduces average willingness to pay, thereby reducing the incentive for producers of high-quality goods to supply these goods to the marketplace. Low-quality goods (lemons) dominate the market. In this context, quality is 'in the eye of the beholder', such that low quality for some consumers may equate to food derived from transgenic crops.

As the previous discussion indicated, there exist potential positive and negative quality reactions to GM food among consumers. Most consumer research suggests that consumers are either indifferent or would choose to avoid GM food if given the choice. A recent stated-preference survey in Canada indicated that some consumers would choose GM over non-GM food, perhaps as a novelty, although these consumers were in the minority with a larger proportion of consumers stating a preference for non-GM food (Hünnemeyer et al. 2003). It is reasonable to assume that firms would expect a negative (rather than positive) reaction to a GM label. However, it is interesting to note initial evidence that the introduction and subsequent removal of a GM-content label from food in The Netherlands did not significantly change consumption patterns for those foods (Marks, Kalaitzandonakes and Vickner 2003). In general, if firms believe that consumers will react negatively to a 'GM content' label, they will not voluntarily label their food as genetically modified as this would signal low quality. Grossman (1981) shows that firms will not voluntarily disclose low quality if quality verification is difficult. The following analysis presents scenarios for the impact of transgenic crops on consumers with and without labelling.

Consumer impacts in the absence of labelling

In the absence of labelling, a pooling equilibrium exists, as consumers cannot distinguish between GM and non-GM food[4]. While some consumers are indifferent, others will suffer an 'adverse quality effect' if they perceive GM food as being of lower quality. This may be for any of the food-safety, environmental or ethical reasons identified earlier. The 'lemons effect' ensues, and *ceteris paribus*, we expect demand for the product to fall, reflecting consumer uncertainty over quality. The extent to which demand falls will depend on consumers' subjective probabilities of consuming GM food. Thus, if the transgenic innovation is drastic and it completely displaces the non-transgenic crop, we can expect a relatively large shift in consumer demand for that specific product. If the innovation was non-drastic and it does not completely displace the non-GM crop, the two crops exist simultaneously, and the fall in consumer demand is smaller.

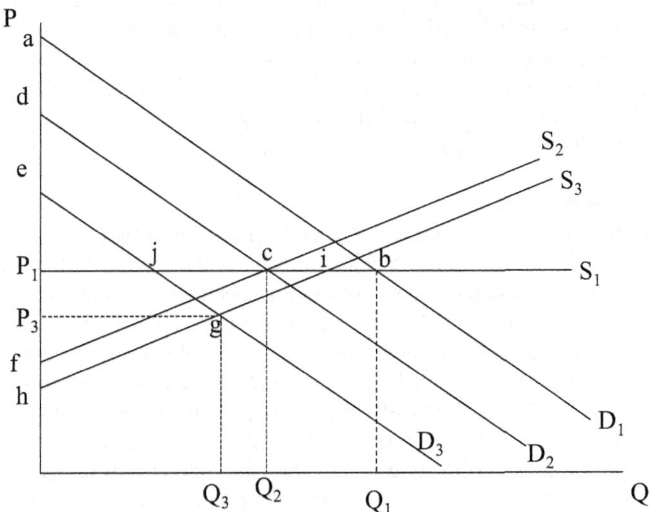

Figure 1. GM-food market effects in the absence of labelling

Figure 1 illustrates these two cases for the GM-food market. The initial equilibrium is represented by point *b* at the intersection of D_1 and S_1, giving equilibrium price and quantity of P_1, Q_1 respectively. A drastic innovation that completely displaces the conventional crop, leading to a consumer expectation that 100% of the food derived from that crop will be GM, is represented by D_3. A non-drastic innovation that only partially displaces the conventional crop results in demand curve D_2. Following Gaisford et al. (2001), initial supply of non-GM output is assumed to be perfectly elastic, reflecting uniform technology and costs across all farms, and for simplicity it is assumed that the farm sector is vertically integrated forwards into the food market. Unequal abilities among farms to adopt transgenic crops (and produce GM food) result in upward-sloping supply functions for GM food given by S_2 and S_3 for a non-drastic and drastic GM innovation respectively.

A number of potential effects from transgenic crops are apparent from Figure 1. In the case of a non-drastic innovation, supply of GM food shifts to S_2, price remains at P_1 and GM and non-GM food co-exist in the product market. Demand shifts to D_2, given consumer uncertainty about the quality of the product, giving a new equilibrium at point *c*. Consumers who are indifferent between GM and non-GM food are unaffected. However, consumers who regard GM food as lower quality, suffer an adverse quality effect equal to area *a-b-c-d* in Figure 1. It is assumed that there are no consumers who actually prefer the GM product. There is a gain in producer surplus for those farms able to adopt the transgenic technology equal to area *P_1-c-f*. Consumers, on aggregate, are unambiguously worse off; however, the net effect on economic welfare depends on the relative size of the loss in consumer surplus and gain in producer surplus.

If the GM innovation is drastic and if transgenic crops completely displace the conventional version of the crop we move to equilibrium point g, at the intersection of S_3, D_3. Price falls to P_3 and quantity falls to Q_3.The complete displacement of the non-GM food product leads to a larger adverse quality effect for consumers with strong anti-GM preferences, equal to *a-b-j-e*. Consumers who are indifferent between GM and non-GM food do not experience a quality effect. However, there is an offsetting beneficial price effect that benefits all consumers. The drop in price from P_1 to P_3 yields a gain in consumer surplus relative to the pre-innovation equilibrium of *P_1-j-g-P_3*. The producer surplus gain is larger at *P_3-g-h*. If there are beneficial price effects, consumers who are indifferent about GM food are better off, while the effect on those with negative quality perceptions is now ambiguous.

Labelling as a quality signal

Labelling the presence or absence of GM-food content provides a quality signal to consumers. An important policy debate surrounds the labelling of GM food. A number of countries (e.g. member states of the European Union, New Zealand, Japan) favour mandatory labelling of GM content, with various threshold proposed from 0.9% to 5% of allowable GM content before a 'GM' label must be applied. Other countries, notably the US and Canada, prefer voluntary labelling by firms wishing to differentiate their food as not genetically modified. Proponents of mandatory labelling argue that it enshrines in law consumers' right-to-know whether they are consuming GM food. They fear that, left to its own devices, the market will under-identify and over-provide GM food relative to consumer preferences. Opponents of mandatory labelling argue that it is misleading, implying a safety or quality difference that has not been substantiated by scientific evidence and as such is inconsistent with a regulatory trajectory that has approved GM foods as safe for consumption. They argue that the market will self-identify non-GM foods if there is a sufficient demand for a non-GM assurance. However, there is a clear incentive for cheating and the mislabelling of GM food as non-GM, so in either case, monitoring and enforcement costs will arise. The following analysis examines voluntary labelling of non-GM food, followed by a discussion of a mandatory GM-labelling policy.

Voluntary labelling of non-GM food

Labelling leads to a separating equilibrium, wherein we can identify separate markets for non-GM and GM food. In the absence of cheating, consumers can accurately and costlessly distinguish between the two types of food. This enables those consumers with quality concerns about GM food – either for food-safety or environmental reasons – to avoid GM products. Thus, labelling enables consumers to express their existence values, or environmental preferences, through the marketplace. For now we will assume that these are negative preferences, i.e. that consumers perceive transgenic crops to be a potential cause of environmental harm and wish to signal this through their food purchases. In this case voluntary labelling will be undertaken by the non-GM rather than the GM food-producing sector.

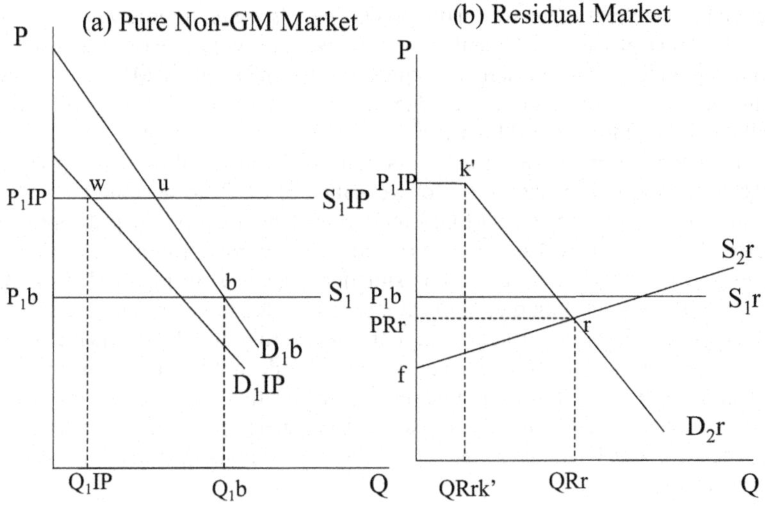

Figure 2. Non-GM identity preservation and labelling

The separating equilibrium under voluntary non-GM labelling is illustrated in Figure 2. Abstracting from the problem of cheating, a voluntary identity preservation and labelling system for non-GM food emerges, as depicted in panel (a) (Gaisford et al. 2001). A second 'residual' market emerges, consisting of GM food and any non-GM food that does not participate in identity preservation and labelling. Equilibrium in the non-GM food market begins at point *b*, with price P_{1b} and quantity Q_{1b}. With the introduction of transgenic crops, a voluntary non-GM identity-preservation and labelling system emerges in response to consumer demand for non-GM assurances based on consumers' environmental and food-safety preferences. The costs of the identity-preservation system are borne by the non-GM food sector, shifting the supply curve in panel (a) to S_1IP. The labelling of non-GM food gives rise to the residual market in panel (b), where S_2r is the supply of GM foods and S_1r is the supply of non-GM food to that market. Following Gaisford et al. (2001), the demand curve D_2r is drawn conditional on a price of P_1IP for the non-GM food. If the prices of the two foods are identical, consumers who are indifferent between GM and non-GM foods will consume in either market, up to quantity QRrk'. At prices below P_1IP, the indifferent consumers and some of those who only weakly prefer non-GM food will switch to the cheaper residual market, giving rise to the downward portion of demand curve D_2r. Given the non-GM price of P_1IP, the residual market is in equilibrium at r, with price of PRr and quantity at QRr. The introduction of substitute GM food at a price of PRr shifts the non-GM market demand curve in panel (a) to D_1IP. This is a pure substitution effect, rather than an adverse quality effect. The new equilibrium on the non-GM market is at *w*, with a price of P_1IP and a quantity of Q_1IP.

In the situation depicted in Figure 2, there is a loss of consumer surplus in panel (a) equal to P_1IP-*u*-*b*-P_1b resulting from the increase in the price of non-GM food 'before' consumers switch to the residual market[5]. With the adjusted non-GM price, there is a gain in consumer surplus on the residual market of *P_1IP-k'-r-PRr*. Some of the consumers who only weakly prefer non-GM food will be better off, while those who remain in the non-GM market will be worse off than prior to the innovation. Consumers who are indifferent between GM and non-GM food are better off due to the beneficial price effect in the residual market. There is a gain in producer surplus in the residual market of *PRr-r-f.* Overall economic welfare will increase if and only if the gain in producer and consumer surplus on the residual (GM) food market outweighs the loss in consumer surplus due to the incidence of identity-preservation and labelling costs on the non-GM market[6].

If there is a price premium for GM food, and if the probability of being caught and/or the penalties from cheating are low, cheating is likely to become a problem. There is an incentive for producers of GM food to mislabel their products as non-GM, particularly if these firms have not incurred identity-preservation costs, unlike 'honest' non-GM supplying firms. In this situation, panel (a) in Figure 2 no longer represents a 'pure' non-GM market. If cheating is suspected to be present, the demand curve in panel (a) will shift further to the left representing an adverse quality effect and resulting in a loss in welfare for consumers who do not want to consume GM food (not shown in graph). The economic rents from cheating are in the form of a gain in producer surplus for GM producers who mislabel their products. In the long run, rampant cheating would weaken the credibility of a non-GM labelling system, rendering it unsustainable.

Mandatory labelling of GM food

If policymakers believe that a significant market failure exists in enabling consumers to express their environmental (food-safety or ethical) preferences for avoiding GM food, they may mandate GM-content labelling. Mandatory GM-content labelling results in a separating equilibrium consisting of an identified GM market and a market for non-GM food identified by default. However, cheating or non-compliance can still be a problem, so that the default 'non-GM' market may contain food that is genetically modified but that has not been labelled. Figure 3 illustrates a mandatory GM-food identity-preservation and labelling system with non-compliance, resulting in an impure non-GM market.

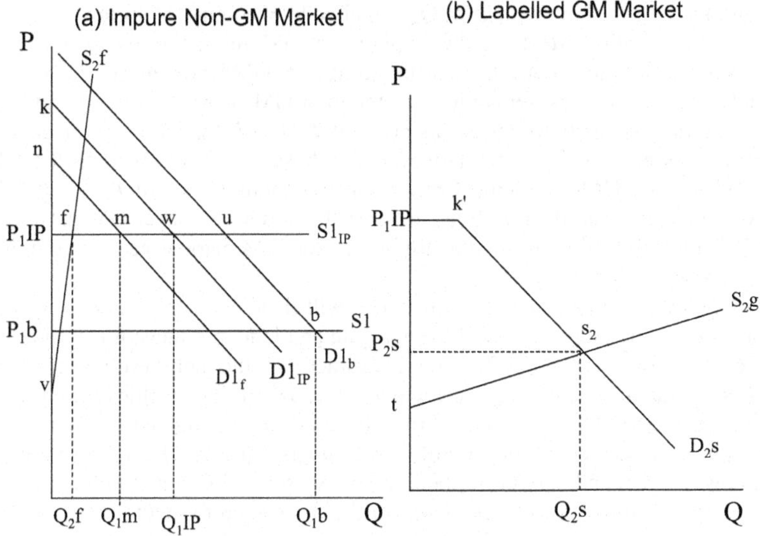

Figure 3. Mandatory GM labelling with non-compliance problem

From the initial equilibrium at point *b*, a GM innovation with mandatory labelling has two effects. First it imposes segregation and identity-preservation costs on non-GM supply chains that wish to continue to market their products as non-GM. It is expected that it will be much more costly to substantiate the absence of GMOs than to acknowledge their possible presence with a 'may contain GMOs' type label. As consumers who choose to purchase GM food will not care if their food is 'contaminated' with non-GM food, it will not be necessary for supply chains providing GM products to segregate their products and incur the costs associated with segregation. This will not be the case for those wishing to sell non-GM food because their consumers will care if their purchases are 'contaminated' with GMOs, hence these supply chains will bear the cost of segregation. Thus, the non-GM supply curve shifts from S_1 to S_1IP reflecting the higher costs segregation indirectly imposed on the non-GM sector. The demand curve for labelled GM food is represented by D_2S in panel (b) and is conditional on the price of P_1IP in the non-GM market. The supply curve of labelled GM food is given by S_2g, yielding an equilibrium price and quantity in the labelled GM market of P_2s, Q_2s[7]. As in the previous case, the cheaper GM-food price entices those consumers who are indifferent toward GM content, and any consumers with only weak preferences for non-GM food, to shift to the labelled-GM market. The departure of these consumers from the non-GM market causes the demand curve to shift to D_1IP in a pure substitution effect.

To this point, the story is similar to that outlined for voluntary labelling with no cheating. However, we now acknowledge the incentive for GM-producing firms to avoid labelling their products in order to take advantage of the relative price differential between unlabelled (assumed GM-free) and labelled GM products. The supply curve for the GM content 'fraudulently' supplied in the (now impure) non-GM market is given by S_2f. At a price of P_1IP, quantity Q_2f of GM food is fraudulently supplied into the non-GM market. Recognition, or even suspicion, among consumers that this is occurring causes the demand curve in the non-GM market to shift left to D_1f, with the result that the equilibrium quantity of food transacted in the (impure) non-GM market falls to Q_1m, with Q_1m-Q_2f supplied by the non-GM sector.

The adverse quality effect on the impure non-GM market leads to a loss in consumer surplus for those consumers with strong anti-GM preferences of *k-w-m-n*. Furthermore, all consumers remaining in the non-GM market suffer a loss in consumer surplus due to the increase in price of P_1IP-*u-b*-P_1b. There is a gain in consumer surplus of P_1IP-*k'*-s_2-P_2s, however, for those consumers who shift to the labelled-GM market. GM-producing firms receive economic rents of P_2s-s_2-*t* in the labelled-GM market given their abilities to adopt the new technology. GM producers who successfully sell GM food fraudulently in the non-GM market receive additional producer surplus o f P_1IP-*f*-*v*[8].

The magnitude of the effects on consumers will depend on the increase in food costs as a result of the industry's need to segregate and label products; this will affect the size of the adverse price effect on consumers in the non-GM market. The effectiveness of monitoring and enforcement and the size of the penalties for cheating will determine the propensity of GM firms to avoid complying with a mandatory labelling policy. If enforcement is rigorous, the adverse quality effect may be quite small. If, however, a mandatory GM-labelling policy proves difficult to enforce due to the complexity of testing foods (particularly further processed foods) for GM content, then widespread cheating is likely, and the adverse quality effect for non-GM consumers will be much higher.

An alternative to a performance-based standard for GM-content labelling based on end-product testing is a process-based approach involving supply-chain audits and documentation to verify GM presence or absence. Since the onus is likely to be on non-GM supply chains to verify the purity of their products, this will magnify their identity-preservation and segregation costs. The intention of a mandatory GM-labelling policy may be to increase the economic welfare of consumers who wish to avoid GM food for food-safety, environmental or ethical reasons. Paradoxically however, the outcome could be a reduction in consumer welfare if negative price effects are substantial and an adverse quality effect remains.

Reactions to environmental benefits

The preceding analysis was predicated on the assumption that the identification of non-GM or GM food through either voluntary or mandatory labelling would provide consumers with the ability to express their environmental (or food-safety, ethical) preferences for non-GM food. As the earlier discussion indicated, however, there are potential environmental benefits from transgenic crops. Thus it is conceivable that some consumers, if aware of these potential benefits, might choose to consume GM food. This introduces counteracting pressures on the demand curves for both non-GM and GM food products. In the absence of labelling, the adverse quality effect on a pooled GM/non-GM market (Figure 1) would be mitigated if a third group of consumers existed that preferred GM products for environmental reasons. With labelling (Figures 2 and 3), positive environmental attributes of GM products could result in both a price *and* a quality substitution effect out of the non-GM market into the GM market. If positive environmental preferences for GM food were sufficiently strong, there would be an incentive for a voluntary labelled-GM market. While the balance of consumer research to date suggests that consumers are either indifferent or prefer non-GM food, the case of positive attitudes towards GM food products should not be discounted.

Risk, uncertainty and the consumer decision-making problem

The potential environmental benefits of transgenic crops have not been convincingly communicated to consumers. This may be because the life-science companies who have developed the crops are not credible sources of objective information about the potential environmental impacts. Also, consumers may perceive the potential downside environmental risks to outweigh any potential environmental benefits. Events with a low probability of occurring but with large negative impacts may be weighted more heavily than events with a higher probability of occurring but with smaller impacts.

Uncertainty over the ability of scientific analysis to assess accurately the long-run implications of a new technology makes some consumers sceptical of the risk-analysis process. The regulatory approval process for new varieties includes an assessment of the risk of outcrossing between GM varieties and other plants and other environmental risks. As with food safety, there remains an element of uncertainty over long-run impacts on the ecosystem and on biodiversity, and for some consumers a lingering concern that science can never have all the answers. Widely publicized contradictions in scientific evidence serve to increase the uncertainty and make it difficult for the general public to discern between sound science and sensationalist science. There has been intense media focus on scientific studies that claim to have found evidence of health or environmental harm from genetically modified organisms. For example, the apparent finding that *Bt* corn harmed monarch butterflies, and the apparent evidence of harm caused to rats fed genetically modified potatoes from a study conducted at the Rowatt Institute in Scotland. In both cases a subsequent review of the research by other independent scientists called into question the validity of the initial reported findings. Media coverage of scientific repudiation, however, tends to be less intense. There is some evidence that negative media coverage has a disproportionately larger impact on public opinion than positive coverage. Liu, Huang and Brown (1998) found that negative news about a food-safety incident related to milk in Hawaii in 1982 had a more immediate effect on consumption than positive news.

Even if consumers are aware that a food is genetically modified, their decision-making process is complicated by the presence of uncertainty. The distinction between decision-making under risk versus under uncertainty is important. A situation involves risk if statistical probabilities can be attached to the randomness facing an economic agent. If it is not possible to attach statistical probabilities to the likelihood of an event occurring, the situation is characterized by uncertainty (Knight 1921; Eatwell, Milgate and Newman 1987). Isaac (2002) distinguishes between recognizable risks, hypothetical risks and speculative risks. Recognizable risks are those where there is sufficient information to attach probabilities. This describes consumer concerns about specific food-safety risks from GM food, and may describe some of the more tangible environmental risks that can be evaluated through short-term field trials. Hypothetical risks yield sufficient information upon which to base a testable hypothesis, but the research to evaluate the risks has not been done. Some concerns over the long-run health or environmental consequences of transgenic crops can be characterized as hypothetical risks. Speculative risks involve uncertainty, and it is not possible to devise a testable hypothesis to evaluate these risks based on current science. Speculative risks represent a fear of the 'unknown' and by definition, we cannot attach statistical probabilities to the unknown occurring. For a portion of consumers, uncertainty over the long-run environmental impacts of transgenic crops prevails and a risk-analysis framework for approving new transgenic crops based on risk assessment, risk management and risk communication will not reduce that uncertainty.

If consumers can form subjective probabilities regarding the recognizable risk they may choose to purchase the product if the perceived benefits from consuming the food outweigh the costs, including the costs of a negative outcome occurring with some probability, p. In cases for which the recognizable risks outweigh the potential benefits for an individual consumer, or for which environmental or food-safety risks are hypothetical or even speculative, consumers may respond by avoiding the product. In the case of environmental risks, which do not have a direct consumption effect on individual consumers, they may choose to boycott genetically modified products, thereby sending a market signal through product avoidance. Additionally, consumers may lobby for removal of the product from the market. In both cases, product avoidance requires that a food be identifiable as genetically modified through an effective and enforceable labelling system.

Conclusions

Consumer response to genetically modified food and transgenic crops is multi-faceted and complex. Initially consumers have either been indifferent or have expressed concerns about agricultural biotechnology related to perceived food-safety or environmental outcomes, or ethical objections to the technology or the technology provider. While this paper has focused on reactions to environmental attributes of transgenic crops, it is extremely difficult to disentangle these reactions from the other consumer concerns. It is difficult to determine the extent to which a negative consumer reaction is due to environmental reasons versus other concerns. Labelling of GM food, if effective, enables consumers to avoid foods about which they have food-safety concerns. It also allows consumers to signal their environmental preferences through food-consumption decisions. However, labelling alone does not remove the environmental 'threat' that may be perceived. This has led some groups to lobby for stronger measures such as a ban on the production or importation of transgenic crops.

New biotechnology innovations that offer direct consumer health benefits through genetic modification of functional traits in food may have interesting implications for consumer acceptance. Similarly, proven (and credible) environmental benefits from transgenic crops could also lead to an interesting dichotomy in consumer markets where different groups of consumers have positive or negative perceptions of the environmental impacts of transgenic crops. The ability of scientific analysis to verify the environmental benefit or damage from transgenic crops will become even more critical, as will the acceptance of that scientific evidence by an often-sceptical public.

Regardless of whether there are perceived environmental benefits or costs from transgenic crops, the credence nature of the GMO attribute means that without labelling a pooling equilibrium emerges. The aggregate effects on consumers are determined by the extent to which adverse quality effects for some consumers are mitigated by a beneficial drop in food prices, or by a counter-acting positive quality effect among other consumers who believe that transgenic crops have environmental benefits. While labelling is often posited as a solution to the consumer information problem with respect to GM food, it is not a simple solution. Segregation and identity preservation costs are likely to have a disproportionate effect on the non-GM producing sector, with potential pass-through of cost increases to consumers. Both a voluntary non-GM label and a mandatory GM-labelling policy will only be effective if the label is credible and backed by sufficient monitoring and enforcement to deter cheating. If substantial mislabelling occurs, consumer benefits are weakened and information asymmetry is not mitigated.

The debate over labelling of GM food and the regulatory approval of existing or new transgenic varieties across jurisdictions is likely to continue for some time. Different regulatory processes have resulted in different rules for labelling and product approval, inevitably leading to international trade tensions. Consumer (and public) concerns over long-run environmental and food-safety implications are often the justification for more restrictive rules with respect to product approval or labelling requirements. It is important to note that in most countries (including the European Union, the US and Canada) the regulatory approval process for new transgenic crops or new foods derived from transgenic crops includes the three-tier scientific risk-analysis approach, including risk assessment, risk management and risk communication. This approach has apparently failed to reassure many consumers. In part this reflects a lack of confidence in the regulatory approval process and in the ability of science to predict the potential extent of environmental or food safety hazards accurately.

The challenge will be to allow technological advances in agriculture that increase yields, reduce costs and improve product quality, while respecting consumer preferences. Regulatory oversight to ensure product safety and environmental soundness remains critically important. Improved communication with respect to proven environmental and quality benefits is also important. Future research could assist in this process in a number of ways. First, by improving our understanding of the consumer decision-making process. Second, by evaluating how consumers react to new information and to different sources of information about transgenic crops or GM food. Finally, by assessing how consumers will trade-off GM content with positive attributes such as improved quality or environmental benefits.

References

Akerlof, G.A., 1970. The market for lemons: quality uncertainty and the market mechanism. *Quarterly Journal of Economics,* 84 (3), 488-500.

Eatwell, J., Milgate, M. and Newman, P. (eds.), 1987. *The new Palgrave: a dictionary of economics*. Macmillan, London.

Gaisford, J.D., Hobbs, J.E., Kerr, W.A., et al., 2001. *The economics of biotechnology*. Edward Elgar, Cheltenham.

Grossman, S.J., 1981. The informational role of warranties and private disclosure about product quality. *Journal of Law and Economics,* 24 (3), 461-483.

Hünnemeyer, A., Veeman, M., Adamowicz, V., et al., 2003. *Consumer preferences for foods containing genetically modified ingredients, Department of Rural Economy, University of Alberta. Paper presented at the workshop "The Economics of Food and Health", Canadian Agricultural Economics Society, Vancouver, May 2-3-2003*.

Isaac, G.E., 2002. *Agricultural biotechnology and transatlantic trade: regulatory barriers to GM crops*. CABI Publishing, Wallingford.

Knight, F.H., 1921. *Risk, uncertainty and profit*. Houghton Mifflin, Boston.

Liu, S.P., Huang, J.C. and Brown, G.L., 1998. Information and risk perception: a dynamic adjustment process. *Risk Analysis,* 18 (6), 689-699.

Marks, L.A., Kalaitzandonakes, N. and Vickner, S., 2003. Evaluating consumer response to GM food: some methodological considerations. *Current Agriculture, Food and Resource Issues* (4), 80-94. [http://cafri.usask.ca/j_pdfs/marks4-1.pdf]

Nelson, P., 1970. Information and consumer behaviour. *Journal of Political Economy,* 78 (2), 311-329.

---

[1] Definitions of genetic modification, genetic engineering and transgenics often differ. For simplicity, food derived from transgenic crops will be referred to as "genetically modified" food throughout this paper.

[2] The authors make no claim as to the scientific merit of these concerns; they are presented as a backdrop to an economic analysis of how they affect consumer decisions.

[3] Intense media attention arose following a scientific study that appeared to show harmful effects on monarch butterflies from *Bt* corn. Subsequent scientific analysis called into question the conclusions of the first study.

[4] The following analysis draws on Gaisford et al. 2001.

[5] There may be special equilibrium circumstances where a pooling equilibrium exists in the residual market if the S_2r GM supply curve intersects the D_2r demand curve above the S_1r non-GM supply curve. Alternatively, if the intercept of D_1IP in the non-GM market lies below P_1IP no consumers would be willing to pay the costs of the identity-preservation system. These cases are discussed in more detail in Gaisford et al. (2001).

[6] This abstracts from any gains to technology providers. See Gaisford et al. (2001) for a more complete discussion of impacts on input markets.

[7] S_2g is drawn conditional on the price of P_1IP in the non-GM market. Technically this means that S_2g represents the supply of GM food in the GM-labelled market, net of any amount fraudulently supplied in the non-GM market.

[8] This surplus gain is net of any loss in producer surplus from shifting out of the labelled GM market, given the specification of S_2g as outlined in the previous note.

15b

Comment on Hobbs and Kerr: Will consumers lose or gain from the environmental impacts of transgenic crops?

Sara Scatasta[#]

Hobbs and Kerr (2004) highlight three key issues to be taken into consideration when analysing the impact of genetically modified (GM) food on consumer welfare. The first issue is the direct impact that GM food will have on consumers through consumption activities in terms of market price and quantities. Assuming GM crops are associated to lower production costs and higher yields, these advantages could be passed on to the consumer in terms of lower market price and higher marketed quantities. The second issue is the indirect impact that GM will have on consumers through changes in quality attributes of related food products as perceived by the consumer. Potential environmental and health impacts related to the consumption of GM food and single consumers' perceptions of these impacts may lead some consumers to be against GM foods. The third issue highlighted by Hobbs and Kerr is the impact of introduction or absence of mandatory GM and voluntary non-GM labelling (here referred to as GM labelling).

On the one hand, since consumer preferences have already been altered by media coverage of topics related to GM food, and since GM is a credence attribute, the introduction of GM labelling may have a positive effect on welfare for consumers who prefer not to consume genetically modified (GM) food. At the same time GM labelling may impose additional costs to producers of GM or non-GM depending on how GM labelling is implemented. For example, in the case of voluntary non-GM labelling, non-GM producers will incur higher costs of demonstrating that their products are non-GM. This may result in higher market prices of non-GM products. On the other hand, absence of GM labelling would result in lower market prices but have a negative effect on welfare for those consumers who do not want to consume GM food.

Several studies have been carried out to investigate the impact of eco-labels on consumption behaviour with respect to environmentally friendly goods (see, for example, Nimon and Beghin 1999; and Wessells, Johnston and Donath 1999). The conclusion was that the consumer might be willing to pay a premium for products that are perceived to be more environmentally friendly than others. Thus, consumers who associate GM food with environmental damages might be willing to pay a premium to avoid consuming those products. If the introduction of GM labelling does not raise the price of non-GM products above this premium, then GM labelling should be introduced. The paper by Hobbs and Kerr supports this conclusion, noting that another important factor in establishing the impact of GM labelling on consumer welfare is consumer expectation about the rate of adoption of GM crops by food

[#] Environmental Economics and Natural Resources Group, Wageningen University, Hollandseweg 1, 6708 KN Wageningen, the Netherlands. E-mail: sara.scatasta@wur.nl

J. H. H. Wesseler (ed.), Environmental Costs and Benefits of Transgenic Crops, 263–264.

producers. The loss in consumer welfare will be directly related to the rate of adoption of GM crops.

In concluding, it should be noted that there is an additional factor that needs to be taken into consideration when analysing the impact of GM labelling on consumer welfare: being non-GM may not be the only attribute differentiating two food products. Other attributes such as taste and colour may be of interest to the consumer regardless of the preference for GM foods. An increase in product price due to the introduction of GM labelling makes all other product attributes more expensive to the consumer. Thus, GM labelling may have a negative impact on welfare for those consumers who do not care about consuming non-GM food if they have strong preferences for some other attributes of non-GM products. In this context it becomes essential to study the impact of GM labelling on consumer welfare on a case-by-case basis.

References

Nimon, W. and Beghin, J., 1999. Are eco-labels valuable? Evidence from the apparel industry. *American Journal of Agricultural Economics,* 81 (4), 801-811.

Wessells, C.R., Johnston, R.J. and Donath, H., 1999. Assessing consumer preferences for ecolabeled seafood: the influence of species, certifier, and household attributes. *American Journal of Agricultural Economics,* 81 (5), 1084-1089.

16

Conclusions

Justus Wesseler[#]

The contributions cover the wide and heterogeneous field of the environmental costs and benefits of transgenic crops. There are several important conclusions that emerge out of the contributions and comments with respect to the overall environmental costs and benefits. First of all, spatial aspects in planting transgenic crops are important. Second, rules and regulations governing the introduction and planting of transgenic crops have important implications for adoption over time and space. Third, the system of intellectual property rights can govern the direction of public and private-sector research incentives including research in transgenic crops with environmentally friendly traits. Fourth, public concern and consumer preferences will guide rules and regulations on transgenic crops. Fifth, rules and regulations have a global dimension and cannot be discussed in isolation. These conclusions lead to the following research themes for further investigation.

Theme 1. Spatial and structural effects of adoption with respect to agricultural biotechnology

The introduction of transgenic crops will have an impact on the scale and scope of crop production, which will have an impact on the quantitative environmental impacts and on the up- and downstream sectors, changing the structure of the agricultural sector. This demands an interdisciplinary analysis by natural and social scientists. Important sub-themes include product differentiation, such as soybeans with specific processing traits, long-term implications for the environment, such as an accumulation of glyphosate use in herbicide-tolerant crops, implications for private-sector marketing as a result of spatial changes in production, impact of spatial changes for agricultural policies such as the reform of the Common Agricultural Policy, the multi-functionality of agriculture and the assessment and management of risk.

Theme 2. Governance management performance under heterogeneity, complexity and uncertainty: Integrating science and politics

The introduction of transgenic crops is regulated by national governments. Governments can choose out of a portfolio of options to govern the introduction and planting of transgenic crops. The chosen management scheme will have an impact on the economic benefits and their distribution. Important aspects that need to be considered are regulation and self-governance, the supply chain and traceability, risk-sharing arrangements, power relationships, the political economy of regulating transgenic crops, monitoring and enforcement, incentive compatibility, learning and adaptive management, risk assessment and management, market power and anti-trust.

[#] Environmental Economics and Natural Resources Group, Wageningen University, The Netherlands. E-mail: justus.wesseler@wur.nl

J. H. H. Wesseler (ed.), Environmental Costs and Benefits of Transgenic Crops, 265–266.

Theme 3. Research management and intellectual property rights

The management of investment in research on transgenic crops at the national and international level and the protection of intellectual property rights (IPRs) is another area where stakeholders have an impact on the quantity and distribution of economic gains from transgenic crops. Analysing the impact of research management and IPRs one should consider the interaction between IPRs, bioprospecting and genetic resources, look at the dynamic aspects of public and private roles, the management of existing IPRs, the design of new IPRs, innovation and evolution of new technologies and the global implication of IPRs (TRIPs), and consider the limits of intellectual property in research management, the access to and the sharing of benefits from transgenic crops.

Theme 4. Public concern and consumer preferences

An important topic in the EU but also other places is to deal with public concerns about and consumer preferences towards transgenic crops. Important to consider here are consumer reaction to different traits of transgenic crops, risk-communication strategies, quality and risk perceptions, values, beliefs and trust, knowledge on consumer-preference dynamics, sources and distribution of heterogeneity, and the media and information industry.

Theme 5. Global dimensions

Stakeholders managing research on transgenic crops have to consider the global dimension the introduction of such crops will have. Important issues that need further investigation include the potential for developing countries, institutional requirements for successful introduction in developing countries, biosafety regulations, capacity building, the global implications of IPRs, national and international regulation and market access, biotechnology in the context of rural development, the role of the WTO and IEA, public access to genomics, improvement of global knowledge base.

List of authors

Beckmann, Volker	Humboldt University, Berlin, Germany
Claessen, David	Biomathematics Unit, Rothamstead Research, UK
De Wit, Pierre	Wageningen University, Wageningen, The Netherlands
Demont, Matty	Catholic University of Leuven, Leuven, Belgium
Den Nijs, Hans	University of Amsterdam, Amsterdam, The Netherlands
Ervin, David	Portland State University, Portland, OR, USA
Gilligan, Chris	University of Cambridge, Cambridge, United Kingdom
Goeschl, Timo	University of Wisconsin-Madison, Madison, WI, USA
Graff, Greg	University of California, Berkeley, CA, USA
Groot, Mirella	University of Amsterdam, Amsterdam, The Netherlands
Hanson, Meira	The Hebrew University of Jerusalem, Jerusalem, Israel
Heijman, Wim	Wageningen University, Wageningen, The Netherlands
Hobbs, Jill	University of Saskatchewan, Saskatoon, SK, Canada
Hogeveen, Henk	Wageningen University, Wageningen, The Netherlands
Hurley, Terrence	University of Minnesota, St. Paul, MN, USA
Kerr, William	University of Saskatchewan, Saskatoon, SK, Canada
Kleter, Gijs	Wageningen UR, RIKILT Institute of Food Safety, Wageningen, The Netherlands
Kuiper, Harry	Wageningen UR, RIKILT Institute of Food Safety, Wageningen, The Netherlands
Laxminarayan, Ramanan	Resources for the Future, Washington, DC, USA
Matoušek, Jaroslav	Czech Academy of Science, Česke Budejovice, Czech Republic
Michalopoulos, Tassos	Wageningen University, Wageningen, The Netherlands
Roland-Holst, David	University of California, Berkeley, CA, USA
Scatasta, Sara	Wageningen University, Wageningen, The Netherlands
Schubert, Jörg	Federal Centre for Breeding Research on Cultivated Plants, Aschersleben, Germany
Simpson, David	Resources for the Future, Washington, DC, USA
Soregaroli, Claudio	Università Cattolica del Sacro Cuore, Cremona, Italy

Stiekema, Willem	Wageningen UR, Plant Research International, Wageningen, The Netherlands
Supp, Patrick	Federal Centre for Breeding Research on Cultivated Plants, Aschersleben, Germany
Tollens, Eric	Catholic University of Leuven, Leuven, Belgium
Van de Wiel, Clemens	Wageningen UR, Plant Research International, Wageningen, The Netherlands
Van den Belt, Henk	Wageningen University, Wageningen, The Netherlands
Van den Bosch, Frank	Biomathematics Unit, Rothamstead Research, UK
Van Ierland, Ekko	Wageningen University, Wageningen, The Netherlands
Weaver, Robert	Pennsylvania State University, University Park, PA, USA
Welsh, Rick	Clarkson University, Potsdam, NY, USA
Wesseler, Justus	Wageningen University, Wageningen, The Netherlands
Zilberman, David	University of California, Berkeley, CA, USA

Wageningen UR Frontis Series

---

1. A.G.J. Velthuis, L.J. Unnevehr, H. Hogeveen and R.B.M. Huirne (eds.): *New Approaches to Food-Safety Economics.* 2003
ISBN 1-4020-1425-2; Pb: 1-4020-1426-0
2. W. Takken and T.W. Scott (eds.): *Ecological Aspects for Application of Genetically Modified Mosquitoes.* 2003
ISBN 1-4020-1584-4; Pb: 1-4020-1585-2
3. M.A.J.S. van Boekel, A. Stein and A.H.C. van Bruggen (eds.): *Proceedings of the Frontis workshop on Bayesian Statistics and quality modellin.* 2003
ISBN 1-4020-1916-5
4. R.H.G. Jongman (ed.): *The New Dimensions of the European Landscape.* 2004
ISBN 1-4020-2909-8; Pb: 1-4020-2910-1
5. M.J.J.A.A. Korthals and R.J.Bogers (eds.): *Ethics for Life Scientists.* 2004
ISBN 1-4020-3178-5; Pb: 1-4020-3179-3
6. R.A. Feddes, G.H.de Rooij and J.C. van Dam (eds.): *Unsaturated-zone modeling.* Progress, challenges and applications. 2004 ISBN 1-4020-2919-5
7. J.H.H. Wesseler (ed.): *Environmental Costs and Benefits of Transgenic Crops.* 2005 ISBN 1-4020-3247-1; Pb: 1-4020-3248-X

---

9 781402 032486